MOLECULAR EMBRYOLOGY

MOLECULAR EMBRYOLOGY

By
Dr. D.R. Khanna
Reader in Zoology
Gurukul Kangri University
Haridwar (Uttaranchal)

DISCOVERY PUBLISHING HOUSE
NEW DELHI-110002

First Published-2004

ISBN 81-7141-773-6

Published by

DISCOVERY PUBLISHING HOUSE
4831/24, Ansari Road, Prahlad Street,
Darya Ganj, New Delhi-110002 (India)
Phone: 23279245 • Fax: 91-11-23253475
E-mail:dphtemp@indiatimes.com

Printed at:

Tarun Offset Printers, Delhi-53

Preface

The fundamentals of Molecular Embryology are presented within the framework of scientific discovery. Researches in Embryology have made almost increadible strides in the past few decades. Consequently, existing concepts of the Molecular Embryology have expanded. There has been a revolution indeed in this direction.

The text integrates the descriptive, experimental and biochemical approaches into a conceptual framework for the analysis of development. All important points are illustrated diagramatically. The title is not intended to be comprehensive nor could it be at length, but it concentrates as putting across the basic principles of the subject as briefly and lucidly as possible. It does this with the aid of careful selected examples—some recent and other classic of the field and with numerous illustrations. The aim is to enthuse the reader with this active and exciting area of research and to lay a solid foundation on which further study of its various facets may be based.

The present title is designed primarily for undergraduate students though it may also serve an introductory text to those preparing for their master's degree. The title is not intended to be comprehensive, nor could it be at this length, but it concentrates on putting across the basic principles of the subject as briefly and lucidly as possible. It does this with the aid of carefully selected examples, some recent and other classic of the field, and with numerous illustrations.

In the preparation of this book large number of books and research papers have been consulted. So no authenticity is claimed.

Text book can not be written without the support and professional contributions of many people. This book is no exception. The author is grateful to those teachers and colleagues whose stimulating discussions have clarified certain vague points for him, but all errors, omissions, and corrections needed are solely his responsibility.

The author tried hard to be accurate and upto date in statement and realises the impossibiiity of completely avoiding errors therefore, the author will greatly appreciate having his attention called to any questionable statement.

The author expresses his gratitute to Mr. Wasan and staff of M/s Discovery Publishing House for their whole hearted co-operation in the publication of this book.

Author

CONTENTS

1

GENE IN EMBRYONIC DEVELOPMENT

Embryonic Adaptations

The evolutionary history of a species determines what structures and processes are available on which selection may act. History and adaptation are intertwined. The result has been a marvelous diversity of larval morphologies. In some groups similar patterns of development and larval stages are retained despite very dissimilar adult morphologies. Thus, barancles and other Crustacea retain the presumably primitive nauplius larva, and mammalian embryos posse's structures resembling the gill arches of fish. These are the cases that led to Von Baer's law and later to Haeckel's recapitulation doctrine. But the opposite also occurs. Certain insects in which the adults are very similar produce quite dissimilar larvae, owing to divergent larval specializations. These cases and the existence of specialized structures associated with development, such as the extraembryonic membranes and placenta of mammals, represent purely larval adaptations that may be totally divergent from ancestral of even related forms. Embryonic or larval stages, while having to meet the inescapable requirements of being integrated and viable organisms, are nonetheless stages in a dynamic process involving differentiation and growth within the embryo.

If we agree that changes in these processes provide the mechanistic basis for achieving morphological evolution, then the nature of the adaptations that underlie the expression of genes controlling embryonic stages of development are of central interest. That genes indeed control morphological development was documented. In this chapter, we consider two major aspects of this expression: the genetic cost of development in terms of the proportion of the genome devoted to development, and specializations in genomic organization to support development.

How Many Genes are Necessary for Development?

. Fortunately, there are methods of estimating the amount of genetic information available to higher organisms. One of the finest of these tools is classic Mendelian genetics. The major difficulty with this method is that surprisingly few organisms have been subjected to sufficiently detailed genetic analysis to yield the desired data. In fact, only one organism, the common fruit fly *Drosophila melanogaster*, presently fulfills the criterion of sufficient genetic familiarity. Since *Drosophila* was first taken into the laboratory by T.H. Morgan and his students in 1910, a monumental amount of information has been gathered on this single modest insect. Thousands of mutations scattered throughout the genome of *Drosophila* have been induced by use of chemical mutagens and X-rays. The current catalog of mutations in *D. melanogaster*, a testament to the assiduousness of *Drosophila* geneticists, is now over 500 pages in length. This vast collection of mutations, as well as the ability to recover more, makes possible a dissection of the relationship between genes and the morphological, metabolic, developmental, or behavioural characteristics of the organism.

Mutations can be used to gain two basic kinds of information about development. Analysis of the ways in which mutations therefore with individual ontogenetic processes can provide major insights into the control and function of such processes. The second kind of information, with which we are concerned here, is the numbers of genes controlling the developmental processes, which in turn allows us to estimate the fraction of the entire genome devoted to ontogeny *per se*. Because any single gene can yield a number of different mutant alleles exhibiting different phenotypes, the number of genes involved in any process cannot be estimated by the appealingly simple procedure of counting up mutations.

Consider two mutations that both affect a particular phenotypic trait. The decision as to whether the two mutations in question are in one gene or two is made by means of a standard complementation test. The number of complementation groups regulating development of *Drosophila* can be estimated by taking a sample of a particular class of mutation and by extrapolating the total number of genes in that class in the entire genome. The extrapolation is based on two assumptions. The first is that genes of similar function are randomly distributed either in a particular chromosome or in the entire genome; that is, there is no clustering of genes of related function in development. With certain important exceptions., this assumption appears reasonable. The second assumption is that genes of like function are equally mutable, and that these mutations are equally recoverable. This assumptions is violated by some genes that turn out to be more mutable than most, but the assumption appears valid if applied to a large sample of genes.

Mutations can be recovered that block development in a variety of ways and can be grouped into general categories. As many mutations is possible are gathered for each category, but an exhaustive sample in which the genome is saturated for the desired class of mutants is, of course, impossible. Thus a non-saturating sample is taken and must be extended statistically. This requires that the genes actually sampled by mutation represent a random sample of the class of interest. What one chooses to consider as developmentally significant classes of genes depends on the particulars of development in the organism being studies.

It is important to remember that *Drosophila* is a holometabolous insect and that the larva represents a morphological, physiological, and behavioural stage completely distinct from the adult. Larval structures do not transform directly into equivalent adult structures. The fly is something of an insect phoenix. Most larval tissues are broken down and absorbed by the adult tissues that develop from the imaginal discs present within the larva.

There are then actually two developmental systems in this organism in which gene function crucial to adult development can occur. The first is during embryogenesis when the component parts and morphology of the larva are produced. The second is the establishment, maintenance, and proliferation of the imaginal discs within the larva, and their differentiation to produce the adult. The first of these systems, larval development, actually represents two temporally distinct systems of gene action. Information present within the egg that is necessary to early development represents the products of gene action during oogenesis.

Subsequent to its period of dependence on the gene products of oogenesis, the embryo begins active transcription to provide the information required for the remainder of development. Thus in *Drosophila* a tally of developmentally important genes will have to include genes that exhibit a maternal pattern of inheritance (i.e., function during oogenesis), genes whose function is vital during embryonic or larval development, and genes that specifically affect the development of imaginal discs. Estimates of all three classes have been made for *Drosophila*. Both Gans and co-workers and Mohler have isolated a large number of maternal effect mutations in genes located on the X chromosome, and similarly, Rice and Garen have isolated maternal effect mutations located on the third chromosome.

There are actually two classes of such mutations. In the first, females produce morphologically abnormal eggs (e.g., with defects in the egg shell), whereas females of the second class produce ostensibly normal eggs that fail to complete development. It is only this latter class that is of interest to us. The tally presented in Table 1.1 indicates that Gans recovered 42 mutations of the interesting class; Mohler, 146; and Rice and Garen, 6. Subsequent genetic complementation tests showed that all of these maternal effect mutations fell into 30,60 and 5 separate complementation groups, respectively. That is, some of the mutations represented repeat hits within the same gene. Before embarking on the analysis that follows, it is worth recalling Thomas Huxley's admonition that "Mathematics may be compared to a mill of exquisite workmanship, which grinds you stuff of any degree of fitness but, nevertheless, what you get out depends upon what you put in ..."

Table 1.1. Estimation of Total Maternal Effect Genes in the *Drosophila* Genome

Investigator	*Chromosome*	*Total No. of Mutations*	*No. of Genes Mutated*	n_o	*Estimated No. of Genes*
Gans et. al.	X	42	30	53	83
Mohler	X	146	60	38	98
Rice and Garen	3	6	5	8	13
Our extrapolation	All				117

Making use of the assumptions of a random arrangement of genes and their equal mutability, the number of unmutated genes of any class can be estimated by use of the Poisson distribution.

This procedure assumes that the probability is high that most of the genes of a class have not yielded any mutations at all, that a smaller number have yielded a single mutation, and that two mutations per gene is even less probable. The quantitative relationship between the no-hit, one-hit, and two-hit groups is given by the expression

$$n_o = \frac{(n_1)^2}{2n_2}$$

where the terms are defined as follows: n_1 is the number of genes that have been mutated once; n_2 is the number that have been mutated twice, and n_o the number of genes for which no mutation has been detected. The complementation data for any class yield numerical values for n_1 and n_2.

The data of Gans and co-workers yield a value of n_o equal to 53 and the data of Mohler yield 38 for total estimates of 83 and 98 maternal effect genes on the X-chromosome. Considering the nature of the estimates, the two numbers are in remarkably close agreement and suggest an average of 90 genes on the X chromosome vital to embryogenesis whose products are supplied as a result of gene action during oogenesis. The same exercise can be carried out using the data of Rice and Garen, yielding an estimate of n_o – 8 and n_{total} =13 for this class of genes located on the third chromosome. The disparity in the totals between the X and the third chromosome are interesting but not readily explicable because the third chromosome contains roughly twice the DNA of the X chromosome and the discrepancy cannot be accounted for by differences in sample sizes examined.

It thus appears that the maternal effect genes are predominantly located on the X chromosome. If we assume that the second chromosome, an autosome similar in DNA content to the third chromosome, is similar to the third chromosome in organization it to would have 13 maternal effect genes located on it. The last autosome, the very small fourth chromosome, can be estimated to have one maternal effect gene, based on its relative size. These estimates taken together suggest that there are a total of about 117 maternal effect genes in the *Drosophila* genome with 90 located on the X chromosome and based on results obtained by examination of the third chromosome, roughly 27 more on the autosomes. In order to estimate the number of genes necessary for embryonic development but not in the maternal effect class, we will have to use a slightly different approach.

It might be possible to collect a random sample of lethal mutations and determine the number of these that are lethal during embryonic development. An extrapolation of the total number of embryonic lethals in the entire genome could then be made by applying the fraction of embryonic among all lethal mutations in the random sample to a determination of the total number of genes in the genome capable of mutating to lethality. Our method will estimately follow this outline except that our random sample of lethal mutations will come from two small but representative regions of the genome that have been saturated with lethal mutations. The first requirement is, of course, that we have an estimate of the total number of genes in the *Drosophila* genome. Judd and his co-workers found in an intensive genetic analysis of a small segment of the X chromosome that there is a nearly 1:1 correspondence between the number of bands in the polytene chromosomes of *Drosophila* and genes as defined by genetic complementation. The total number of bands visible in the polytene chromosomes of *D. melanogaster* is 5,000, and a reasonable estimate of the total number of genes is also 5,000.

Needless to say, this number has been hotly debated by geneticists since, it was originally proposed in 1972, but subsequent studies have continued to substantiate the approximate correspondence between bands and genes. The proportion of these 5,000 genes capable of mutating to lethality comes from two studies; one by Shannon and her co-workers, and the other by Hochman, Shannon et al., analyzed the mutations in a small region of the X chromosome between the genes *zeste* and *white,* a region that, contains 13 bands and 13 identifiable genes.

Hochman's study entailed a similar analysis of the small fourth chromosome, which contains only 50 bands altogether, on which he was able to define 43 genes. Of the 13 defined genes on the region of X chromosome studied by Shannon and co-workers, only two failed to mutate to lethality, whereas six of the 43 genes identified by Hochman on the fourth chromosome yielded no lethal mutations and are thus presumably not vital to the fly. Extrapolating from this limited sample, approximately 85% of the 5,000 genes of *Drosophila* appear to be capable of producing lethal mutations. In developmental analyses of the lethals thus recovered and

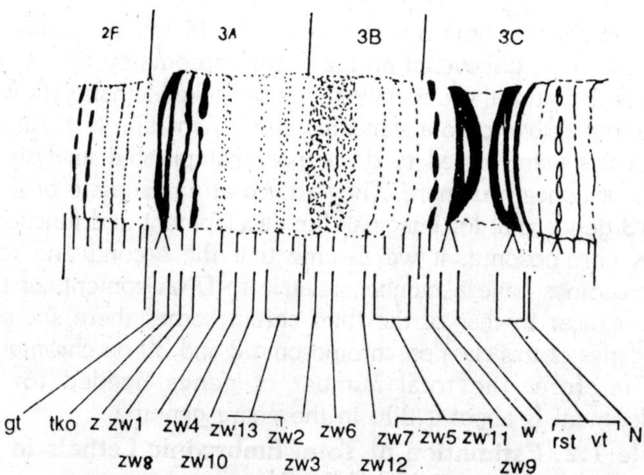

Fig. 1.1. Identification of genes in a small region, between the genes zeste and white, or the X chromosome of D. melanogaster. Thirteen genes, corresponding to the thirteen identificable bonds, have been identified.

mapped, Shannon *et al.*, and Hochman found that 1/10 and 5/37, respectively were embryonic lethals.

If the frequencies of embryonic lethals to all lethals are an accurate reflection of the total in the genome then, there are 425-550, or approximately 500 genes capable of yielding mutations that are not only lethal but are lethal specifically during embryonic development. Combining this estimate with that for maternal effect lethals, there are 617 genes necessary for early embryogenesis and for the formation of a functioning larva. An estimate of the number of genes necessary for the development and maintenance of the imaginal discs in the absence of any larval defects can be calculated from the results of a study by Shearn and Garen. These workers selected for lethal mutations that permitted ostensibly normal embryogenesis and larval development, but in which death occurred during metamorphosis. These mutations were found to cause imaginal discs either to be defective or lacking altogether. A total sample of 57 mutations of this type was isolated on the third chromosome. Complementation tests of the 57 mutations revealed 52 complementation groups or genes.

By the same procedure used in the calculation of maternal effect genes, n_o=384 for this class of genes on the third chromosome. The total theoretical number of imaginal disc function genes on chromosome 3 is therefore 436. Mutations of this class have also been discovered on the X chromosome by Kiss and co-workers. Because these mutations are sex-linked lethals with which it was impossible to do a standard complementation test, Kiss and co-workers were forced to do a more complicated analysis that need not concern us here. They arrived at an estimate of a total of 118 genes vital for imaginal disc development and function on the X chromosome. If we assume that the second and fourth chromosomes have a number, relative to DNA content, of these genes similar to that of the third chromosomes there should be 436 genes of this class on chromosome 2 and 20 on chromosome 4. Therefore, the total number of genes needed for disc development is about 1,000 in the entire genome.

Table 1.2. Estimation of Total Embryonic Lethals in the Drosophila Genome

Investigator	*Region*	*No. of Bands*	*No. of Genes*	*No. of Lethals*	*No. of Emb ronic Lethals*	*No. of Emb ryonic lethal in Genome*
Shannon et al.	zeste white	13	13	10	1	425
Hochman	Fourth Chromosome	50	43	37		550

If we now consider all of these developmentally important genes there are 1,617 genes necessary for normal embryogenesis and metamorphosis, or roughly 30% of the total genes, as defined by Mendelian genetics. What must be stressed is that certain types of genes are entirely omitted from the estimates. These are the genes that exist in multiple copy, such as those for histones, ribosomal RNAs (rRNAs) and transfer RNAs which are discussed later in this chapter.

The nature of the screens for the mutations we have been considering makes it unlikely that genes existing in more than one identical copy would be detected. Further, the estimates probably omit genes whose functions are strictly devoted to metabolic house-keeping through the life of the fly. Thus, for example, a mutation in cytochrome c would not show a temporal specificity of action like the mutations considered before. If these assessments are correct, then the genes comprising 30% of the *Drosophila* genome that have been determined from screens for developmentally important genes probably represent a reasonable estimate of the genetic cost of development from egg to fly.

Genetic Cost of Development at the Molecular Level

Two problems arise in applying estimates of the number of genes required for the development of *Drosophila* to other organisms. *Drosophila* appears, at least in some major respects, to have a genome organized rather differently from that of most other animals. It is entirely possible that genetic inferences drawn from *Drosophila* may be misleading when applied elsewhere. The second difficulty lies in the definition of a gene. The classic geneticist's definition is largely operational, with a gene being defined by its effect on viability or visible phenotype.

Unfortunately, genetic analysis has been made in only a most rudimentary form for most organisms studied by embryologists. For these animals an alternative molecular approach to estimates of numbers of active genes has been used. To molecular biologists an operational definition of a gene has generally meant a sequence of DNA that encodes the amino acid sequence of a protein. Although techniques exist for separating and estimating the number of relatively common protein species present in a cell, it is nearly impossible experimentally to estimate the number of species of rare proteins. Fortunately, it is feasible to determine the number of different mRNAs present and thus yield an estimate of the number of functioning genes.

The molecular biologist's definition of gene will be ultimately congruent with the geneticist's definition in many instances. However, it is not at all clear that all genes necessarily produce mRNAs, nor does a mRNA contain all of the information present in the gene from which it is transcribed. Embryos have two sources of mRNAs available: those stored in the egg during oogenesis, and those synthesized by embryonic cells during the course of development. The major functional role for the large and diverse store of mRNAs of the egg is to allow the newly fertilized egg to begin a massive programme of protein synthesis to provide the proteins needed for assembly of nuclei, membranes, and other sub-cellular structures needed during the period of rapid cleavage. Once cleavage has produced sufficient nuclei, the embryo itself can sustain a sufficient rate of mRNA synthesis to provide for the protein synthetic requirements of further development.

Each individual species of mRNA present in the embryo represents the coding sequence portion of a particular structural gene. Reference to Table 1.1 leaves little doubt that such messenger sequences provide reflection of only portion of the genome. It is also evident that many of the regulatory genes important in morphogenesis may not produce mRNAs at all. Thus a crucial class of genes that manifests itself to the Mendelian geneticist because mutations in it result in altered phenotypes or in developmental lethals of the sort tallied in the previous section of this chapter will simply be missed if we use mRNA diversity as an index of the number of genes required for developed.

It is not possible to gauge accurately the quantitative effect of this discrepancy, but fortunately there is not a total incongruity between regulatory genes and genes represented as mRNA copies. Genes that function as regulators of morphogenesis are, at least in some cause transcribed. The RNAs detected by Kalthoff in eggs of the insect *Smittia*, seem to represent transcripts of the class of gene because these RNAs provide the determinant for differentiation of the anterior end of the embryo. Further, many structural genes do have important roles in development and morphogenesis. The chorion proteins of the insect egg shell, which will be discussed in some detail in the next section of this chapter, serve as useful examples of this category of genes. Individual chorion genes are switched on and off in a concerted pattern as morphogenesis of the egg shell proceeds.

The functions of chorion proteins are of course as structural components, and not as regulatory molecules, yet failure of these genes to switch or presence of a defective gene can result in assembly of a modified or defective final structure. There is another source of ambiguity in using mRNA diversity to reveal gene numbers. Purely genetic methods allow the isolation of mutations that affect specific developmental processes or periods. These mutations reveal genes specific to ontogeny rather than genes basic to the metabolism or structural maintenance of cells at all stages of development.

Messenger RNAs of embryos contain both stage-specific and housekeeping sequences, and both groups of sequences contribute to the overall diversity of the mRNA population. A distinction between groups of sequences can only be made by experiments in which mRNAs from embryonic stages are compared to those from adult tissues to reveal the proportion of sequences in the embryo common to all stages. These can be assigned to the house-keeping category. Measurements of mRNA diversity are made by use of nucleic acid hybridization techniques. However, instead of annealing complementary strands of DNA of each other, RNA is annealed to the coding strand of genomic DNA. Therefore, the proportion of the DNA that hybridizes to mRNA from any stage can be used to calculate the number of genes represented. This kind of estimate has been made with varying degrees of completeness and success for several organisms. Values of mRNA diversity, and thus numbers of genes active in production of proteins in eggs and embryos of several protostomes and deuterostomes, are presented in Table 1.3. Estimates of mRNA diversity in this table are derived from either total cytoplasmic RNA or, preferably, from RNA associated with polysomes and thus presumably actually engaged in directing protein synthesis. Polysomal RNA is more likely to include only; *bona fide* mRNA than is total cytoplasmic RNA, but a note of caution, is necessary in considering even studies made with polysomal RNA.

Table 1.3. Diversity of Cytoplasmic RNAs Present in Eggs and Embryos

Organism	*Stage*	*Estimated No. of Different Genes rep.*	*Source of complexity Values*
Protostomes			
D. melanogaster	Egg	8,000	Hough-Evans et. Al., 1975
(fruit fly)	Larvae	3,100	Bishop et. Al. 1975
	Larvae	5,400	Levy and McCarthy 1975
	Larvae	14,500	Zimmerman et al., 1980
Musca domestica (house fly)	Egg	16,000	Hough-Evans et al., 1980
Urechis caupo (echiurid worm)	Egg	21,000-31,000	Davidson, 1976
Deuterostomes			
Xenopus laevis	egg	18,000-27,000	Davidon, 1976
(frog)	tadpole	20,000	Periman et. Al., 1977
Arbacia punctulata (sea urchin)	Egg	20,000	Davidson, 1976
Strongylocentrotus	Egg	24,000	Davidson, 1976
Purpuratus	16-cell	18,000	Hough-Evans et al., 1977
(sea urchin)	Blastula	15,000	Hough-Evans et al. 1977
	Gastrula	11,000	Galau et al., 1976
	Pluteus (larval stage)	10,000	Galau et al., 1976

Most of the RNA sequences revealed by hybridization studies are too rare to be identified by their presumptive protein products: They are mRNAs by inference only. The numbers of active genes estimates for *Drosophila* eggs and larvae based on nucleic acid hybridization data have generally been found to be high with respect to genetic estimates for genes specifically needed for larval development. This is not surprising, because house-keeping as well as development specific genes are expected to be expressed. On the average, published mRNA diversity values have been similar to genetic estimates of the total number of genes in *Drosophila*. But, one recent study by Zimmerman et al., using methods designed to detect all classes of mRNA, revealed that there are approximately 14,500 mRNA, sequences present in larvae. This is over twice the number predicted by the genetic analysis of Judd and his co-workers.

Resolution of the discrepancy is difficult without more precise data on the specific identities of the presumptive genes counted by these two very different approaches. The other organisms list in Table 1.3 generally exhibit diversities of genes expressed as mRNAs in development higher than those of *Drosophila*. It is not clear what this really means, but differences in diversity may be in an aspect of the C-value paradox (the discrepancy between morphological complexities and amounts of DNA in the genomes of organisms). The most detailed studies of mRNA diversities in development have been performed by E.H Davidson and his collaborators with the sea urchin *Strongylocentrotus purpuratus*. This group, in a paper by Galau et al., compared the diversity of genes expressed in the mRNAs of developmental stages and adult tissues. They also addressed a perhaps more significant question than mere gene numbers; that is, how many genes are stage-specific and how many are expressed in several stages and in adult cells? Their results, summarized in Table 1.3, reveal a curious fact.

The number of genes being expressed as proteins actually declines during sea urchin development, despite a pronounced increase in morphological complexity and differentiation between early cleavage and the larval pluteus stage. Thus, egg mRNAs share the sequences present in the gastrula and possess an equal number of sequences not present in gastrula at all. Blastula and pluteus stages share most of the gastrula sequences, but have other sequences as well. Three adult tissues were also examined, which is no mean feat because an adult sea urchin is essentially a limestone box filled with gonads and little else. Adult tissues exhibit a lower diversity of mRNAs, with values ranging from about 2,000 to 4,000 genes expressed as mRNAs in any tissue. Even though these tissues have a much lower mRNA diversity than gastrula, they nevertheless share about 1,500-2,000 sequences with gastrula. Interestingly, these appear to be the same subset of gastrula sequences in all three tissues.

Do these represent house-keeping genes whereas the rest of the diversity is required for embryonic development or maintenance of adult cell types? That this is not an improbable conclusion is suggested by measurements of mRNA diversity in organisms of extremely simple morphology, such as bacteria or fungi. In these forms it is reasonable to expect that very little gene activity is devoted to morphological organization, but rather to synthesis of proteins involved in metabolism, maintenance of cell structure, and cell replication. The bacterium *E. coli* has been found by Hahn and his colleagues to contain a mRNA diversity of 2,300 sequences, essentially the fully capacity of the genome, assuming that only one strand of DNA is transcribed.

Hereford and Rosbash similarly have found that yeast, one of the most simple of eucaryotes, expresses 3,000-4000 mRNA sequences. Another simple eucaryote, the fungus *Achlya ambisexualis*, studied by Timberlake and his collaborators, has a similar mRNA diversity of about 2,000-3,000 sequences. These are other eucaryotes of equally simple morphology that exhibit higher mRNA diversities and differences in gene expression in the small number of different cell types they possess. The fungus *Neurospora crassa*, studied by Dutta and Chaudhari, expresses about 10,000 genes in mycelial growth, but only about 5,000 in conidial cells.

Similarly, Firtel observed that of the approximately 16,000 genes expressed in the life cycle of the slime mold *Dictyostelium discoideum* about 11,000 were specific to stage of differentiation, and about 6,000 were expressed at all stages. These results contrast with the finding of Zantinge and co-workers that of the roughly 10,000 genes expressed in the fungus *Schizophyllum commune* over 90% are shared between morphologically different mycelial types. Taken together these observations of diversities of genes expressed in simple organisms are difficult to interpret.

Part of the high diversities observed probably results from the fact that these stuc'ies examined total cell RNA. That much of this RNA diversity is never translated into proteins is suggested by recent observations by Firtel and his collaborators on *Dictyostelium*. Only half of the RNA sequences measured by Firtel appear to serve as mRNA. If this finding also applies to the fungi *Neurospora* and *Schizophyllum*, it will serve to bring estimates of numbers of genes expressed closer to the mRNA diversities of yeast and *Achlya*.

The most meaningful approach; may be to regard the lowest gene numbers as representing the minimum numbers of genes required for house-keeping functions. In some forms, large changes in numbers of genes expressed accompany a rather simple morphological differentiation, but in others not greatly different in their morphogenetic abilities very small gene numbers seem to be sufficient. Again, in the absence of any solid data on the functions of the genes expressed, we suggest that the cases in which smaller numbers of genes are expressed may have more value in revealing the minimum number of genes actually involved in morphological differentiation. The comparatively low mRNA diversities of adult sea urchin tissues suggest that a rather small number of genes represented by proteins are required for maintenance of the differentiated state in animals.

This conclusion also follows from studies made with differentiated tissues of higher organisms, such as those or the mouse by Hastie and Bishop and on the chicken by Axel and co-workers. Mouse kidney, liver, and brain, like chicken oviduct and liver, were found to contain about 12,000 different mRNAs. In both organisms only about 10-15% of the mRNA species was unique to any tissue: The rest were shared. That only 1,000-2,000 genes expressed as proteins are sufficient to define individual tissues agrees with the result for sea urchins, but the question of why there should be such a large number of sequences, about 10,000 common to several very disparate tissues, remains open.

The situation may in fact be even more complicated than is suggested by these studies. Hahn and his collaborators have examined the diversity of mRNAs in mouse brain by a hybridization technique capable of detecting rarer mRNA sequences than those observed by Hastie and Bishop, and Van Ness et al., detected the presence of as many as 170,000 different mRNA species. It should be kept in mind that brain is an incredibly complex tissue composed of perhaps hundred of cell types and sub-types. Thus, high mRNA diversity may reflect a high degree of cellular diversity in this tissue. Large numbers of genes have also been found by Kamalay and Goldberg to be expressed as mRNAs in the tissues of a higher plant, tobacco. Roughly 25,000 sequences are present in leaf, stem, root, petal, anther and ovary. About two-thirds of the sequences are shared between tissues, leaving 6,000-10,000 sequences tissue-specific. But our concern here is with the genetic cost of ontogeny *per se*.

However, it may be risky to generalize from the only two organisms for which estimates have so far been made. In both the fly *Drosophila* and the sea urchin *Strongylocentrotus*, a relatively large proportion of the genes expressed at some time during the life cycle are expressed in a specific manner during ontogeny. The crucial question of how many of these genes control morphogenesis is simply unanswerable at present. The overall proportion of genes concerned with morphogenesis may be great, but paradoxically the number of genes that actually regulate morphogenesis may not be.

Many structural genes required for morphological entities could not be assembled. Yet these genes provide little in the way of regulatory information. They are instead regulated in their action. Genes of this type should not be thought trivial, however, because the products of some of them, as for example, tubulins, actins, or cell surface proteins, provide the actual machinery for cell shape-change and cell movements directly underlying morphogenesis. Much of the control exerted by regulatory genes, those genetic gray eminences, must be devoted to orchestrating the expression of ontogeny-specific structural genes. If regulatory genes were very large in number, interactions between them would be so complex as to render viable evolutionary changes nearly impossible. That large numbers are not involved, at least in several instances for which quantitative estimates of regulatory gene numbers are available comes from evidence in which it was shown that 10 or fewer genes determine head shape in two species of Hawaiian *Drosophila*.

Gene Switching and Multigene Families

Simply tallying the number of genes whose activity is required for normal development provides an estimate of the complexity of ontogeny. However, such an estimate will be a misleading one unless it is understood that structural genes specific to development are not expressed everywhere in the embryo at all times. Both spatial and temporal controls regulate gene expression. The patterns of localization are crucial to the establishment of the initial distribution of groups of determined cells within the embryo. Although such localization phenomena are a *sine qua* non for ontogeny, the process of development is fundamentally one of a series of cascades of events of ever-growing complexity. The initial localization patterns established during cleavage only serve to rough out a simple early embryonic morphology that generally changes dramatically with the morphogenetic events initiated by gastrulation and subsequent organogenesis. The cells whose fates are determined during cleavage, or later, are distinguished by their distinct locations within the embryo, and by their unique patterns of gene expression.

The diversity of mRNAs extracted from a whole gastrula or larva represents the sum of mRNA diversities in several cell types. Increasing complexity in ontogeny requires that the process producing differentiation and morphogenesis have two characteristics: Events must occur in correct temporal relationship to one another, and differentiating regions of embryos must interact with one another. Many structural genes expressed only at specific stages in development are subject not only to temporal control, but also have the interesting property of belonging to multigene families.

Naturally, not all genes switched on or off during development are members of such families; many must be genes represented only once in the genome. However, a surprisingly large number of multigene families exists, and their expression, in almost all cases, involves developmentally regulated gene switching. Multigene families provide the embryo with a means of meeting its changing needs for proteins with similar but not identical functions as development proceeds and metabolism as well as cellular and embryonic architecture change. Because many multigene families include genes whose products are quantitatively important and easily isolated for study, these families have already contributed a great deal to our understanding of gene switches. Our definition of a multigene family is slightly modified from that of L.E. Hood and his co-authors.

A multigene family is a group of genes that exhibits close sequence homology, and has related or overlapping phenotypic functions. The degree of multiplicity can vary from a few copies, as in the case of globins, to several hundred copies, as in the case of histones and structural RNAs. The multiplicity and other characteristics of several well-documented multigene families are presented. A large number of multigene families are known. Some of these, the highly repeated satellite sequences, are not transcribed at all. Others, including many moderately repeated families, are transcribed to yield RNA sequences that apparently are not translated to produce proteins. The function of these RNA sequences is still to be determined. However, many multigene families consist of *bona fide* structural genes of well defined function, which include the genes for the structural RNAs, ribosomal, 5s and tRNAs, and the genes for a wide spectrum of proteins.

Proteins encoded by multigene families include species important in cell motility and shape, such as actins and tubulins; structural proteins important to morphogenesis, such as collagens, keratins, and chorion proteins; some serum proteins; the oxygen carrier proteins, haemoglobins; some membrane proteins; the histones important in chromosome structure; the storage proteins of yolk; and antibodies. The arrangements of several of this multitude of multigene families on chromosomes are known, and there are several quite distinct multigene arrangements. The genes for several structural RNAs are arranged in the tandem repeat pattern, in which a series of identical genes for a particular product are linked together as gene—spacer—gene—spacer. The regions between the structural genes are sometimes transcribed, and sometimes not: They serve no known codogenic function, and are thus reasonably thought of as spacers between tandem structural genes.

Genes in a multigene family are not necessarily identical. For instance, the family of genes coding for the related but not identical β-globins expressed in an orderly sequence during the development of some mammals are linked in the second pattern. A cluster of related genes can itself be the basic unit of a set of tandem repeats. The genes for the histones of sea urchins and other higher organis form a cluster of genes organized. The best-known cluster is that which accounts for the bulk of histone synthesis in the sea urchin embryo. This cluster has the structural genes for the five individual histone species arranged in the order.

—(spacer—H2A—spacer—H3—spacer—H2B—spacer—H4—spacer—H1—spacer)—

The major histone types are very distantly related to one another, and each is in fact composed of several subtypes, making five histone multigene families. Thus, there are at least four or five distinct genes each for histones belonging to the H1, H2B, H3 and H4 families, and seven or more members of the H2A family. Some members of the major histone families are organized in clusters as above, but it is clear that not all histone genes are organized in this way. Multigene families have their evolutionary origins in duplications or high-order replications of single-copy genes. The initial duplication or multiplicative replication yields tandem genes.

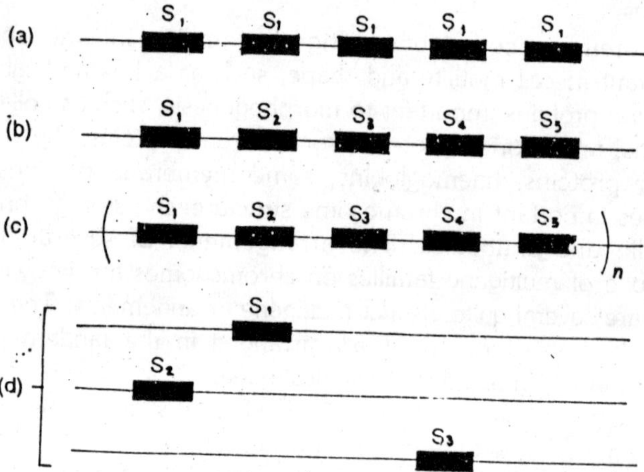

Fig. 1.2. Patterns of organization of multigene families. (a) Identical genes linked in tandem, as seen in ribosomal RNA genes; (b) related, but non-identical genes linked in tandem, as seen in globin genes; (c) a gene cluster of non-identical genes, as seen in sea urchin embryonic histone genes in which the clusters are tandemly linked; (d) related genes dispersed over several chromosomes, as seen for actins or tubulins.

In some cases a large number of identical tandem genes is retained. The correction mechanism that keeps identical such genes as the 18 and 28 S ribosomal RNA genes probably depends on the maintenance of a tandem arrangement. In some of the smaller tandem families duplication has been followed by gene divergence to produce related but non-identical genes. In the case of the β-globin genes these have been maintained over considerable evolutionary time. In other cases, related genes, for example, those for tubulins or actins, are organized. These genes are scattered over one or several chromosomes. The diversity of multigene families thus far recognized suggests that membership in a multigene family has little to do with the function of the gene product *per se*, only with the control of the expression of the gene.

There appear to be two primary reasons for the existence and function of these families in development. The first, and less intrinsically interesting reason, is that some gene products are needed in a short time and in enormous quantities. In these cases, the multigene family consists of a larger number of identical gene copies, usually linked in tandem. The genes for ribosomal RNAs are organized in this manner, and are transcribed to produce the huge amount of ribosomal RNA needed for assembly of the ribosomes used by cell in protein synthesis. A good, if extreme, example of the demand for the function of large numbers of ribosomal RNA genes is provided by the work of D.D. Brown and I.B. Dawid on the ribosomal genes of the oocyte of the frog *Xenopus laevis*.

The embryo makes no new ribosomes until after gastrulation, and so depends on the store is considerable: The *Xenopus* egg contains about 10^{12} ribosomes. Somatic cells of *Xenopus* contain 450 copies of the ribosomal RNA genes per haploid DNA complement. This is sufficient to provide for the needs of the relatively small somatic cells, but even this multiplicity of genes is inadequate to produce the ribosomal RNA needed by the egg. Brown and Dawid found that these genes were amplified a further 4,000 times in oocytes. A somewhat different strategy is used in oogenesis to provide the 5S RNA also required for ribosome assembly. There are 24,000 copies of the major oocyte specific 5S RNA gene per haploid DNA complement in *Xenopus laevis*. A different, and smaller, 5S family is transcribed in somatic cells. The second function of multigene families is to provide gene switching in development.

Essentially, multigene families containing related but not identical genes produce similar products specifically required in distinct cell types or at different times in development. The best understood example is provided by the small multigene families containing the genes for the globins. The evolutionary relationships of human globins, based on protein sequences. The ancestral haemoglobin diverged from myoglobin near the time of the origin of chordates late in the Precambrian. The ancestral β-globin in turn diverged from the ancestral α-globin about 500×10^6 years ago, early in the history of vertebrates in the early Paleozoic, and the foetal γ-chain diverged from the β-chain at most 200×10^6 years ago, early in the history of mammals.

Finally, the δ-chain, a variant of β-globin found as a minor component of normal adult haemoglobins, diverged from the β-chain about 40×10^6 years ago. The genes for α-chains comprise a small multigene family of three members, and the genes for β-chains a family of seven members. The functional haemoglobin molecule is a tetramer composed of two subunits of the α-type and two of the β-type, thus α_2, β_2. In humans there are seven clustered genes of the β-globin type, arranged. Two of these code for the A_γ and G_γ chains found in the foetal haemoglobin $\alpha_2\,\gamma_2$: Two genes, δ and β code for chains expressed after birth in major ($\alpha_2\beta_2$) and minor ($\alpha_2\delta_2$) haemoglobin species.

Humans also possess another β-type globin, ε and an α-type globin, δ which are expressed only in the embryonic haemoglobin $\zeta_2\varepsilon_2$. Two other members of the β family, $\psi\beta_1$ and $\psi\beta_2^*$, are unexpressed pseudogenes. The time course of haemoglobin switching in human development reveals changes in which haemoglobin genes are expressed, and in site of expression. In the early embryonic portion of development, ζ-and ε-chains are synthesized by nucleated megablast cells produced in the yolk sac. This synthesis falls rapidly, and by the sixth week of development is replaced by a pattern of synthesis in which non-nucleated red cells derived from stem cells in liver and spleen produce the α and γ globin chains characteristic of the foetus.

During late foetal development bone marrow becomes the preponderant site of globin synthesis. Shortly following birth there is second switch in globin synthesis, and the adult pattern is assumed. The switch exhibits a very significant characteristic. The transition is one involving gene regulation within individual stem cells rather than a replacement of γ-producing stem cells by β-producing stem cells, because during the switch single red cells produce both γ and β-globin chains. While haemoglobin gene switching is the rule in vertebrates, there is a surprising diversity of switch patterns, even within the mammals.

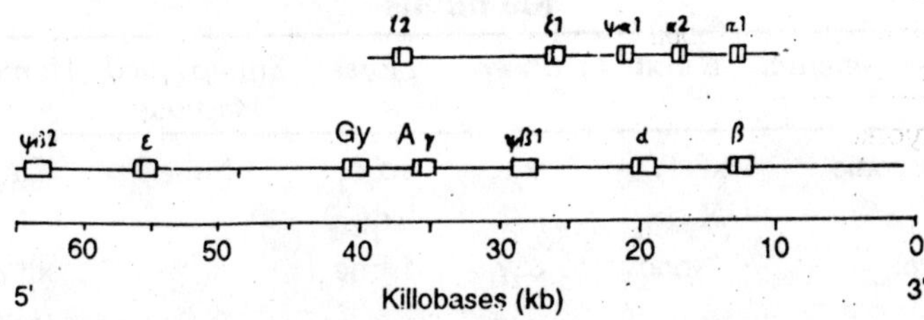

Fig. 1.3. Arrangement of human α-like and β-like globin genes. The scale indicates lengths of chromosomal DNA in kilobase units. Globin genes are shown as boxes. Black segments indicate coding sequences, whereas white segments indicate introns. Pseudogenes (ψ) are indicated as entirely open boxes. The 5′ to 3′ direction of transcription is from left to right.

Humans express distinct embryonic, foetal, and adult globins. Foetal haemoglobins have been shown to have a higher oxygen affinity than the adult haemoglobins of the mother, thus facilitating oxygen transfer across the placenta to the foetus. However, not all mammals possess distinct foetal haemoglobins. Rodents, carnivores, and horses, for example, pass through a direct transition from embryonic to adult globins in the fetus. In these cases, the affinity for oxygen of the adult haemoglobin in the foetal red cells appears to be modulated by small molecules in the cytoplasm so that the foetal blood has a higher oxygen affinity than the maternal blood. Organization of the β-globin genes may be very similar, even in organisms exhibiting significant differences in switching patterns.

Table shows that rabbits produce the embryonic β-like globins called ε(Y) and ε(Z), but have no foetal β-globin equivalent to human γ-chains. Instead, the fetus synthesizes the adult β-chain. Lacy and her collaborators and Hardison *et al.*, have found that the β-globin genes of the rabbit are organized in a cluster very similar to that of humans, and that the individual genes in the cluster undergo developmental switches. Two of the genes corresponding in position to the γ-genes in the human β-gene cluster are expressed in the embryo, presumably to produce the ε-chains. The gene corresponding in position to the human δ-gene appears not to be expressed at any stage, and the gene corresponding to the human adult β gene serves the same function in the rabbit.

Table 1.4. Haemoglobin Gene Switches in Some Mammals

Developmental Stage	*Rabbit*	*Sheep*	*Horse*	*StumpTailed Macaque*	*Human*
Embryonic	$x2\varepsilon(Y)_2$	$\alpha_2\varepsilon_2$	${}^{s}\alpha_2\varepsilon_2$	None	$\zeta_2\varepsilon_2$
	$X_2\varepsilon(z)_2$		${}^{F}\alpha_2\varepsilon\ _2$		
Foetal	None	$\alpha_2\gamma_2$	None	$x_2{}^{1}\gamma_2$	$x_2{}^{A}\gamma^2$
				$\alpha_2{}^{2}\gamma_2$	$\alpha_2{}^{c}\gamma_2$
Adult	$\alpha_2\beta_2$	$\alpha_2\beta_2{}^{s}$	${}^{s}\alpha_2\beta_2$	${}^{1}\alpha_2\beta_2$	$\alpha_2\beta_2$
			${}^{F}\alpha_2\beta_2{}^{2}$	$\alpha_2\beta_2$	$\alpha_2\delta_2$

While the globins illustrate the role of switching in the expression of a succession of genes directly involved in metabolic functions, there are more complex multigene systems whose products are directly involved in gene expression or morphogenesis. Some of these families include a large number of members under developmental regulation. The histone genes, best studied in sea urchins, nicely illustrate the adaptation of using multigene families to satisfy both the necessity of producing a large amount of protein in early development and the need for switching to express a sequence of related proteins in development. During cleavage the embryo doubles its nuclei and the chromosomes they contain as rapidly as every 10-20 minutes.

Sufficient histones to support the assembly of chromosomes during rapid cleavage can come either from stores of proteins within the egg, as Woodland and Adamson have shown for the frog *Xenopus*, or from massive synthesis of histones in the cleavage-stage embryo, as has been found by Kedes and his collaborators to be the case in sea urchins. These two strategies make different quantitative demands on histone genes. The histones of the frog egg are accumulated slowly during weeks or months of oogenesis, whereas those of the sea urchin embryo are synthesized over a period of a few hours. The sea urchin egg contains only about 25% of the histone mRNA it will need to produce histones during cleavage; the remainder is transcribed during cleavage.

As a consequence, the main histone gene family of sea urchins is much more highly repeated than that of frogs (or for that matter humans). Whereas sea urchins possess 300-1,200 copies of these genes, Birnstiel et al. and M.C. Wilson and co-workers have found *Xenopus* and human histone genes to be repeated only 10-20 times. The histones of sea urchins have been found to be the subjects of a very elaborate set of switches, including both temporal and tissue-specific changes. Newrock and his collaborators have documented in intricate detail a sequence of changes, first observed by Ruderman and Gross, in which of the five major histones, three (H1, H2A, and H2B) are represented early in cleavage by a short-lived synthesis of cleavage-stage-specific sub-types. This synthesis is followed by synthesis of a sequence of other histone subtypes in a stage-specific manner.

The cleavage-stage subtypes, *cs,* of each histone type are succeeded by α-subtypes during cleavage, then by β- γ, and other subtypes in the blastula. Experiments by Newrock and co-workers, Kunkel and Winberg, and Childs *et al.*, all show that the histone protein switches result from a gradual succession of changes in mRNA synthesis. Each subtype is the product of a distinct gene related to other subtype genes in a family. The switching of histone subtypes results in a change in chromosome protein composition as development proceeds. Such changes may result in a "remodeling" of chromatin, potentially of importance to differentiation of cells within the embryo. Switches are not limited to proteins functioning, like globins or histones, primarily in the internal economy of cells. Some are intimately involved in morphogenesis. For example, the microtubules so central to cell movement and cell shape are composed of α-and β-tubulins, which are the products of small multigene families.

The tubulins have been found by E.C. Raff and her co-workers to be regulated by switches during the development of *Drosophila*. Some tubulin species are synthesized throughout development, but at least one β-tubulin is switched on and then off during a restricted period of embryogenesis, and Kemphues *et al.*, have demonstrated that there is a tissue-specific β-tubulin expressed only in the testis. The testis-specific β-tubulin is required for the assembly of a very specialized microtubule structure, the axoneme of the sperm tail. The study of the role of switches of multigenes in morphogenesis has been developed by F.C. Kafatos and his collaborators in their studies of the chorion proteins that make up the shell of the silkmoth egg. Seemingly a humble object, the egg shell can be seen with use of scanning and transmission electron microscopy to have an elegant structure and surface geometry, which would certainly have pleased D'Arcy Thompson.

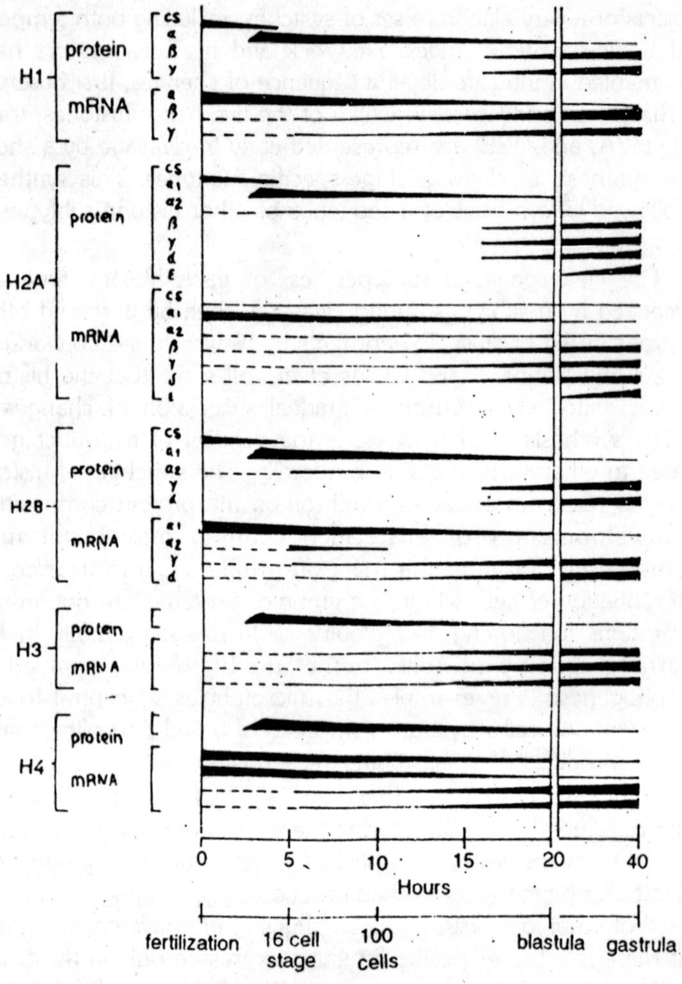

Fig. 1.4. Expression of members of histone families during the development of sea urchin embryos. Thickness of bars is a schematic indication of level of synthesis of protein or amount of mRNA present. Note that translational regulation as well as mRNA synthesis control histone protein synthesis because mRNAs stored in egg are present in early development but not immediately translated.

The predominant surface features are the hexagonal pavement, which marks the former sites of follicle cells, and the tall, chimneylike respiratory structures evocatively called *aeropyles*. The pit in the foreground is the *micropyle*, through which the sperm enters at fertilization. In cross-section the shell can be seen to be mechanically strong but light, with internal vaulting. The layers of the shell are secreted by a sheath of follicle cells that synthesize the chorion proteins, and there are close to 200 different chorion protein species synthesized by the silkmoth. These fall into five broad classes defined by molecular weight and amino acid sequence relationships. Each class comprises a family of related genes, but as C.W. Jones et al. have shown there is also a considerable degree of relatedness between restricted regions, or domains, of proteins belonging to different families within the chorion superfamily.

The relatedness of the gene for these proteins is further indicated by the finding of Marian Goldsmith and her collaborators that in the silkmoth *Bombyx mori* a large number of the genes

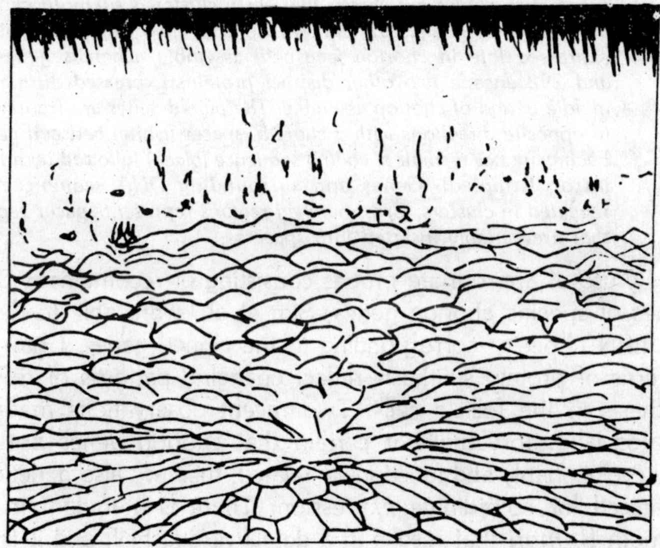

Fig. 1.5. Surface structure of the eggshell of the egg of the silkmoth Antheraea polyphemus. This view shows the micropyle (site of sperm entry) surrounded by a pavement of concentric cell imprints in the foreground and the aeropyles (respiratory structures) standing in the background.

coding for members of these families are organized into three clusters of genes on a single chromosome. Expression of chorion genes is controlled by a series of switches. Sim et.al., have found that synthesis of members of the three higher-molecular-weight families predominates early in choriogenesis, with synthesis of members of the two lower-molecular-weight families predominating throughout middle and late choriogenesis.

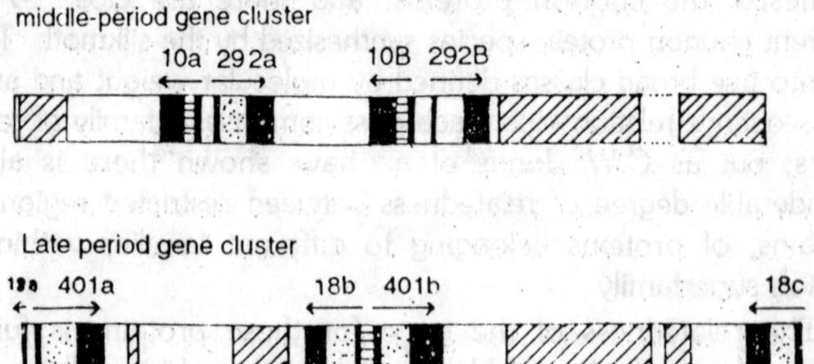

Fig. 1.6. Arrangement of two sets of co-ordinately regulated chorion protein genes. Two clones are shown that carry clustered silkmoth chorion protein genes. Genes 18 and 401 encode two different proteins expressed late in chorion (eggshell) assembly, whereas genes 10 and 292 encode two other distinct proteins expressed during the middle period of chorion assembly. The paired genes are transcribed in opposite directions with a short 5′ spacer (white) between genes. Each gene has a small 5′-coding sequence (black) followed by a large intron (stippled). Genes and surrounding DNA sequences are repeated in clusters. Cross-hatched regions represent spacer regions that are variably inserted into spacers.

By use of appropriate probes consisting of recombinant DNA clones of specific chorion genes, Sim et al., were able to detect the mRNA species corresponding to the cloned genes. Changing patterns of protein synthesis reflect changing patterns of mRNA synthesis in the follicle cells. Two recent observations made in Kafatos laboratory make it certain that chorion genes are not only evolutionarily and functionally related; they are also genetically organized for co-ordinate expression. There is a mutant in the silkmoth *B. mori* that results in a defective eggshell, and it is the consequence of a deletion of section of DNA containing about half of the chorion genes.

The deleted genes are primarily late-expression genes, suggesting a clustering of genes by time of expression. Use of recombinant DNA clones of fragments of DNA containing more than a single specific chorion gene has allowed Jones and Kafatos to examine gene organization from the point of view of temporal control of expression. The two clones contain different genes, but both contain two copies of each of two distinct genes. The genes belong to different chorion gene subfamilies; but they are co-ordinately expressed. Those in the upper fragment are utilized midway through choriogenesis, whereas those in the lower fragment are expressed late in the process.

In both cases physically contiguous pairs of genes appear to be linked in transcription, with a common control element lying between each pair. Eggshell structure has undergone a variety of evolutionary modifications in related species of silkmoths. This can be seen in the eggshells of *A. polyphemus* and *A. pernyi*. These species may have diverged as long ago as 10-30 × 10^6 years, although the not surprising dearth of a fossil record for these big moths makes this a somewhat uncertain estimate. Has the change in egg shell morphology resulted from changes in the structural genes for chorion proteins, or from changes in control of expression? Recent experiments performed in Kafatos laboratory indicates that changes in chorion structural genes have occurred, but have been limited in extent. Morphological differences primarily reflect differences in gene expression.

In light of the importance that changes in relative timing of processes have in modifying development during the course of evolution, it is interesting that the relative timing of expression of chorion genes is the same in both species of *Antheraea*. Differences in chorion gene expression are instead quantitative, with some chorion mRNA species differing greatly in amount. There are mutations in chorion protein genes that begin to reveal the function of these switched genes in morphogenesis. *D. melanogaster* chorion proteins comprise a much smaller family than that of silkmoths, perhaps 20 genes. A mutation in one of the chorion protein genes in this species recently discussed by Digan and her collaborators produces a morphologically deranged eggshell, indicating that at least one of the chorion proteins has a role in the organization of eggshell structure. Chorion genes bring us almost full circle to the question of gene numbers considered in the previous sections of this chapter.

The numerous genes involved in assembly of this one structure and their intricate switching suggests that there may be a demand for the expression of large numbers of related genes in many of the morphogenetic processes of development. This possibility is certainly borne out by the histones, tubulins, actins, and other proteins synthesized by embryos. As these proteins, once thought to be the products of one or at most a very few genes, have been studied in more detail it has become obvious that their expression in development in reality involves whole families of structurally related and functionally co-ordinated genes. The expression of large numbers of structural genes is necessary for development. Many developmentally regulated structural genes are members of evolutionary-related multigene families derived from an ancestral gene by duplication and divergence.

Multiple, related genes allow fine-turning of structural gene expression, with individual members expressed during precisely controlled periods or groups or cells in development. All of the β-globins, for example, serve the same general function, but in somewhat different ways and with different efficiencies with respect to the foetal versus the adult environment. The possession of multigene families provides organisms with evolutionary flexibility in that changes in time or location of expression of a member of a multigene family will not affect expression of other members of the family. Dissociation of development processes may thus be facilitated.

2

GENES AND EMBRYOGENESIS

To understand the relation between an organism's Genotype. Its genetic constitution, and its phenotype, the result of the activities of its genes, it one of the chief aims of developmental biology. The phenotype may be considered at any level, from the cellular and subcellular up to studies on organs or whole organisms; in any case it is not a static thing but something which changes in the course of development as the cells move down the pathways of the epigenetic landscape—or sometimes, as we have seen, back up them and into different ones.

Development and Evolution

Ultimately the evolution of organisms must be considered in these terms, since natural selection acts upon the phenotype but produces its results through changes in the genotype, and any evolutionary change must be such as the developmental mechanisms will generate through modifications of existing developmental processes. Some modifications will be much more likely to occur than others, and some will set up systems where other modifications are an almost inevitable consequence, just as one human invention suggests another leading to an appearance of directed evolution along particular pathways. The changes of body form produced as a result of differential growth and analyzed by D' Arcy Thompson in 1917 in his classic study *On growth and Form* by the method of Cartesian transformations, are of this type.

The form of the structure is defined by superimposing a grid and deforming the co-ordinates of this grid (how this is done in

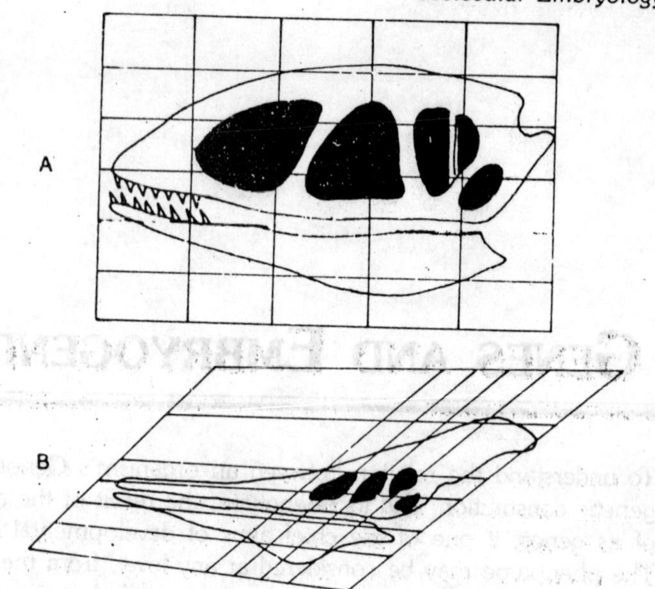

Fig. 2.1. Cartesian transformation showing how the skull of Pteranodon (B) may be related to that of Dimorphodon (A).

systems consisting of many individual structures is quite unknown and is one of the major problems of growth control): it produces a series of topologically identical but otherwise different forms, the extremes of which e.g., the skulls of the primitive reptile *Dimorphodon* and the extinct flying reptile *Pteranodon* appears to be totally distinct. One especially interesting aspect of development biology in this evolutionary context is the problem of accounting for adaptive features of the type which are produced generally as a reaction to an environmental stimulus, but which in particular cases appear as inherited characteristics. For example, as we saw in chapter, calluses—extreme thickenings of the stratum corneum—appear whenever the skin is subjected to rubbing; but on the sole of the foot and—more unexpectedly—in patches on the breast and rump of ostriches in areas which press against the ground when the birds are squatting, these calluses appear during embryonic development. Lamarckian explanations are nowadays discounted, yet random mutations for development of calluses in exactly the right place are almost as hard to accept.

In 1957 Waddington proposed an explantation in terms of genetic assimilation, based upon his epigenetic landscape model and the canalization of development which it implied, i.e., the adjustment of the genotype to produce one definite result, regardless of minor genetic and environmental variations. His suggestion was that in the ostrich the response to the environmental stimulus in the ancestral ostrich was facilitated and made more uniform by selection, establishing a branch pathway which still required an environmental stimulus to shunt development into it: but the pathway being established, a wide variety of minor environmental or genetic stimuli would do it, so that it is very probable that a suitable random mutation would arise and be selected, so that the whole process would come under genetic control.

Waddington showed that genetic assimilation of a comparable type could be demonstrated in the laboratory, using a stock of the fruit fly *Drosophila* in which a small proportion of the population responded to heat shock by forming a wing lacking the usual crossvein. Selection for response to the heat shock produced a stock in which some flies showed the crossveinless condition without being exposed to the stimulus and further selection from these individuals led to a stock in which crossveinlessness occurred spontaneously in most of the flies.

Genes in Individual Development : Pleiotrophy

In order to discover how genes act, it is necessary to substitute mutant genes for their normal wild-type alleles and observe what effect this substitution has upon development: if embryos consisted of a mosaic of autonomous parts and processes, and if each gene affected only one part or process, it would then be easy to define the activity of the gene by simple subtraction. But we have seen that this is not the case; the embryo is a highly complex interacting system, and a genetic alteration in one part of the system is almost certain to have repercussions in other parts. A spurious suggestion of simplicity often arises from the names which mutants are given, referring to the character or structure which is most obviously identifiable anatomically in the fully developed embryo or adult. For example, a number of genes have been given the name "wingless" in the fowl because, indisputably, the birds have stumps or no wings at all, but the gene *wg* leads as well to absence of the lungs and air sacs, and also the metanephric kidney, suggesting that some widespread cell function is affected which is manifested only in regions of some specific sensitivity; it is not a gene concerned specifically with controlling the development of the wing.

No simple "one gene-one character" relationship exists; generally a mutant gene will produce a multiplicity of different effects and this phenomenon is known as *Pleiotropy*. Therefore, the first stage in analysis of gene action is to establish what Gruneberg, who surveyed a long-series of studies on mouse mutants in 1963; has called a *pedigree of causes*, tracing each terminal manifestation back through its embryological development to its origin in some primary effect, form which all the abnormalities will also have arisen and diverged. Frequently this includes the analysis of inductive interactions which having failed to occur, or occurring in abnormal locations through displacement of the interacting tissues, lead to a cascade of further developmental effects. One of the mutants in Gruneberg's survey illustrates how difficult such an analysis may be: mice homozygous for the gene *congenital hydroephalus* (*ch/ch*) show widespread skeletal abnormalities and also bulging cerebral hemispheres.

The primary effect of the gene appeared to be on the formation of cartilage condensations in the mesenchyme, producing the skeletal defects, including deformities of the skull which led mechanically to abnormalities of the brain which in turn interfered with the normal distribution of cerebrospinal fluid. But in 1970 Green showed that there were a number of other effects, especially, urinogenital defects arising from an enormous excess of mesonephric (embryonic kidney) tubules which filled up the whole region between the normal mesonephros and the kidney, and there was also a striking deficiency of cells in the neighbouring coeliac nerve ganglion. These two are likely to be related, since neural-crest cells, from which the ganglion cells are derived, inhibit the development of mesonephric cells in normal mice.

Green suggests, a tentative unitary hypothesis which combines all of these aspects, that the *ch* gene may have an effect upon the mesenchyme which inhibits the movement of neural-crest cells through it and also interferes with cell interactions involved in the mesenchymal condensation process. Hypotheses of this sort have only a limited value until they are supported by experimental evidence of the sort which Mayer and Green in 1968, and Mayer in 1973, produced in analyses of the mutants *steel (st)* and dominant white spotting (*W*) in the mouse.

The pleiotropic effects of the *steel* gene are anaemia, sterility and white coat colour caused by the absence from the skin of melanocytes derived from the neural crest; the retinal pigment cells are pigmented so that these white mice do not have pink eyes. In all these cases the effect is not directly upon the cells which fail to develop, but upon the tissues in which their terminal differentiation normally takes place; primordial germ cells are present but very few arrive at or survive in the gonads; the precursor cells of the erythropoietic systems are present and can be used to repopulate the bone marrow of irradiated normal hosts; the neural crest produces melanoblasts which differentiate into melanocytes when combined with normal skin, but normal melanoblasts *st/st* skin produce no melanin, either in the epidermis or dermis, even when one or the other component is from a normal mouse.

In the *dominant white spotting* mutant, which is also sterile and white, the reverse in the case; the effect is directly upon the cell, and its skin supports normal differentiation of melanoblasts from *wildtype* or from *steel* mice. Whether the inhibiting effect of the *steel* tissues is upon the migration of the cells through them, or upon their survival or their differentiation when they arrive at their destinations, is at present unknown.

Mutations and Changes in Ontogeny

If morphology is the manifestation of a complex set of developmental processes so too are these processes the manifestation of the actions of a constellation of genes. This has been the essential assumption of this book and we have assembled at least a *Prime facie* demonstration that a portion of the metazoan genome is specifically involved in control of ontogeny and evolves by a mode distinct from that of structural genes. Thus far, however, we have only dealt with isolated specific examples of gene control of morphogenesis and have not attempted to answer the central question of how genes control the course of development. In a very real sense we return to Roux's programe for developmental mechanics, but instead of deleting cells or other embryonic structures to determine their roles in development, as was done by the classical experimental embryologists, the developmental geneticists exploits mutation as a delicate and precise scalpel to delete or modify individual genes. The genetic paradigm for studying any system theoretically under genetic control is the following.

In order to analyze a process, in this case ontogeny one recovers mutations that alter that process. Once recovered, individuals carrying the mutation are compared in phenotype with normal individuals. This comparison should give a clue as to how the gene in question affects normal development. However, before continuing with the manner in which the comparison is made, it should be noted that there are two basic ways in which mutations can be seen to affect ontogeny. First are disruptive changes, in which the path of normal development is altered to produce morphological abnormalities (e.g., missing structures). In their most cataclysmic form these mutations cause lethality. Second, there are homoeotic changes. This type of mutation causes an alteration in development such that one structure of the organisms is replaced by a homologous organ or limb. We will forego any further discussion of this latter type of mutation until the following chapter and concern ourselves here mainly with disruptive changes. In the analysis of the defects produced by a disruptive mutation it is rarely a simple matter of observing the terminal phenotype of a lethal individual in comparison to the normal, because development is a complex and highly integrated process.

The vast majority of events takes place in relation to other events, and indeed, are dependent on one another. This is most clearly seen in the fact that many mutant conditions have pleiotropic effects, in which several morphological alterations are caused by a single genetic defect or change. An example of this is the defect observed in individuals suffering from what is called *Pelger (Pg)* anomaly. This trait is inherited as a simple autosomat dominant in humans. *Pg/+* heterozygous individuals are clinically normal but show abnormal nuclear segmentation in one of their white blood cell types. The polymorphonuclear neutrophil leukocytes at maturity normally have four to five lobes on their nuclei. In Pg/+ individuals there are usually only two, and rarely, three lobes. This same trait also exists in the rabbit and shows the same pattern of inheritance and blood picture in *heterozygotes*.

By crossing rabbit heterozygotes it is possible to obtain homozygous *Pg Pg* individuals. The leukocytes of these individuals have no lobes at all and the genotype has low viability. In addition to the leukocytes phenotype the rare survivors demonstrate an extreme dwarfing phenotype in which the limbs and rib cage do not develop fully. The pertinent question to ask at this point is: What is the causative relationship, if any between these two pleiotropic defects? It is conceivable that both phenotypes are

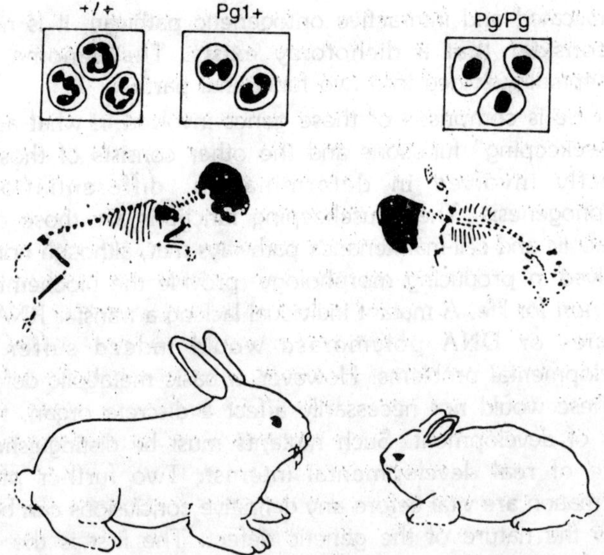

Fig. 2.2. The Pelger anomaly in the rabbit. The polymorphonuclear neutrophil leukocytes of normal (+/+), heterozygous (Pg+) and homozygous (Pg/Pg) rabbits are shown at the top of the figure. Below are the skeleton and adult morphologies of Pg/+ (left) and Pg/Pg (right) rabbits. Note the severe stunting of the limbs in the homozygous individual.

actually the result of a third as yet unknown lesion caused by the *Pg* allele. A wide range of pleiotropic effects can also be seen in another inherited defect of the blood. This is sickle-cell anemia.

It too is inherited as a simple autosomal dominant in humans. The difference between this trait and *Pelger* is that we know the precise biochemical lesion that produces the defect. Individuals suffering from this disease have a single amino acid substitution at position 6 of their β-haemoglobin chains. This single change alters the conformational properties of the resultant haemoglobin tetramer under conditions of low oxygen tension. Red blood cells containing this type of mutant haemoglobin deform to a characteristic "sickle" shape in the peripheral and venous portions of the circulatory system. This shape-change has two direct consequences. The body recognizes the abnormal shape and destroys the sickled red blood cells resulting in anemia. Also, the sickled cells clump and tend to block capillaries, and thereby local blood supply, which is of course necessary for normal growth and function of the organs. When one is presented with a complex phenotype at the end of a complicated and interactive ontogenetic pathway, it is necessary to consider that a dichotomy exists. The genome can be conceptually divided into two functional parts.

One is comprised of those genes involved in what are called "housekeeping" functions and the other consists of those genes directly involved in determination, differentiation, and morphogenesis. The housekeeping functions are those common metabolic and cell-maintenance pathways that, although not directly involved in producing morphology, provide the biochemical *sine qua non* for life. A mutant individual lacking a transfer RNA (tRNA) species or DNA polymerase would indeed suffer severe developmental problems. However, a basic metabolic defect such as these would not necessarily affect a discrete organ, tissue or time of development. Such mutants must be distinguished from those of real developmental interest. Two further pieces of information are vital before any definitive conclusions can be drawn as to the nature of the genetic defect. The first is the primary site of action of the gene. That is, there a specific tissue or organ in which the gene is active? Moreover is the gene autonomous in its action? The question relates to the fact that there are actually two types of pleiotropy.

In the first, relational pleiotropy, as seen in the example of sickle-cell anemia, there is one primary site of action for the gene (i.e., red blood cells) and all of the other observed defects are related to, or occur; because of that single defect. The second is direct pleiotropy. In this case all of the diverse defects in different tissues and or organs are caused by the direct action of a single gene. Grunberg has argued based on his studies of mutations in the mouse, that relational pleiotropy is predominant. However, there are cases of direct pleiotropic effects. The second piece of information necessary to any meaningful discussion of genetic control of development is the time at which the gene is active. What is the time of onset; is the time of action continuous; does it occur in a single discrete interval; or is gene action required at several discrete intervals. This complex question relates of course to analysis of the nature of the product of the gene being studied.

Analysis of Time and Place of Gene Action

The techniques used to determine the primary site of gene action are similar to and indeed have been borrowed from classical embryology. In their simplest form, an organ or piece of tissue is transplanted from a mutant individual to a normal recipient. The reciprocal graft is also performed. This operation is usually accomplished before any mutant defect is apparent. One can then determine the fate of the developing organ or tissue in its new environment.

If the genetic defect of the organ or tissue in question is autonomous to this structure, that is, it is the primary site of action of the gene, it would be expected that the mutant tissue would produce the abnormal phenotype, even in a normal host. Similar kinds of experiments can be performed in tissue or organ cultures in a manner similar to that described in mouse, lizard and chick dermis. Instead of cross-species associations mutant and non-mutant tissues or organs are mixed. The creation of mosaic individuals has been accomplished on an even grander scale by B. Mintz and her collaborators. These workers have been able to fuse entire mouse morula stage embryos *in vitro*. These "hybrid" embryos are then implanted into pseudopregnant females. These fused or tetraparental mice are-comprised of a mixture of cells of two different genotypes, both of which are active. This technique can also be used to analyze autonomy of mutant expression by fusing mutant and normal morulae. One further technique of this gene should be mentioned; parabiosis. This is the fusion of entire animals rather than just organs or tissues. However, it should be noted that the result in this case is not a true integrated mosaic.

All of these techniques require that transplanted tissues, organs or fused embryos are compatible. In lower vertebrates, such as amphibians, this causes little problem; however, mammals have the added complication of transplant rejection and care must be taken to ensure that the mutant and normal individuals are immunologically compatible. The remaining two techniques are peculiar to developmental genetics and are used almost exclusively in *Drosophila melanogaster*. These are the production of gynandromorphs and the induction of mosaic individuals by mitotic recombination. Gynandromorphs are adult flies whose bodies are made up of both male and female tissue.

By taking advantage of the type of cleavage in the diptera and a special ring-X chromosome, this type of mosaic individual can be produced regularly in laboratory cultures. The normal *Drosophila* X chromosome is a rod with the centromere at one end. A mutant form of this chromosome exists in the form of a closed ring. This ring has the interesting property that it is unstable in the first few cleavage divisions of the embryo. Specifically, this instability can result in the loss of the ring chromosome from one of the two daughter nuclei of the first cleavage division. When loss occurs at this stage, the remaining cleavage divisions yield a population of nuclei in which half have the ring-X and half do not. If the zygote begins its division as a rod-X/ ring-X heterozygous female, then after loss half the nuclei will be ring X/ rod-X female and half will be rod-X/ null-X male (sex is determined in *Drosophila* by the ratio of X chromosomes to autosomes, not by the Y chromosome, as in mammals). After eight syncytical divisions the egg is filled by a cloud of nuclei. This cloud, however is not a random mixture of the XO/ XX types. These two classes are present as two spatially separate groups, the position of which is determined by the plane of the first cleavage division. Therefore, when this population of nuclei migrates to the peripheral cytoplasm to form the cellular blastoderm, it does so as two adjacent and contiguous groups of either male or female nuclei. When an adult fly is derived from this gynandromorphic embryo it too will be mosaic.

The amount and position of the adult tissue that is male or female is not constant. This occurs because the plane of the first cleavage division is random with respect to the axes of the egg.

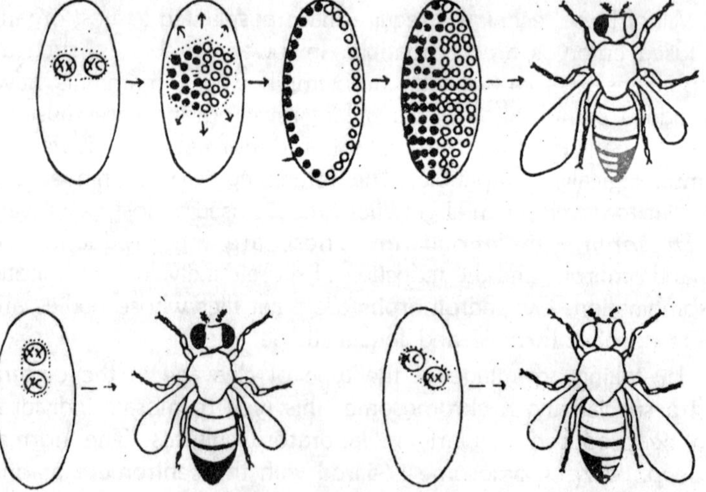

Fig. 2.3. The production of gynandromorphs in D. melanogaster.

Therefore, if the plane of division is perpendicular to an anterior-posterior midsagittal plane in the egg, a bilateral gynandromorph will result. Variation from this simple case will result in a greater or lesser portion of male tissue, depending on how many XO nuclei populate those regions of the blastoderm destined to form adult tissues. If a mutant gene is resident on the X chromosome, its effects can be analyzed in a mosaic simply by using a rod-X chromosome containing the mutant allele in the above scheme. The resultant gynandromorph will be normal XX female in part and mutant XO male in the remainder of its cells. The final technique, induced somatic recombination, is not restricted to the X chromosome and therefore has a somewhat wider applicability.

In a fly that is heterozygous for a mutant gene and its normal allele, normal mitotic cell division will ensure that every nucleus of that individual is identical and also heterozygous. However, if the developing organisms is irradiated with X-rays, recombination can be induced between homologous chromosomes (similar to meiotic exchange). When cell division takes place subsequent to the

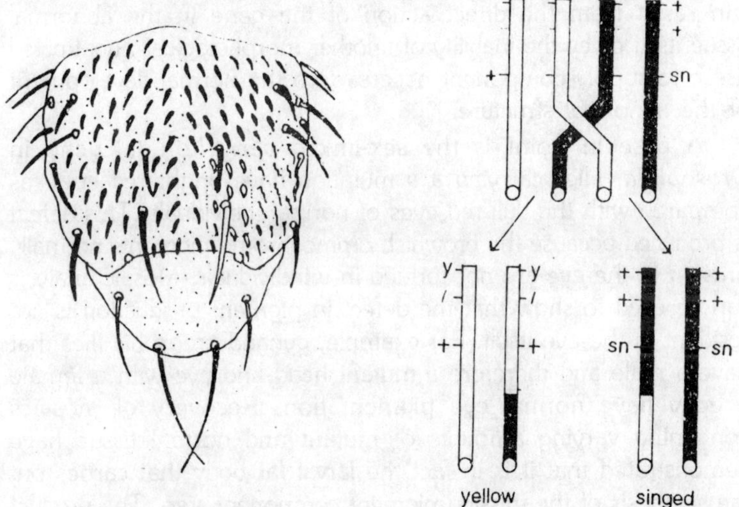

Fig. 2.4. A test for cellular autonomy of lethality by X-ray-induced somatic exchange. On the right is the genotypic X chromosome constitution of the heterozygous cells that genes rise to the two homozygous daughter cells after the exchange event. y=yellow body colour. l=lethal, and sn=singed bristles. To the left is a dorsal view of the thorax of D. melanogaster.

exchange event, the two cross-over products can segregate from each other. The daughter cells resulting from this process will no longer be heterozygous for the mutant gene, but one will be homozygous for the mutation and the other homozygous normal. The subsequent daughter cells of each of these two reciprocal types will form a clone and, dependent on where in the animal and when in ontogeny, will form mosaic spots of differing position and size in the adult.

It should be noted that in reality what is produced is an organisms of three mosaic cell types because of two homozygous clones will be produced in a background of heterozygous tissues. In this situation it is also possible to determine the survival of the mutant versus normal tissue, or their relative growth rates, because the single cross-over event should produce twin spots if the markers were appropriately place in the original heterozygous cell, that is, adjacent patches of the two types of tissues. The above methods of producing mosaics can yield information as to the site-specificity of gene action for any observed mutationally produced defect. That is, a morphological abnormality produced by a mutant gene can result from the direct action of the gene in the abnormal tissue itself or by the inability of another morphologically unaffected tissue to supply component necessary to the normal development of the abnormal structure.

A case in point is the sex-linked *vermilion (v)* gene in *Drosophila*. Flies carrying a *v* mutation have bright red eyes, as compared with the dull red eyes of normal individuals. This defect is produced because the brownish ommochrome pigments normally present in the eye are not formed in *v* individuals. Mosaic analysis can be used to show that the defect in pigment production is not resident in the eye itself. For example, gynandromorphic flies that have a male and therefore *v* mutant head and eye with a female *v* body have normal eye pigmentation. Recovery of mosaics containing varying amounts of mutant and normal tissue have demonstrated that it is in fact the larval fat body that carries out the synthesis of the missing pigment component step. The product of this step is then apparently transported to the developing eye where it is utilized in eye pigment formation. In this manner it is indeed possible to distinguish between the direct and relational pleiotropic effects of any single genetic lesion. By coupling these observations with a determination of the earliest observable defect in the mutant organism, it is sometimes possible to deduce probable causative relationships and begin to understand the nature of the genetic defect. However, if the observed lethal phase occurs very early in development before discrete structures or organs are formed it becomes difficult to determine a precise focus of action of the gene in question. Early lethality can result from two different causes. The embryo may lack some indispensable biochemical functions (e.g., an element of the protein synthetic machinery), or some early but discrete morphogenetic activity is aberrant. How then can we distinguish these two types of defects?

One way is to determine the time of action of the gene. An indispensable function necessary to all cells should cause lethality at all stages and in all tissue types. A specific morphogenetic gene should be more discrete in its time as well as its place of action. Ideally, what one would like to do is bypass the early lethal phase, and subsequent to that point, reinstate the genetic defect. This can be accomplished in two ways. The first is by the technique of somatic recombination, but with the added variable of including the mutant clones at various times in ontogeny. If the gene under analysis does indeed code for some indispensable metabolic cellular function, mutant clones should fail to survive or produce a normal phenotype, irrespective of the time or place of their production. If, however the gene has a discrete time of activity only clones produced after this time will survive or be normal morphologically.

Analogously, mutant genes that function in a specific tissue or organ will not produce surviving or wild-type clones in these structures. Both discrete time and specific site of action of a single gene are also possible. A mutation in this type of gene would result in the inability of clones to survive or produce normal structures in a specific tissue up to the time of gene action. Subsequent to this time in development, however, the cells should be unaffected. Finally, although limited in applicability, the most informative method is to recover conditional mutations, specifically, temperature sensitive lesions. This type of mutation allows one to control the time of onset of the mutant effect to any point in development simply by altering the temperature regimen. (This class of mutation is of course, not of much use in homeothermic animals). As an example, let us consider the case of a heat-sensitive lethal in *Drosophila* that is inviable at 29°C the non-permissive temperature. If raised at this temperature, mutant individuals succumb in the pupal stage, whereas at 20°C development is normal. Cultures of this mutant are grown first at the high temperature and then shifted down to the low temperature, and the fate of the "shifted" animals is observed. Likewise, reciprocal shift-up experiments are performed.

The earliest shift-down experiment at which the lethal syndrome is observed denotes the beginning of what is called the Temperature-Sensitive Period (TSP). In the shift-up one determines the latest time at which the mutant phenotype is no longer expressed. This indicates the end of this sensitive interval. If a discrete TSP is found, this can be confirmed by applying pulses at restrictive temperatures to cultures of mutant individuals. Further, it is possible to do pulse shifts for only portions of the TSP to determine if the mutant phenotype can be partially ameliorated or if certain pleiotropic effects can be obviated.

The relationship of the TSP and the actual time of lethality can also be informative, especially if the TSP precedes the lethal phase by a significant time interval. By relating this type of result to early lethals and indispensable functions, it can be seen that if a gene and its product are necessary at all times, a continuous rather than discrete TSP will be realized. If, however, the early period is past, shifts to the non-permissive temperature should have no detrimental effect. A vital step in the irreversible process of development has been completed "The Moving Finger writes; and, having writ, moves on." Individually, the techniques outlined above can be used to answer specific questions about the nature of developmentally important genetic lesions.

Moreover, when several of them are used to consort, truly significant information can be obtained as to how the normal genes are involved in the developmental process. The utility of this approach can be seen in examples, such as the results obtained by D.T. Suzuki and his collaborators in their analysis of two temperature-sensitive lethals in *Drosophila melanogaster*. The first of these was originally recovered as a simple recessive sex linked lethal that was also temperature-sensitive. If grown at 29°C, death occurred in the pupal state, whereas if grown at 22°C the flies were normal. Shift studies demonstrated that the TSP immediately preceded the lethal period. During these shift studies, it was noted that some of the surviving adults exhibited a mutant eye colour.

Subsequent genetic tests show that the TS lethal was actually an allele of a previously described locus called *raspberry (ras),* by virtue of its mutant eye colour. This particular mutation is known to have pleiotropic effects on pigmentation in that not only is the eye colour changed but the pigmentation of the adult testes and larval Malpighian tubules is also altered.

Further temperature shift studies on the TS *ras* mutation allowed Grigliatti and Suzuki to determine that the temperature sensitive period for pigmentation of the Malpighian tubules occurred during the early larval stages, whereas the TSP for both the eye and testis pigmentation took place in the late pupal stage more than four days later in the life cycle. Therefore, the action of this particular gene is necessary at two distinct times in development. However, the question still remains: is this a relational pleiotropic effect or direct pleiotropy? This question was answered by determining the autonomy of the pigmentation defect. If the lack of pigmentation was caused by the inability of a single tissue to manufacture a common pigment molecule which was then transported to the testes, Malpighian tubules, and eyes, the defects would be relational. However, it was possible to show by using the ring -X method to produce gynandromorphs that the eye and Malpighian defects are specific and autonomous to those two tissues; that is, the tissue must carry the mutant allele in order to express the mutant phenotype. Therefore, it would appear that the same gene, *ras* in necessary to produce pigment in three different tissues that two discrete periods in the ontogeny of the fly.

This result is in contrast to that seen in the relational pleiotropy of the sickle-cell-anemia syndrome and demonstrates that both direct and relational pleiotropy exist. This point is even more dramatically demonstrated by a second mutation analyzed by Suzuki and his collaborators. This mutation is called *shibire* (meaning paralyzed in Japanese); it was originally isolated as a sex-linked, temperature paralytic lesion. Adult male and female flies carrying this mutation have normal ability at 22°C. If these adults are shifted to 29°C, they immediately fall to the bottom of the culture vessel totally paralyzed. If the flies are subsequently returned to 22°C they begin to move in a few minutes and soon appear normal in all respects.

The physiological basis of this adult defect was determined by Ikeda and his co-workers who, by implanting microelectrodes in the flight muscles of normal and *shibire* flies, measured the junctional and action potentials evoked by stimulation of the motor nerve that innervates this muscle. On warming to 29°C, both the junctional and action potentials were lost in *shibire* fibers but not in wild-type flies. However, if the muscle was stimulated directly rather than through the nerve it could be made to contract in *shibire* flies, even at 29°C.

Moreover, it was possible to demonstrate transmission of an impulse along a *shibire* mutant nerve fiber at 29°C. Therefore, it would appear that the paralytic defect resides in the neuromuscular junction, which cannot transmit a stimulus at 29°C. However, this is not the only defect observed in *shibire* mutants. In order to determine if there was any paralysis of larvae and possible developmental defects. Poordry, Hall, and Suzuki performed temperature shifts on developing embryos and larvae. They found that a shift up to 29^{0}C during any stage of development resulted in paralysis and death; therefore, the *shibire* normal function was apparently indispensable to the fly. By performing pulse experiments, it was shown that a shift of >18 hours to 29°C was sufficient to kill developing larvae and embryos.

Heat pulses for shorter periods (two, four and six hours) revealed unexpected developmental defects. Six-hour heat pulses revealed six critical periods in development when the *shibire*, gene or its product is needed, or death occurs. One period of high sensitivity occurs at gastrulation, when a two-hour heat pulse is sufficient to kill the animal. This lethality does not occur as in the 18-hour heat pulses from irreversible paralysis, but must result from some other defect. The heat pulse experiments also revealed temperature-sensitive periods for several visible defects in the adult fly. The most striking of these was the production of a vertical "scar" on the eye facets caused by a disruption of these structures. The scar is produced by a heat pulse of three to six hours administrated during a period from about 48 hours before the pupal stage until just after the pupa is formed.

By a judicious spacing of these pulses, it is possible to create a fly with a double scar on the eye and demonstrate that the position of the scar moves across the eye in a posterior-to-anterior direction as the temperature-sensitive period proceeds. Interestingly, in this period direction of movement and position in the eye corresponds to a wave of cell divisions that moves across the developing eye in a manner analogous to the scar. The formation of bristles and hairs on the thorax and head of the fly is also affected by heat pulses, but the sensitive period for these structures is slightly later than that for the eye.

Early heat pulses produce duplication of bristles, whereas later pulses result in the deletion of the same bristle types. As with the eye scar, these effects show a posterior-to-anterior movement with progressively later pulses. The ubiquity of defects produced in *shibire* flies demonstrates that the function of this gene is far less specific than a lesion in the neuromuscular junction. This is perhaps best shown by the fact that gastrulation, an event that takes place long before nerves or muscles are evident or functioning, is extremely sensitive to temperature in *shibire* embryos. The tentative conclusion from these studies is that the *shibire*' gene product is a membrane component that is necessary for some types of cell-to-cell interaction or communication. This membrane component would be necessary for different process that all relate to its primary function in cellular communication. This membrane component would be necessary for different processes that all relate to its primary function in cellular communication, but each different temperature-sensitive period with its corresponding defect would reflect the manner in which a specific function is important at that point in development.

Here again we have a case of direct pleiotropy and the demonstration that a single gene can affect several seemingly disparate developmental events. The other point to be made from these results is the usefulness of temperature-sensitive alleles. A non-conditional *shibire* mutation would have produced simply a dead embryo with very little cytodifferentiation and therefore no clue as to the nature of the gene product or its ubiquity. There are however certain limitations that must be kept in mind when interpreting the results of an analysis of a temperature-sensitive mutation.

The TSP is generally thought to reveal the time at which the gene product is utilized. That is in the case of an enzyme the period in which its metabolic function is needed. This, however is not necessarily always the case. These are also instances in which temperature sensitivity results from an abnormal synthesis of the protein during translation. The protein thus formed is inactive, even at the permissive temperature. Additionally, the protein, once

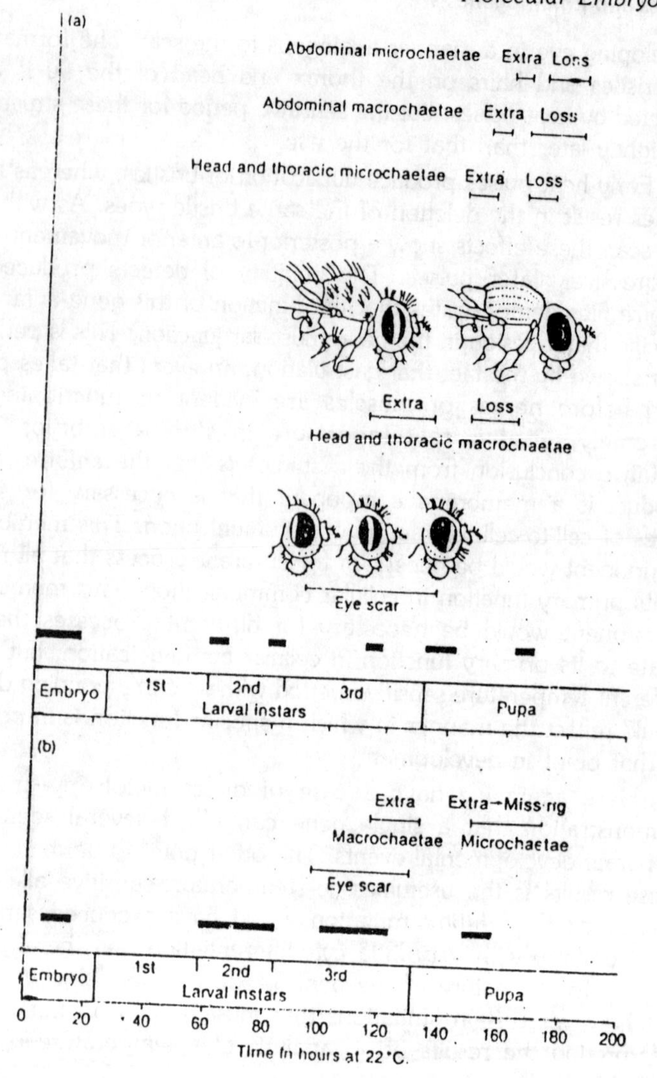

Fig. 2.5. Temperature sensitive periods for the various effects caused by the shibire[s] (a) and Notch[ts] (b) Mutations. The black bars below the line bars indicate the times at which mutant animals are particularly sensitive to pulses to the non-permissive temperature. Pulses longer than those necessary to produce the phenotypic anomalies listed result in lethality.

synthesized under permissive conditions, in no longer sensitive to an elevation in temperature. Therefore, the TSP for this lesion reflects not the time of gene action, but rather the time of synthesis of gene product. The relationship of the time of gene transcription, the time of gene product utilization, and the TSP of a temperature-sensitive lesion can only be determined by a more intimate knowledge of the nature of the gene product or an independent assessment of the time of transcription of the gene in question.

In order to gain a full appreciation of the role of a gene in ontogeny, it is therefore necessary to adopt several experimental strategies taken from a variety of fields. Utilizing the methodology of developmental genetics, it is possible to demonstrate the wide range of observable genetic defects with respect to their temporal distribution in ontogeny. In the following discussion, we will produce a picture of the repertoire of genes that is available to the evolutionary process in order for it effect morphological changes.

Maternal Effect Mutations

In organisms as diverse as sea-urchins and frogs, the early cleavage events, and indeed the majority if not all, of development prior to gastrulation do not rely on the zygotic genome. The information necessary for performing these initial and crucial steps of ontogeny is specified instead by the maternal genome during production of the egg. Utilizing the example of shell coiling in *Limnaea*, this conclusion is supported by the existence of what are referred to as maternal effect genes in a wide variety of organisms. Maternal effect genes, when mutated have a unique pattern of inheritance. If one were to cross two individuals heterozygous for a recessive trait, it would be expected that one-quarter of the progeny would express that trait. However, in the case of *maternal (mat)* mutations the *mat/ mat* individuals develop normally. Moreover, males of this genotype are fertile and produce normal offspring when mated to normal females.

The homozygous females, on the other hand, produce abnormal offspring. This production of defective progeny results from the fact that these females produce abnormal oocytes that do not complete normal development. The *mat/mat* female survives because she was produced by a heterozygous (*mat/+*) mother that is capable of producing normal eggs. What one would like to conclude about this type of mutation is that these genes are producing necessary "morphogens" that are placed in the developing oocyte as "instruction" for early development. However, it is also possible that the egg is unable to develop simply because

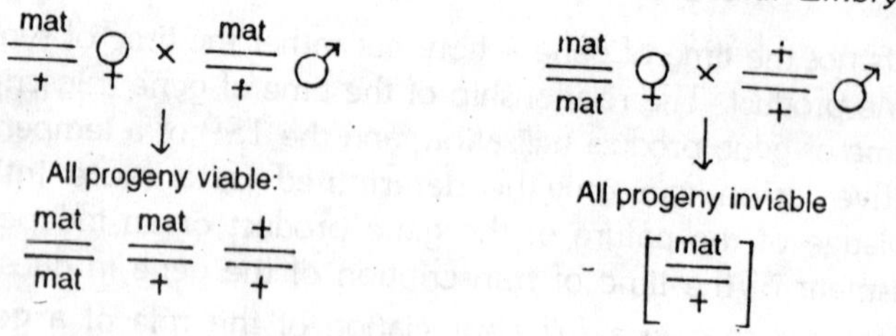

$\frac{mat}{+}$ ♀ × $\frac{mat}{+}$ ♂ → All progeny viable: $\frac{mat}{mat}$ $\frac{mat}{+}$ $\frac{+}{+}$

$\frac{mat}{mat}$ ♀ × $\frac{+}{+}$ ♂ → All progeny inviable $\left[\frac{mat}{+}\right]$

Fig. 2.6. Pattern of inheritance of a maternal effect lethal mutation. The homozygous mat condition is not itself lethal as long as the female parent is heterozygous (mat–). However, if a homozygous female is mated all of her progeny die, irrespective of their genotype.

it suffers from some general metabolic defect. A case in point is present by a group of five different maternal effect lesions on the X chromosome of *Drosophila melanogaster : cinnamon (cin) deep orange (dor), almondex (amx), fused (fu),* and *rudimentary (e).* These all produce in addition to their maternal effects, visible adult morphological aberrations, from which their colourful names are derived.

Hemizygous males carrying any one of these mutations are viable and fertile, as are heterozygous females. By crossing mutant males to heterozygous females, homozygous females can be obtained, which when mated to mutant males, are completely sterile. For example *dor/dor* females produce eggs that arrest development at gastrulation. The remaining four mutations also result in embryonic lethality, the lethal phase varying somewhat from that of *dor.* All five of these mutations have an added perturbation in their pattern of inheritance. If homozygous mutant females are mated to normal males, some progeny are produced. These consist entirely of heterozygous females, that is, eggs fertilized by X-bearing sperm. No males survive.

Apparently, the presence of the wild-type allele of the deficient gene can ameliorate the egg's defect, even if it is supplied by the sperm. This of course implies that at least a portion of the zygotic genome is active during gastrulation. It has been shown by Gehring that injection of wild-type egg cytoplasm into *dor* embryos will rescue the deficient animals. In a further analysis Kuroda developed an *in vitro* test for the *dor'* substance. Using cultures of embryonic cells isolated from deficient embryos, he noted that the mutant cells demonstrated abnormalities, such as failure of muscle cells to fuse, in contrast to cells from normal embryos. If the culture mutant cells were exposed to extracts of normal eggs, the defects were prevented.

Moreover, by using cytoplasm for rescued embryos of various stages, Kuroda was able to show that the *dor'* substance was produced at and after gastrulation, but not before, by *dor'* embryos rescued by *dor'* sperm. With this assay, he was also able to show that the rescuing substance is heat labile, and therefore, may be a protein. The question still remains, however: What is the normal function of the *dor⁺* gene? We may address this question by considering first the pleiotropic effect of *dor'*, an eye pigmentation abnormality, and second, the nature of the product of one of the other genes, *rudimentary*, which shows an identical pattern of inheritance. The eye pigmentation defect of *dor'* involves the synthesis of pteridines, a group of heterocyclic compounds related to the nucleic acids.

The *rudimentary* gene has been shown by Norby, Jarry and Falk, and Rawls and Fristrom to encode the first three enzymes in the biosynthetic pathway for pyrimidines. An involvement of *dor* in nucleic acid metabolism is also indicated by the fact that the sterility of *dor/dor* females can be partially alleviated by supplementing the growth media of these flies with extra purines. Therefore, it would appear that the *dor* mutation and possibly other similar mutations are producing their defects by causing a deficiency in nucleic acid synthesis or degradation. Such mutations illustrate the problem of distinguishing between nonspecific genetic defects and those specifically affecting mechanisms unique to morphogenesis. A more likely candidate for a morphogen deficiency is the *oocytedeficient (0)* mutation in the axolotl. By transplanting ovaries from *0-mutant* -bearing individuals to normal hosts. Humphrey demonstrated the defect to be autonomous to the mutant ovary. This result is not expected if the defects is caused by the deficiency of some low-molecular-weight diffusible metabolite. However, despite the extensive work of Briggs and his co-workers which resulted in the identification of a proteinaceous rescuing substance in normal embryos and oocytes, the precise nature of the defect is not known.

One intriguing observation, however, is that the rescuing protein is localized to the germinal vesicle (the oocyte nucleus), implying that the protein in some way interacts with genetic material. In light of this possibility, it is of interest to consider a newly discovered mutation in *Drosophila*, Mortin and Lefevre have described a sex-linked (X chromosome) dominant mutation that they have called *Ultrabithoraxlike (Ubl).* In females heterozygous for the mutation *(Ubl/+),* the halteres of balancer organs common to all dipteran insects are enlarged relative to normal specimens. These halters are found on the third segment of the thorax and are, in evolutionary terms, vestigial wings that act as stabilizing organs during flight. The *Ubl* mutation mimics another dominant mutation on the third chromosome, called *Ultrabithorax (Ubx),* which is homoeotic in nature. The enlargement of the haltere is indicative of a transformation of that organ toward a fully differentiate wing. This becomes clear in the phenotype of female flies heterozygous for both *Ubl* on the X chromosome *(Ubl/+),* and *Ubl* on the third chromosome (*Ubl*+). These flies posses a haltere that has the marginal hairs and bristles characteristic of a wing as well as wing veins, two structures not normally found on the small bulbous haltere.

A further discovery made by Mortin makes *Ubl* germane to the discussion at hand; that is, it shows a maternal effect (*Ubx* does not). *Ubl* is lethal in homozygous females and hemizygous males; however, it is possible, by supplying the fly with a duplication containing the normal allele of *Ubl,* to obtain adult female flies of the genotype *Ubl/ Ubl/+*. If these females are mated to normal males, they are fertile and produce heterozygous *Ubl/+* offspring. If, however, they are mated to *Ubl/Y* + males (viable because of the same duplication), they are virtually sterile and unable to produce even *Ubl/+* progeny. What makes this result intriguing and different from the similar rescuable sterility demonstrated by *dor* females is the nature of the gene product of the *Ubl* locus.

Greenleaf and co-workers have recovered an additional mutation at this locus that confers resistance to α-amanitin, which is known specifically to inhibit RNA polymerase II from eucaryotic organisms. This polymerase is the enzyme that transcribes those genes that code for RNA (mRNA), that is, structural genes. Moreover, like the *Ubl* lesion at the locus, the α-amanitin-resistant allele interacts with *Ubx* to produce a winglike haltere. Garcia-Bellido and his collaborators have shown that activity of the *Ubx* locus is necessary as early as the blastoderm stage, where it is involved in the initial determinative events controlling segmentation. This fact, coupled with the demonstration of the maternal effect of *Ubl,* indicated the possibility that a maternally supplied polymerase is necessary to these initial determinative events.

It remains to be shown whether the interaction of the altered polymerase with the homoeotic genes is specific to these loci, or if the seeming specificity of the polymerase defect is more apparent than real. Alterations in this subunit of polymerase II may affect binding of the holoenzyme to the promoters of many or all genes; thus, the interaction may be revealing differential promoter strength, with *Ubx* having a week promoter extremely sensitive to enzyme alternations. Nonetheless, what has been shown is that the function of a gene, *Ubx,* whose correct function in early embryogenesis is crucial to normal morphogenesis may be dependent in large part on a maternally supplied enzyme important to gene transcription. Although a direct and unique role of the previously described genes in morphogenesis remains problematical, Rice and Garen have isolated on the third chromosome of *D. melanogaster*, a group of maternal effect mutations with a seemingly specific morphological involvement. Three of these produce very characteristic and specific defects at the blastoderm stage in embryos derived from homozygous mothers. The first, *mat (3)1,* does not form a normal cellular blastoderm after the initial syncytial cleavage divisions. However, the pole cells do form at the posterior end of the embryo. In the normal embryo, this group of cells is destined to form the gametes, and thus forms the germ line.

Therefore, it would appear that germ-line cells and few somatic cells develop in this mutant. Interestingly, despite the lack of a normal blastoderm, the pole cells attempt to invaginate as they would normally do during gastrulation, indicating that at least some of the early cell movements of the embryo do not require cell-interactions or adhesions. A second mutation, *mat (3)6,* also forms only a partial blastoderm. The syncytial nuclei migrate into the cortical cytoplasm as in normal embryos; however, cellularization takes place only in the anterior and posterior ends of the embryo. The distributions of cells in these two mutants at the blastoderm stage. Again, in *mat (3)6,* like met (3)1, the pole cells attempt to invaginate and the embryo goes through an abortive attempt to invaginate and the embryo goes through an abortive attempt at gastrulation.

In further work on these embryos, Rice and Garen assayed the potentialities of those cell formed by the *mat(3)6* embryos.

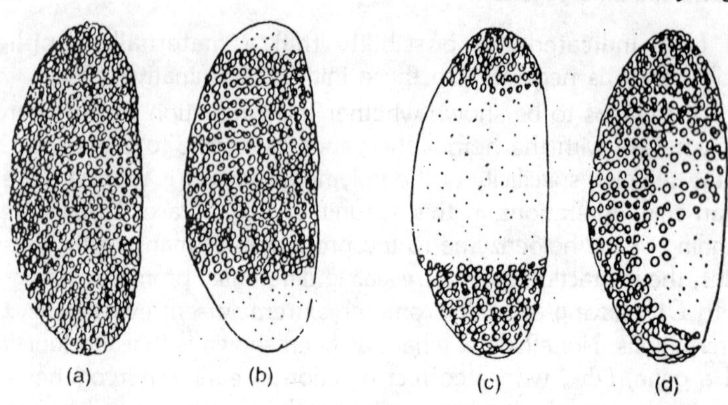

Fig. 2.7. The pattern of cellularization in the blastoderm of progeny of three maternal effect mutation-bearing females of Drosphila. The anterior end of the embryo is at the top, dorsal to the right. The pole cells at the posterior end of the embryo are larger than the somatic cells. (a) Normal, (b) Grandchildless, (c) mat (3)6, (d) mat (3)1.

Previously, Gehring had shown that if a blastoderm-stage embryo is cut in half and these half embryos are injected into the abdomens of adult female flies, cell cultures can be established. These cell cultures can then be assayed for their ability to produce adult structures by a subsequent transplantation from the female's abdomens into a metamorphosing larva. As the larva changes into an adult, so too will any injected cells capable of forming adult or imaginal tissues. By this technique, Chan and Gehring were able to demonstrate that, at the blastoderm state, the anterior end was already committed to produce only anterior adult structures, and the posterior end only posterior adult structures. Rice and Garen applied this technique to their mutant embryos and found that only the most anterior head and posterior abdominal structures were formed. They found no thoracic elements.

It would appear, therefore, that only very specific portions of the cellular blastoderm are formed in these mutants and that maternal gene products are involved in the cellularization of certain portions of the early embryo. A third maternal mutation of this genre was studied by Mahowald and his collaborators in another species. *D. subobscura*. This mutation, *grandchildless (gs),* is inherited as a maternal effect lesion but with a slight difference from those of Rice and Garen, Homozygous recessive *gs/gs* females are fertile and produce viable progeny. However, their male and female progeny are sterile. Thus, the original *gs* female can have no grandchildren. The basis for this phenomenon is that the embryos produced by *gs/gs* mothers do not form pole cells and thereby result in agametic adults. The precise reason for this failure is that the nuclei that normally migrate to the most posterior tip of the cleavage-stage embryo do not do so at the normal time.

It has been shown by Illmensee and Mahowald that this region of the embryo contains the "determinants" instrumental in specifying the fate of these pole cells destined to become the germ line. The *gs* mutation apparently prevents this nuclear-cytoplasmic interaction from taking place. Interestingly, at the time of blastoderm formation in *gs* embryos, not only is there a failure of cellularization at the posterior tip of the embryo but at the anterior end as well. This is shown where a comparison of the three blastoderm defect mutations can be made. An intriguing fact is that the spatial distribution of cellularization in *gs* is reciprocal to that in *mat(3)6*. The final, and perhaps the most striking of the maternal effect lesions is the *bicaudal (bic)* mutation of *Drosophila melanogaster*. This lesion, an autosomal recessive was first found by Bull and has been more recently analyzed by Nusslein-Volhard.

Homozygous *bic/bic* females produce embryos which at gastrulation, are seen to have two posterior ends and no anterior. Amazingly, whereas progeny of bicaudal animals fail to hatch from the egg case, these two-tailed monsters complete embryonic development and reach the larval stage before dying. All of the normal larval cuticular structures for the posterior are formed at both ends of the embryo. This phenotypes is strikingly similar to the one produced experimentally by Kalthoff in the lower dipteran *Smitia*. It is therefore, reasonable to conclude the *bic* may be the locus responsible for the synthesis of a product whose localized activity is similar to that of the morphogen that was destroyed experimentally by Kalthoff. The preceding examples demonstrated the involvement of the maternal genome and its products in many of the early events of development. The existence of genes like *dor* and *r* show the influence of maternally supplied enzymes to the metabolism of the early embryo, whereas the defects observed in the *mat* and *bic* studies indicate that positional and organizational cues are also supplied during oogenesis. Finally, the effects of the *Ubl* mutation in *Drosophila* and the *0* mutation in the axolotl reveal the role of maternal information necessary for the proper activation of the zygotic genome.

Mutations Affecting Organogenesis

The early events of development have been shown to be highly dependent on maternally supplied information. However, at about the time of gastrulation, the embryo's own genetic information becomes important to further development and the organism gains control of its own destiny. Developmental events subsequent to the formation of blastoderm require the synthesis of RNA and its translation into protein. We can also demonstrate the necessity of embryonic genetic information by virtue of the large number of mutations that affect events after gastrulation and thus reveal the presence of genes that control these events. These mutations show no pattern of maternal inheritance. A case in point is the *Notch* locus of *Drosophila melanogaster*. *Notch (N)* is a sex-linked dominant mutations that is also recessive lethal.

Homozygous females (*N/N)* and hemizygous males (N/Y) die in the embryo at about six hours after fertilization. This time corresponds to a point just subsequent to gastrulation about one-quarter of the way through embryogenesis. Histological and morphological observations by Poulson on these lethal embryos showed that the ventral and lateral ectoderm, which normally gives rise to both epidermis and neural cells, produces exclusively neuroblast-like cells and no epidermis. Therefore, it would appear that the *Notch* locus is necessary for the differentiation of neural versus epidermal tissue from embryonic ectoderm. This however turns out to be a somewhat simplistic interpretation. Shellenbarger and co-workers isolated and characterized a temperature-sensitive allele of the *Notch* locus. Flies carrying this lesion grown at 22°C are normal, whereas at 29°C the embryonic lethality described previously is expressed. In shift experiments similar to those on *shibire* presented earlier, it was found that his early embryonic period was not the only developmental stage sensitive to a deficiency of the *Notch* locus and its product.

Pulse shifts revealed three additional periods in which normal activity of the *Notch* locus was vital to the mutant organisms. Shifts to the non-permissive temperature during the second or third larval instars or the pupal stage produced lethality. Moreover, shorter pulses during specific periods of the third larval instar and pupal periods produced the same eye scarring and bristle defects seen in the *shibire* mutants. Therefore, like *shibire*, *Notch* is much more pervasive in its effects than might have been expected from observations on its primary phenotype. All of the structures seen to be affected are ectodermal, and in further experiments utilizing gynandromorphs Shellenbarger was able to show that the observed morphological defects were autonomous to this cell type.

Therefore, it would seem that again like *shibire*, *Notch* makes a product that is common to and necessary for the functioning of ectodermal cells at several discrete periods throughout development. What should also be noted is that two distinct genes, *Notch* and *shibire*, are necessary for the completion of the same set of ontogenic genetic events and that a deficiency either gene results in a strikingly similar set of defects. There exists in the mouse (*Mus musculis*) a complex gene, the *T* locus, that exhibits many analogies to the *Notch* system just described. The first allele of this locus was discovered as an autosomal dominant referred to as *Brachyury* (*T*). In the heterozygous, condition, *T/+*, mice have short tails. The homozygotes, T/T are lethal and die as embryos in the uterus of their mother.

Shortly after the discovery of this dominant mutation it was found that crosses of heterozygotes (T/+) to wild caught mice often produced progeny with no tails at all. These tailless mice were shown to be the result of recessive alleles of the T locus that were quite common in natural populations of mice. The genotype of these tailless mice was therefore T/t. Crosses of these heterozygous T/t mice produced true-breeding tailless mice. This was subsequently shown to be caused by a "balanced lethal system." That is, not only were the T/T offspring expected from the cross lethal, but the t/t were as well. Therefore, only the heterozygous T/t survived to produce the next generation. This intriguing situation has been exploited both genetically and developmentally by L.C Dunn and his students, D. Bennett and S. Gluechsohn-Waelsch, in a series of elegant studies. In genetic studies of the newly recovered balanced lethal strains, it was found that crosses between tailless mice whose recessive t allele came from different populations often produced normal progeny.

Specifically, the cross T/t^a x $T/\ t^b$ produced progeny that were tail-less and normal in a 2:1 ratio. The normal-tailed progeny could be shown to be t^a / t^b genotype. Therefore not only was this genotype non-lethal, but it was normal morphologically. A further attribute of this locus was discovered in the balanced lethal crosses themselves. Normally, T/T^a crossed to a like heterozygote produces only tailless offspring. However, normal-tailed mice are produced at a low frequency (1/500 = 1/ 1000). These normal mice are nearly always produced as rare genetic recombinants in the chromosomal region at or adjacent to the T locus in chromosome 17. These recombinant offspring can be shown to be tailed by virtue of the fact that concomitant with the recombination event the original *t* allele has changed to a new *t* allele that complements t^x in the same manner that certain wild-derived *t* alleles, complement each other.

The difference, of course, is that t^a is in this case directly related by descent to t^a. This transition of one *t* allele to other complementing types has been shown to occur with most of the isolated recessives. Some derived alleles, for example, t^x can in turn, by the same mechanism, produce another complementing allele, for example, t^y. This transition of one recessive *t* to another forms a graded series, resulting finally in the production of what are called *t viable* or t^v alleles. These are all non-lethal and only express a tailless phenotype in T/t^v heterozygotes; t^v/t^v individuals have normal tails. A further class of recessive *t* mutations is that of the semilethals, which range in viability from 2 to 51% of normal. Like the completely lethal alleles, they transform of t^v types.

Complementation crosses (T/t^a x T/t^b) of all of the wild and genetically derived recessive alleles have revealed that the 111 extant mutations fall into eight separate groups, all of which fail to complement T. Each group has a different number of members from a low of one for the t^{w73} group to a high of 66 for the *t* alleles. Of the five dominant *T* mutations one was induced by X-rays, whereas all of the other mutations known at the locus are apparently spontaneous in origin. These are a few of the genetic attributes of this complex series of genetic lesions, all of which are related by either complementation and genetic position or by derivation one from the other. But what are the developmental attributes of this complex locus? Each one of the eight complementation groups produces a different pattern of defects from early to late in development of the embryo. The earliest observable defect is produced in homozygous t^{72} embryos. Fertilization and the cleavage divisions of the zygote result in the formation of a ball of cells called a morula.

The first evidence of differentiation of cell types in the mouse is the transition of this morula into a second stage, the blastocyst, which consists of an external trophectoderm and an inner cell mass. Homozygous t^{12} embryos do not perform this step and the non-differentiated "morulae", die before implantation in the uterine wall, an event which in normal embryos takes place at about four days after fertilization. Moreover, t^{12} cells appear to be autonomous in their lethality. Chimeras, which contain both t^{12} embryonic cells and normal embryo cells, still express lethally, and never develop beyond the point when t^{12} embryos normally expire. Therefore t^{12} apparently affects a locus necessary for the first differentiative step in the mouse embryo, the formation of trophectoderm, which will eventually form the chorion and other portions of the zygotically derived extraembryonic elements common to placental mammals. The next earliest acting allele is t^{w73}. Homozygotes of this type produce a blastocyst.

However, the trophectoderm of these mutant embryos does not form the necessary associations with the uterine wall, and the improperly implanted embryos die shortly thereafter. After successful implantation of embryos in the uterine wall, the inner cell mass begins to grow and further differentiation takes place.

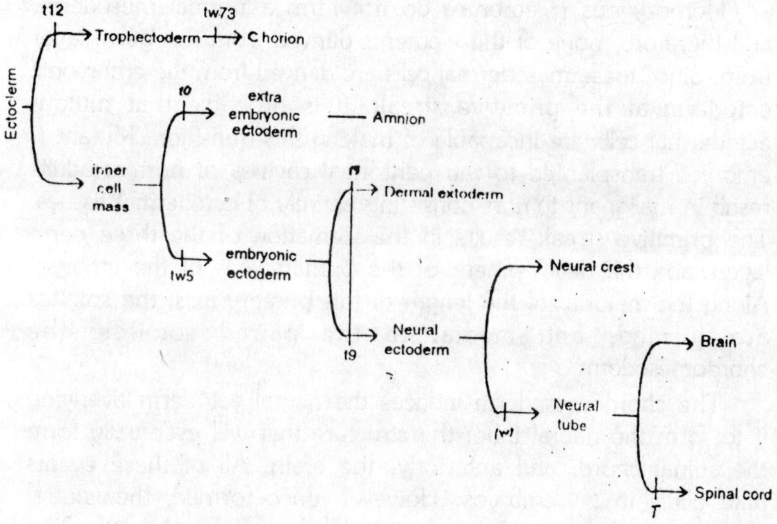

Fig. 2.8. Hypothetical flow of decisions made during the differentiation of the ectoderm and its derivatives in the mouse. Superimposed on this flow are the various t-allele mutations that block the indicated differentiative events.

One of these differentiative events is the formation of extraembryonic ectoderm and the embryonic ectoderm. The former eventually formes the placenta and portions of the extraembryonic membranes. The later will form the embryo proper. The mutation t^o fails to form extraembryonic ectoderm and dies in the early egg cylinder stage. In normal embryos, further growth of the inner cell mass proceeds and produces an elongated ectodermal mass covered by endoderm, which are together called the egg cylinder.

The extraembryonic cells also continue to proliferate and differentiate. Homozygous t^{w5} embryos produce an egg cylinder like stage and then the embryonic ectoderm cell become pycnotic and die. The extraembryonic cells, however, do not seem to be affected and the dead embryo persists in the midst of apparently normal extraembryonic development for several days until the extraembryonic cells die as well. At this point in the development of the mouse, 6.5-7 days post-fertilization, differentiation of the embryo proper begins. This is evinced by the development of the primitive streak on the egg cylinder and the formation of a mesodermal layer of cells between the per-existing embryonic ectoderm and endoderm.

Homozygous t^3 embryos do not form a normal mesoderm, and therefore, none of the elements derived from this germ layer from. Since these mesodermal cells are derived from the embryonic ectoderm at the primitive streak, it is probable that mutant ectodermal cells are incapable of making this transition. Mutant *t*, embryos transplanted to the peritoneal cavities of normal adults result in malignant tumors consisting entirely of ectodermal tissues. The primitive streak results in the formation of the three germ layers and the development of the primary axis of the embryo. Along the majority of the length of this primary axis, the somites are formed, and medial to the paired somites, the chordomesoderm.

The chordomesoderm induces the neural ectoderm overlying it to form the neural tube, the structure that will eventually form the spinal chord, and anteriorly, the brain. All of these events take place in t^{w2} embryos. However, once formed, the ventral portion of the neural tube and the brain degenerate. The remaining dorsal cells spatially replace the dead ventral cells, but apparently cannot replace them functionally because these embryos always show a variety of defects and succumb before birth. The final lethal condition of the *T* locus is found in individuals homozygous for dominant *T*. These individuals have a similar lethal phase to t^{w1} individuals, that is, later than the majority of the other recessives. T/T individuals do not extend their primitive streak

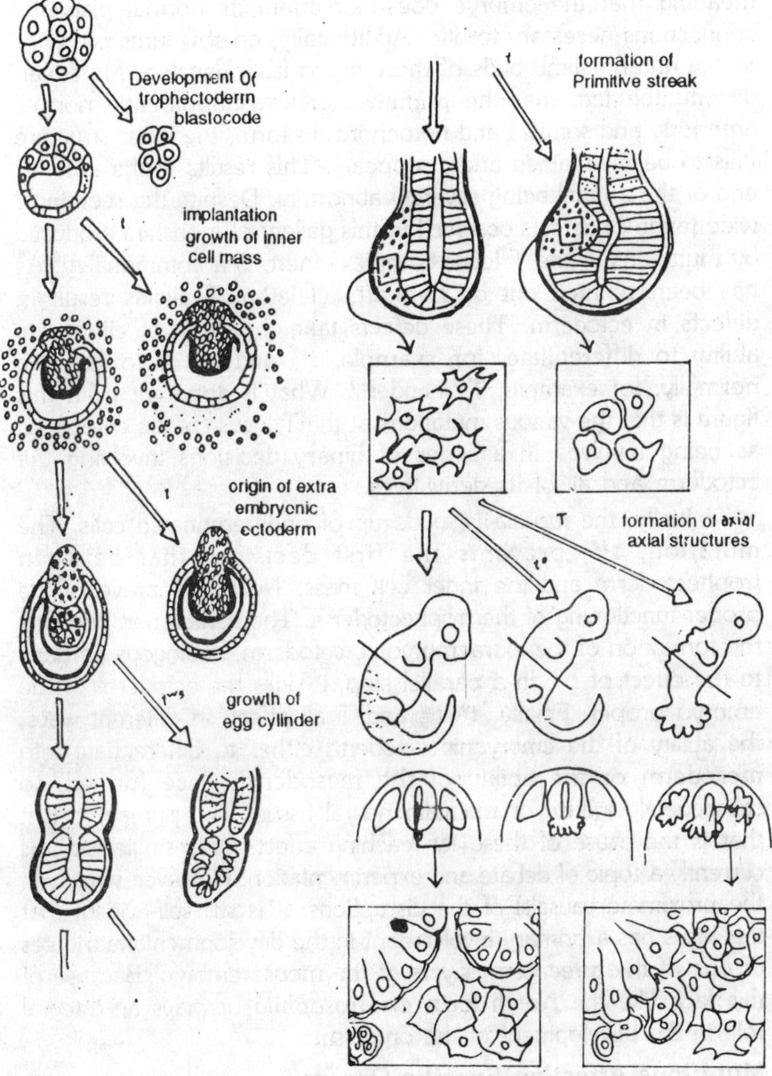

Fig. 2.9. T-locus mutants in the mouse, showing the effects of different t-alleles on a series of stages in embryogenesis.

fully to the posterior end of the embryo therefore, none of the structures dependent on the formation of mesoderm in this region ever develop.

Most notably, the allantoic placental stalk does not form, meaning that the embryo does not attain its normal placental connections necessary for life. Additionally, no structures posterior to the anterior limb buds of the embryo itself develop. Moreover, despite the fact that the primitive streak is apparently normal anteriorly and somites and notochord do form, the latter structure fails to be maintained and disappears. This results in the anterior end of the animal being severely abnormal. Despite the seemingly wide range of defects observed in this gallery of monsters produced by mutations in the T locus complex, there is a commonality. As has been pointed out by Bennett, all lethal T alleles result in defects in ectoderm. These defects take the form of either the ability to differentiate, for example, t^{12} and t^{o}, or to function normally, for example, T^{u73} and t^{w1}. What is apparent from this figure is that the various mutations at the T locus can be envisioned as being involved in a series of binary decisions involving the ectoderm and all of its derivatives.

Initially, the morula is made up of undifferentiated cells. The mutation, t^{12}, prevents the first decision, that between trophectoderm and the inner cell mass. Next, t^{u73} prevents the proper functioning of the trophectoderm. The t^{o} mutation prevents the formation of the extraembryonic ectoderm, analogous perhaps to the effect of t^{12}. In a parallel step, t^{w5} kills the ectoderm of the embryo proper. Finally, $t^{9,}$ t^{wl} and T all affect, in different ways, the ability of the embryonic ectoderm either to differentiate into mesoderm or the ability of the mesoderm, once formed, to subsequently induce or maintain neural tissue. The primary defect that is the cause of these far-reaching effects of a single locus is currently a topic of debate and experimentation. However, whatever, the proximate cause(s) of the disruptions, it is still self-evident that this locus has a primary involvement in the developmental attributes of one of the three germ layers of the mouse embryo. Because of this fact, like the *Notch* locus of *Drosophila*, it plays an integral role in the development of the organism.

Mutations Affecting Specific Organs

While the *T* locus seems to have quite wide-ranging effects on the entirely of ectodermal development, there are also mutations that are more specific in their defects. An example of such a mutation is *cardiac lethal (c)* in the Mexican axolotl, *Ambystoma mexicanum*. This mutation was originally discovered and analyzed by Humphrey. He found that the gene is inherited as a simple autosomal recessive such that if two heterozygous (c/+) individuals are mated, 25% of their offspring die in the early larval stages shortly after hatching. These mutant individuals show normal swimming behaviour but are swollen with fluid and have a poorly developed digestive system and gills. The primary cause of these defects is the failure of the heart to develop and beat. Thus, the animals have no circulation of blood, and respiration probably occurs by diffusion through the skin, thereby allowing mutants to survive for only a limited time period. The autonomy of this heart defect was shown by Humphrey through the production of parabiotic twins of normal and *cardiac* individuals.

Mutant and normal embryos were selected before heart formation, and a block of tissue was removed from the side of each individual. The two embryos were then conjoined at the wound site and allowed to heal together. When such fused individuals were allowed to complete development, it was found that the normal partner ameliorated the swelling and other defects and allowed survival of the c/c individual. However, the heat of the c/c partner never developed beyond a simple non-beating tube, and circulation in the mutant partner was wholly supported by the normal partner. It has been shown that many vertebrate organs, including the heart, develop as the result of certain, inductive interactions during development.

Specifically, Jacobson and Duncan have shown that the heart of the salamander is induced from heart forming mesoderm by the anterior endoderm. The reason that c/c individuals fail to form a heart could be caused by a failure of anterior endoderm to induce the process, or from a failure of cardiac mesoderm to respond to the inducer. In order to determine which was the case, Humphrey transplanted normal heart forming mesoderm into c/c hosts and c/c mesoderm into normal hosts. He found that the c/c mesoderm was capable of forming a beating heart under the influence of the normal anterior endoderm, whereas the mutant hosts did not yield normal heart formation. This result can be interpreted as indicating a defect in the inductive potential of the c/c anterior endoderm. However, it is also possible that the mutant individuals are actively inhibiting heart formation. This latter possibility was made less tenable by Lemanski and his co-workers. Mutant heart mesoderm, as well as normal, was cultured *in vitro*. Under the conditions used, normal cardiac mesoderm will beat vigorously while the mutant heart mesoderm does not.

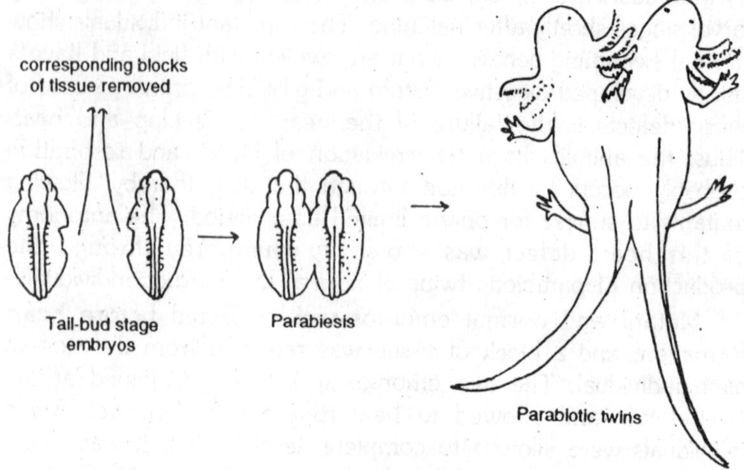

Fig. 2.10. The technique of parabiosis. Two animals of different genotype, indicated by presence and absence of stippling, are fused during early embryogenesis and the ability of the normal individual to rescue the mutant member of the pair is ascertained.

If heartbeat in the mutant tissue was being suppressed *in situ*, then *in vitro* culture should have removed that influence. Furthermore, when the c/c mesoderm was cocultured with normal anterior endoderm, the mutant tissue could be seen to contract, showing that the mutant mesoderm was capable of a normal response if supplied with the proper inductive influence. Thus, the *cardiac lethal* appears to the expressed by the failure of anterior endoderm to provide the inductive signal that triggers the differentiation of the heart from cardiac mesoderm. The results of the studies on another autosomal recessive axolotl mutation, *eyeless (e),* stand in contrast to those seen with *cardiac*. As the name implies, homozygous individuals (e/e) have no eyes. They also exhibit two other pleiotropic defects: dark pigmentation and sterility.

The primary defect occurs quite early in the formation of the eye, which is blocked at or before the formation of the early optic vesicle. Like the heart, the eye is formed as the result of an inductive interaction. The eye is induced from anterior neural ectoderm in the presumptive forebrain (anterior medullary plate) by the chordomesoderm, which comes to lie under its during gastrulation. The induced neural tissue grows laterally as a part of the brain and finally evaginates to form the optic vesicles. Van Deusen investigated the nature of the *e* defect by transplanting prechordal mesoderm from the dorsal lip of the blastopore of mutant and normal embryos.

Transplants of *e/e* mesoderm to normal blastulae resulted in the induction of optic vesicles. The reciprocal transplant (normal mesoderm to *e/e* blastulae), however, did not result in the induction of early eye development. The same result was obtained in reciprocal transplants taken from later stages during early gastrulation but before optic vesicle formation. That is normal ectoderm was capable of forming optic vesicles under the influence of either normal or *e/e* mesoderm, whereas *e/e* ectoderm could not be induced by either type. Therefore, it appears the eyeless defect us caused by the inability of the ectoderm to respond to mesodermal inducer. This conclusion is supported by the fact that morphologically differentiated eyes taken from normal individuals and transplanted to *e/e* hosts survive and function. Thus, one the optic vesicle is induced, the *e/e* genotype is capable of maintaining a fully formed eye.

However, the pleiotropic defects in pigmentation and sterility in the mutant individuals remain to be explained. The excess pigmentation phenotype can be mimicked by simply removing the optic vesicles of a normal individual, which results in heavy pigmentation in the blinded larvae. Conversely, pigmentation can be made normal by transplanting a normal eye rudiment into a developing *e/e* individual. Some of the individuals used in transplantation and eye vesicle operations were grown to sexual maturity.

The experimentally blinded individuals could be shown to be fertile, whereas the *e/e* recipients of normal eyes remained sterile. Therefore, the sterility was not caused directly eyelessness. Moreover, Van Deusen was able to show that genotypically *e/e* ovaries transplanted into normal hosts were capable of oogenesis. The sterility was shown by further transplantation experiments to result from a defect in the hypothalamus and the inability of the mutant organ to induce gonadotropins from the anterior pituitary.

The hypothalamic primordium resides immediately adjacent to the eye primordium in the anterior neural ectoderm. Thus, the non-functioning of this ectodermal organ may be caused by the same inability to respond to induction as the eye, and may be the result of a direct pleiotropic activity of the eyeless gene in both the ectodermal cells that give rise to the eye and the hypothalamus. In both of the previous cases, the existence of the inducer and its effect on the responding tissue is inferred by the developmental attributes of each system.

Unfortunately, little concrete evidence exists as to the nature of the inducer or its mode of action. However, in the case of a mutant gene that affects development of mammalian secondary sexual characteristics, we do possess more definitive information. This is the *Testicular feminization* locus (*Tfm).* This gene is inherited as a sex-linked character in humans, mice and rats. Examples also may be present in dogs and cattle. Females heterozygous for the mutant gene, *Tfml +,* are ostensibly normal, but one-half of their genotypically male progeny (*Tfm/* Y) are phenotypically female, although sterile. In order to understand the mechanism of action of this gene it should be realized that all mammalian embryos begin development indeterminant as to sex. Before development of the gonads, XX and XY embryos have both a Wolffian (male) and Mullerian (female) duct system as well as an undifferentiated urogenital sinus. If the embryo is XX in genotype, the Wolffian ducts degenerate, whereas the Mullerian ducts develop into fallopian tubes, uterus and vagina, and the urogenital sinus into the female external genitalia. This developmental programme is the "ground state", that is, it will take place even in the absence of ovary, for example, in a castrated male.

In XY embryos, on the other hand, the gonadal primordium develops rapidly into testes that begin early on to synthesize and secrete testosterone. This hormone then actively promotes secondary male sexual development, inducing the Wolffian ducts to form vas deferens, seminal vesicles and ejaculatory ducts, and the urogenital sinus to form the external male genitalia. Moreover, the sertoli cells of the testes secrete an anti-Mullerian factor, which causes the regression of the female Mullerian duct. Developmental studies on *Tfm* "male" mice by Ohno Lyon, and co-workers have shown that this mutation affects the ability of all tissues in a male to respond to androgens, and thus, by default female development ensues. *Tfm* / Y mice can be shown to have reasonable levels of circulating testosterone and do indeed produce the anti-Mullerian hormone from their testes. Therefore, these individuals are not simply mimics of castrated males. Moreover, high levels of injected exogenous testosterone do not ameliorate the *Tfm* defect. Ohno and Lyon were also able to demonstrate that the kidney of normal and castrated normal male mice can be induced to produce high levels of the enzyme alcohol dehydrogenase by administered testosterone, whereas the kidneys of *Tfm* "males" could not.

Therefore, the lack of androgen response is not only restricted to sex organs but to non-dimorphic organs as well. This fact is further substantiated by the fact that *Tfm* human "males" do not form axillary or pubic hair at puberty, which is a normal response to rising hormone levels at this point in human development. The basis for this defect has been shown by Ohno as well as Meyer and co-workers to reside in the absence of a specific testosterone receptor protein that is apparently ubiquitous in its distribution in male and female tissues. This receptor is not produced in *Tfm*/Y individuals, it is also the case that *Tfm*/ + females only produce the receptor in half of their cells because of the Lyon X-inactivation effect. What is also apparent is that although females produce the androgen receptor, it is not necessary to their normal sexual development. Lyon was able to create tetraparental male mice by fusing +/Y and *Tfm* / Y blastocysts by the method.

Some of the resulting chimeric males proved fertile and transmitted the *Tfm* mutation-bearing X chromosome. Because of this, Lyon was able to create *Tfm*/*Tfm* homozygous female mice. These females are entirely normal and fertile, thereby demonstrating the lack of necessity of the *Tfm* gene product to normal female sexual development. The cases of *Testicular feminization* in mammals and of *eyeless* and *cardiac lethal* in axolotls demonstrate the integral involvement of genes in the development of specific organs and organ systems both at early and relatively late periods of development. That this type of gene is important to evolution was illustrated by the example of the blind cave fish, a situation analogous to the eyeless condition of the axolotl. That blindness in the cave fish and eyelessness in the salamander do not occur by homologous genetic alterations is illustrative of a further point.

The development of any organ or organ system is dependent not on a single gene but on a group of genes, a conclusion also derived from the case of the *Notch* and *shibire* mutations in *Drosophila*, which produce nearly identical defects at similar points in ontogeny. Further, genes of this sort that cause large effects can be modified in their expression by a variety of genetic alterations, including changes in the genetic background and the inclusion of other mutant genes in the developing system. Thus, the production of the organism and its component organs is the result of coordinated sets of gene actions and interactions. Perturbations in these groups at a variety of points are available to the evolutionary process, all of which can lead of morphological change. It should also be pointed out that although the sum of all these sets of gene actions produces a harmonic whole, the sets are not inexorably linked. Mutations can also change one ontogenic process independently of the others, yielding the dissociability that is so necessary to morphological evolution.

Cell Death in Normal Development

The degeneration of the Wolffian ducts in female development and the Mullerian ducts in the male are examples of the initial elaboration of structure and its subsequent necrosis. These processes are, like all the other ontogenetic events we have discussed, under genetic control. A specific example of this fact is demonstrated in the development of tetrapod limbs. In the chicken, limb buds first appear as lateral thickenings of the somatopleure at about 55 hours of development. These buds grow out from the body as protuberances covered by ectoderm and filled with mesodermal tissue. As growth proceeds, the contours of the limb, characteristics of either wing or leg, appear.

The process of contour formation is accomplished by a series of cellular necrosis of the mesodermal regions of the limb. Early in limb-bud formation the necrotic regions can be seen by staining with certain vital dyes and are found at both the anterior and posterior margins of the bud, where it joins the body wall, and in the center of the bud. These three regions are called the anterior and posterior necrotic zones and the opaque patch, respectively. The anterior and posterior zones are responsible for the contouring of the more proximal regions of the limb and the opaque patch for the separation of the tibia and fibula in the leg, and the radius and ulna in the wing. Later in limb development the separation

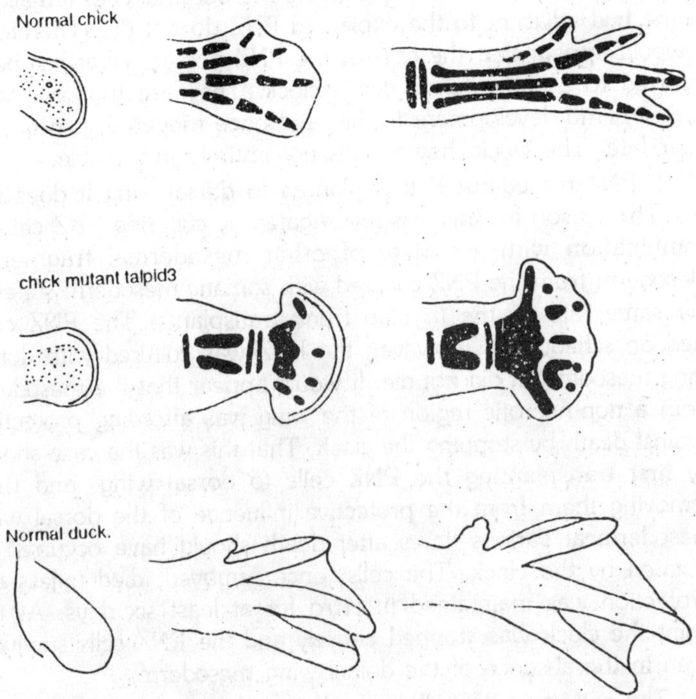

Fig. 2.11. The growth of the chick and duck hindlimb. The stippled regions indicate the position of cell death at the posterior and anterior margins of the limb of both the chick and the duck, as well as the intradigital necrosis observed in the chick but not the duck. The talpid mutation apparently eliminates the majority of necrosis, resulting in the broad syndactylous paddlelike appendage diagrammed in the center of the figure. The dark areas in the normal and mutant chick limbs represent the areas of formation of cartilage.

of the digits is accompanied by interdigital necrosis. Studies on the posterior necrotic zone (PNZ) of the chick wing bud by Saunders and Fallon have revealed several interesting properties of this group of cells. At 96 hours the PNZ is at its maximal extent, with about 1,500-2,000 cells dying. These cells are phagocytized by a population of nearly 150 macrophages.

If the mesoderm of the prospective PNZ is taken from a wing bud 40 hours before necrosis is evident and transplanted to the flank of a host, death still ensues and does so on schedule. This schedule is autonomous to the PNZ cells and is not dependent on the age of the host. However, the mesodermal cells remaining in the bud adjacent to the explanted PNZ do not die. Therefore, it would appear that the cells of the PNZ possess what has been referred to as an internal death clock. They are triggered very early in limb development to die, and once triggered, proceed to their fate. The clock, however, is not entirely irreversible.

If PNZ mesoderm is transplanted to dorsal wing it does not die. The reason for this was investigated by culturing PNZ cells in combination with a variety of other mesodermal fragments. Mesoderm from the PNZ cultured with somatic mesoderm suffered the same fate as the *in vivo* trunk transplants: The PNZ cells died on schedule. If, however, the PNZ was cultured with dorsal wing mesoderm it did not die. It would appear that the mesoderm from a non-necrotic region of the wing was affording protection against death by stopping the clock. That this was the case shown by first transplanting the PNZ cells to dorsal wing, and then removing them from the protective influence of the dorsal wing mesoderm at various times after death should have occurred as dictated by the clock. The cells, once removed, died unless the protection was maintained *in vitro* for at least six days. At the point the clock was stopped entirely and the PNZ cells survived, even in the absence of the dorsal wing mesoderm.

The pattern of the cell death observed in the proximal portions of the wing is therefore a product of two factors: a cellularly autonomous clock that is triggered in a specific group of cells, and the position of those cells with respect to the remainder of the mesoderm of the limb. Analysis of the cell death in the interdigital regions reveals another level of regulation. There is little cell death in these regions of the duck foot. This lack of death results in the webbing present between the second, third and fourth digits. In the interval between the first and second, digits where there is no web, there is necrosis.

Saunders and Fallon constructed chimeras of duck-limb mesoderm with chick ectoderm and the reciprocal. These hybrid leg buds were then grafted onto the flank of host chick embryos, and the pattern of interdigital necrosis observed. In both cases the duck pattern was seen. Therefore, it is possible to conclude that duck ectoderm, like the dorsal mesoderm of the wing, can inhibit the necrosis normally seen in chick foot development. Chick ectoderm, however, does not induce necrosis in duck mesoderm.

Thus, it is possible that the regulation of interdigital necrosis is similar to that in the PNZ; that is, it is an autonomous character of the interdigital mesoderm that may or may not be expressed, based on the environment in which the presumptive necrotic cells come to lie. That the pattern of cell death in limb development is indeed genetically controlled was shown by Hinchliffe and Thorogood in an analysis of the *talpid (ta)* mutation in the chicken. The *ta* mutation is inherited as a simple autosomal recessive and in the homozygous conditions causes osseous polydactyl and soft tissue syndactyl of both the wing and leg.

The observed polydactyly results primarily from a fusion of the radius and the ulna, and of the tibia and the fibula. The resulting limbs, especially the leg, are broad and paddlelike and bear six or seven digits rather than the normal four. The development of these limbs is characterized by an almost total lack of cell death. The absence of the anterior and posterior necrotic zones accounts for the breadth of the mutant appendage because the contouring, which normally narrows this portion of the limb, does not occur. The fusion of the forelimb bones results from the absence of the opaque patch. This failure in separation may also result in the broadening of the distal portions of the appendage and the seeming mirror-image symmetry seen in the distal limb. There is also an absence of the normal interdigital necrotic zones and the subsequent failure in separation of the digits. Interestingly, this same pattern of syndactyly has been observed by Johnson in the *polysyndactylous* mutant of the mouse, but in the absence of osseous of polydactyly. What remains to be determined is whether the *talpid* mutation affects the death clock in the various mesodermal regions, or if it is allowing the rescuing factor produced by dorsal mesoderm to reach the cells destined to die.

However, whatever the proximate cause, it is clear that there is genetic regulation of the process of cellular necrosis, and that cell death is important in the production of the final morphology of the vertebrate limb. With respect to this final point, it is interesting to note as, do Hinchliffe and Thorogood, that the simple bifurcating pattern seen in the limbs of the *talpid* mutants is reminiscent of the pattern of elements found in the paddles of crossopterygian lungfish, such as *Eusthenopteron* and *Sauripterus*. Therefore, it is possible that the complex pattern of development observed in the limbs of higher tetrapod vertebrates may have evolved through a process of genetically regulated and patterned cellular necrosis. There are also mutations that increase regions of necrosis. Thus, mutations in *Drosophila* such as *Bar* eyes or *vestigial* wings, which dramatically reduce the eye and wing, do so by an increase in the regions of cell death normally present in the development of these two structures.

Additionally, the *wingless* and *rumpless* mutations in the chicken seem to act by a similar mechanism, allowing necrosis to excess. One can envision this type of alteration, in a less drastic form, as being important in the reduction or elimination of structures necessary during one stage of development but unnecessary at a later time. A simple example is the tail of the tadpole larvae of frogs. Again, as for the mutations that suppress cell death, there are two conceivable levels at which the process could be regulated. The intrinsic clock could be activated in more cells or extrinsic factors, such as the diffusing, protective substance found in dorsal wing mesoderm may be eliminated.

Genes Involved in Late Events and Growth

It is clear that mutations in genes directly controlling developmental pathways, particularly those functioning early in development, may have cataclysmic effects. However, there are also late-acting genes that, while affecting the overall morphology of the organism often have no obvious deleterious effects. These include genes that control the growth characteristics of the organisms subsequent to the production of basic morphology and organogenesis. Such genes have been discovered as a result of mutations that alter hormone function to produce gigantism or dwarfing.

Alterations in form (e.g., relative size of appendages) can result from changes in patterns of growth introduced as pleiotropic effects of the basic hormonal change. Although this type of alteration has certainly led to evolutionary changes, for example, giantism in European cave bears during the Pleistocene and dwarfing in elephants. These kinds of late developmental changes have not resulted in basic reorganizations in morphology. After all, a pygmy elephant is still undeniably an elephant. There is one size affecting mutation, however that demonstrates the plasticity of the developmental process and thus deserves some special comment. This is the *giant (gt)* mutation in *Drosophila melanogaster*. This sex-linked recessive trait was originally discovered by Bridges and Gaberchevsky in 1928.

The entirety of *Drosophila* development, from fertilization to adult, normally takes place in 10 days. Individuals bearing the *gt* mutation take from two to five days longer to complete the same process. The resulting *gt* individuals are normal in morphology, but are nearly twice the size of normal flies. The manner in which this change is produced is quite interesting. The development of *gt* flies is normal from the embryonic through the late—third larval instar. However, at the point in time when normal individuals pupate and begin metamorphosis, the *gt* individuals continue on as larvae. It is during the larval instars and early pupal stage that those groups of cells, the imaginal discs which are destined to from the tissues of the adult fly, proliferate, and it is the number of these cells that controls the size of the fly. The *gt* larvae, during their period of extended larval development, apparently go through at least one extra cell division. This is implicated by the fact that those cells in the larvae that are polytene (e.g., the salivary glands) participate in at least one, and in some cases, two extra rounds of DNA synthesis. The *gt* larvae, after this 2-5 day extra period of growth, pupate at about twice normal size. Metamorphosis then takes place and a morphologically normal albeit double sized adult ecloses after a slightly protracted pupal stage. Therefore, the animal is able to regulate its development such that no extra elements are formed, despite the apparent proliferative event prior to differentiation.

We have previously encountered this type of developmental plasticity in the evolution of early ontogenic events in the tunicates and lower chordates. In these organisms, as an adaptation to the relative importance of the larval stage in different species, the numbers and the times of the cleavage divisions have changed. This has resulted in differing portions of the early embryo of different species contributing to specialized larval organs. This demonstrates the ability of two quite different developmental systems to take a major alteration into stride and produce a completely integrated organism. Other terminally acting genes include those that control patterns and amounts of pigment production and the effects of genetic variation in these genes are obvious in natural populations of most organism. Although these kinds of alterations are clearly important to selective processes, and thus evolution, they are probably not important to alterations in morphology *per se*. Our basic tenet is that genes control ontogeny, as we have shown, this control is exerted at several levels.

Maternal effect mutations can be utilized to demonstrate the genetic control of the organization of the egg in organisms, such as *Drosophila*, that have a mosaic mode of early development. Other mutations, such as *tailless* in the mouse and the *cardiac* and *eyeless* defects in the axolotl, indicate the existence of genetic controls over subsequent stages of development, just as specific genetic information is necessary for the proper function of the cascade of ontogenetic events resulting in differentiation of the basic germ layers of the embryo and the proper inductive events necessary to organogenesis. Finally, genetic changes can alter later events in development, including growth and pigmentation, and thereby modify that final form of the adult organism.

The evolutionary process may select alterations in expression of the class of developmentally important genes discussed in this chapter to produce new developmental pathways. What must be stressed, however, is the difference between this class of gene and the nature of the mutation (alterations in expression) important in evolutionary change. The mutations discussed in this chapter have been, for the most part, drastic and deleterious in their effects. They are important in revealing the genetic elements underling certain developmental processes. In all likelihood, it is mutations that are more subtle in their effects that contribute to morphological evolution. Mutations that alter timing or duration of events or strengths of interactions will all result in evolutionary modification of developmental pathways. An altogether different class of genes and mutational changes are also crucial to pattern formation and morphogenesis.

3

Expression of Gene in Development

It is a well-known fact that almost all eukaryotic cells retain a complete set of gene sequences, and that many but not all cells retain in intact genetic potential. It is now appropriate to examine how genes are expressed and regulated with the eventual aims of determining whether gene regulatory mechanisms themselves control the process of differentiation, and how the different patterns of gene expression that characterize cells of different types come to selected and stabilized.

In the genomes of eukaryotic cells, patterns of gene expression are very variable; some genes are constitutively expressed, while others are only rarely transcribed.

In eukaryotic cells, it seems that certain classes of genes are transcribed more or less continuously, and only in extreme situations is their activity repressed. For example, genes coding for larger ribosomal RNA (rRNA) or transfer RNA (tRNA) are present as multiple copies. These genes are transcribed uniquely by RNA polymerase I for the larger ribosomal RNA or RNA polymerase III for tRNA and 5S RNA, and they may not be open to the modulation of expression that characterized many other sequences. However, although the products of some of these genes, the ribosomes are used continuously in all cells, it does not follow that all of these multiple copies are continuously transcribed at maximal rate. Electron micrographs of spread chromatin from nucleoli often indicate that some of the repetitious rRNA genes are inactive. It is also true that in the nucleated erythrocytes of lower vertebrates such as *Xenopus*, all genes may be turned of (Maclean, Hilder & Baynes, 1972), including those for ribosomal and transfer RNA. Therefore, it is clear that mechanisms do exist for inactivating sequences, even those deemed to be constitutive in normal cells.

One of the clearest demonstrations that some specific genes are at least *available* for transcription in different kinds of differentiated cells is provided by examining *Drosophila* polytene chromosomes in different larval tissues of these flies.The pattern of bands and interbands in these chromosomes does not vary between different tissues, yet it is now concluded that the interband regions probably represent 'housekeeping' genes which code for essential proteins, that they are expressed in every cell, and that they are retained in a state of permanent decondensation. Thus, although the present interpretation of these chromosomes and their underlying pattern of gene expression is subject to verification, it supports the logical view that 'housekeeping' genes may be 'left on' for much of the life of the cell. Of course, there is none part of the life of the cell when transcription of even the most essential house-keeping genes ceases, namely mitosis, and so again even these genes are regulated to some extend within phases of each cell cycle.

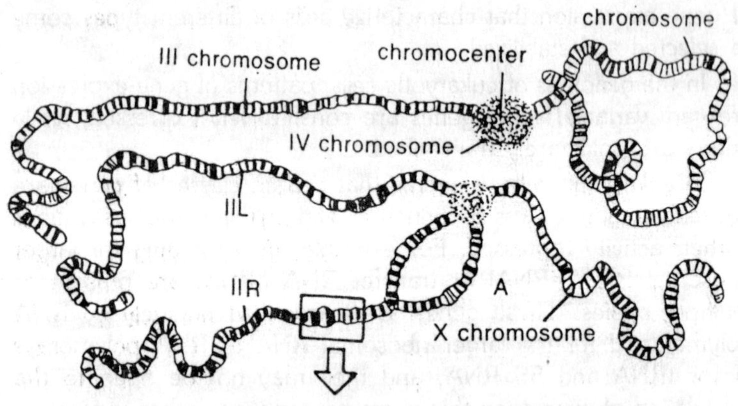

Fig. 3.1. Polytene chromosomes of Drosophila.

Turning our attention to cell-specific genes,which code for the products only found in specialized tissues, it is immediately clear that differential expression is the rule. Whether expression is measured at the level of thc messenger RNA or the protein, genes coding for products such as globin, crystalline, fibroin, ovalbumin, casein, and immunoglobulin give every indication of complete repression in all but the specialized tissue characterized by their presence. Support for this tenet of restricted expression also comes from electron microscopy of spread films of eukaryotic chromatin, in which investigators have sought transcription complexes. These studies underline the comparative rarity of the event of transcription and the fact that most of the DNA within the nucleus of a differentiated cells is not being actively expressed. Indeed it has been calculated that there is only one molecule of RNA polymerase II per 750 nucleosomes-worth of DNA in an average cell nucleus, i.e. one enzyme molecule per 150000 base pairs of DNA.

But discussion in this area can all too easily become a mixture of overgeneralization and oversimplification. Before we proceed to draw firm conclusions on the general pattern of gene expression in differentiated cells, it will be useful to tabulate such evidence as is available and take account of any conflict in the conclusions that emerge. We would urge the reader, wherever possible, to consult the original literature referred to, since what is presented here is itself a considerable generalization.

Situations of Total Genetic Shutdown

(a) During the mitotic phase of the cell cycle, chromatin is highly condensed to form chromosomes, and transcriptional activity of all genes is suspended.

(b) During meiotic division of germ cells a somewhat similar situation to (a) is evident, although in some rare cases, such as the lampbrush chromosomes of meiotic diplotene in vertebrates, transcription proceeds very actively.

(c) The nucleus of mature nucleated erythrocytes of amphibians is transcriptionally inactive. Chromatin in these cells is highly condensed, but not organized into diserete chromosomes and transcription can be partially reactivated in these nuclei by transferred them into new cytoplasm or exposing them *in vitro* to altered environmental conditions.

(d) Sperm cells clearly contain a complete genetic endowment but no transcription occurs until the sperm nucleus is activated within the egg cytoplasm.

(e) Many other situations involving complete suspension of transcriptional activity are known including cells within some

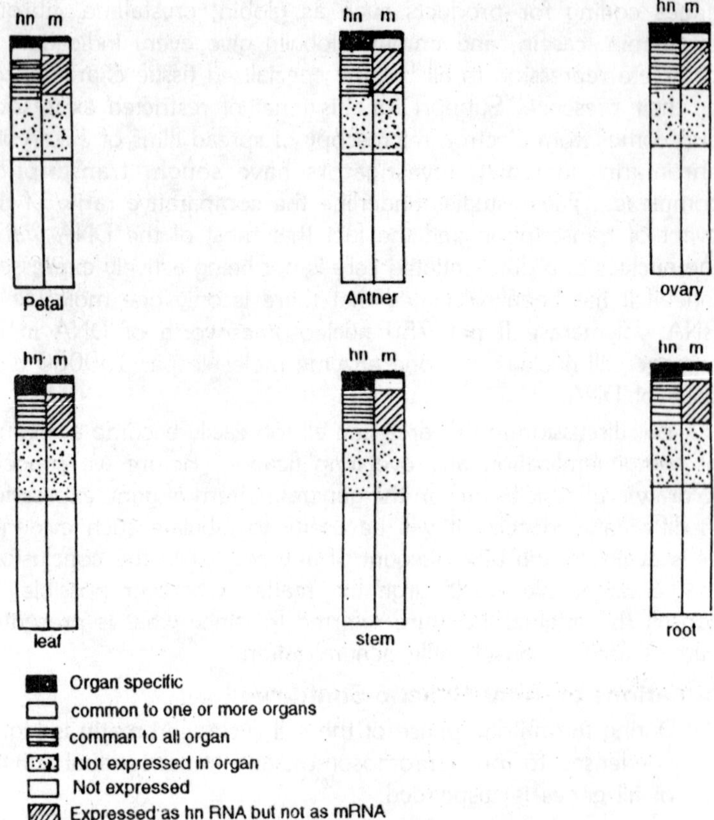

Fig. 3.2. Regulation of gene expression in the tobacco plant. Histograms indicate the quantities of heterogeneous nuclear (hn) RNA (left-hand column) and messenger (m) RNA (right-hand column) as a percentage of the total single-copy proportion of the tobacco genome, itself representing 40% of the genome, hnRNA represents the fraction of single-copy DNA expressed as nuclear RNA, and mRNA represents the fraction expressed as polyadenylated cytoplasmic RNA. The mRNA sector, although drawn to the same block size as the hnRNA, represents only 12% of single-copy DNA, 5% of the genome, while the hnRNA plus mRNA represents 50% of single-copy DNA or 205 of the genome.

plant seeds, cells within diapausing *Artemia* gastrulae, cells within inactive organisms such as desiccated *Tardigrada*, nuclei within bacterial and fungal spores, and nuclei within desiccated amoeba cells, as for example in the slime mould *Dictyostelium*.

Evidence for Constitutive Expression of Some Genes

(a) If the interbands of *Drosophila* polytene chromosomes are correctly interpreted as being loci for 'house-keeping' genes, then the evidence is that such chromatin is permanently decondensed and is transcribed at a low but constant rate.

(b) Electron microscopy of spread films of DNA extracted from nucleoli reveals tandemly arranged sequences coding for the 458 precursor of ribosomal RNA, each gene adorned with a Christmas-tree arrangement of RNA in the process of synthesis. Many, though not all, of these genes seem to be transcribed at maximal rate, with RNA polymerase molecules packed thickly along the coding sequence of each ribosomal sequence.

(c) There is a constant and universal requirement for the products of certain genes in all cells and at all times. These include products such as the three kinds of rRNA, 28S, 18S and 5S (although only the smaller population of somatic-type 5S genes are universally expressed). tRNA of the 20 basic type and a few hundred proteins such as histones, unbiquitin, lactate dehydrogenase, RNA polymerases, and the like.

(d) The interesting analysis done by *Goldberg* (1983) of mRNA in the tissues of tobacco shows that some 8000 different mRNAs, representing some 1.5% of the single-copy DNA, are common to all tissues examined. Many of these mRNAs, judging from their hybridization kinetics, are likey to be products of either single-copy genes, or genes with very low copy numbers.

Many genes are expressed only in certain tissues, often for very short period of the life of an individual cell.

(a) The interesting regulation of 5S genes in *Xenopus*, provides an important example here. *Xenopus borealis* possesses 19000 copies of the oocyte-specific 5S gene, and these genes are active in only the oocyte and no other cell. The few hundred copies of somatic-type 5S genes are, by contrast, active in the oocyte and in all somatic cells.

(b) The enzyme *lactate dehydrogenase* (LDH) is coded by a small family of genes, each gene determining the structure of a subunit. Subunits A and B are expressed in almost all mammalian cells, but one of the genes in the family, coding for subunit C, is active only in spermatocytes within the developing testis. Another gene, coding for LDH subunit E in fish, is active only in cells of the brain and the eye. However, it is true that in these examples, as in many others, the gene activity or inactivity is monitored at the level of the protein product, and it is only assumed that transcriptional regulation occurs. In this case it is probably a correct assumption.

(c) To take account of the caveat mentioned at the end of (b) above, it is appopriate to mention an experimental approach that monitors gene activity at the transcriptional level by detecting specific messenger RNA molecules. Using a radioactive complementary DNA (cDNA) probe, made by nick translation and reverse transcription from a purified mRNA sample, it is possible to assay for messages in specific cells by *in situ* hybridization. Although the method is not sensitive to very small numbers of molecules, it does provide convincing support for the restriction of cell-specific protein mRNAs to the tissues characterized by these products.

(d) The puffing of restricted segments of the *Drosophila* polytene chromosome provides visible evidene of the activity of genes coding for cell-specific products, according to all recent data relating to these structures. Certain puffs, known as heat-shock puffs, can be induced to appear specifically when salivary glands are exposed to heat shock either *in vivo* or *in vitro*. These puffs can also be induced by exposing polytene nuclei from salivary glands to cytoplasm from heat-shocked *Drosophila*

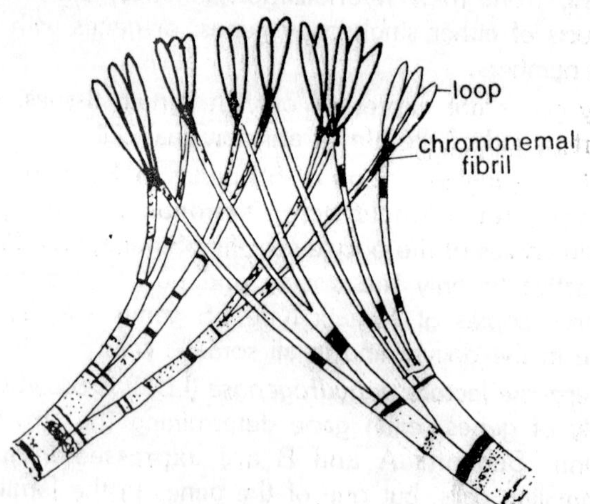

Fig. 3.3. A chromosomal loop of chromosome puff.

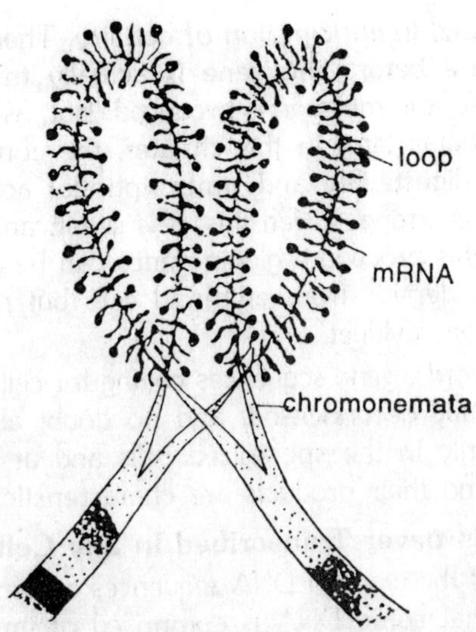

Fig. 3.4. Suggested form of the individual loops in chromosome puffs. Note the tiny fibrils with dense RNA granules attached to the loop.

tissue culture cells. Here then is additional evidence for the decondensation of specific genes. The correlation between such chromatin decondensation and transcriptional activity is readily proved by autoradiography using tritiated precursors of RNA.

(e) An interesting way to detect active gene sequences is to exploit the fact that such sequenes, in contrast to inactive genes, are not condensed. They are therefore amenable to digestion with endonuclease enzymes. An enzyme such as DNAase I will readily cut up DNA in extended chromatin but is relatively ineffective in degrading DNA within condensed chromatin. Exploiting this methodology, it is possible to digest chromatin from various tissues, then to challenge the DNA with specific DNA probes for the presence or absence of specific gene sequences. Absence of the sequence is deemed to imply its digestion and therefore its state of decondensation in the tissue from which such chromatin was obtained. Conversely, the ability of the probe to hybridize will confirm the presence of the sequence and thus indicate its earlier state of condensation. It must be concluded that decondensation of chromatin in the vicinity of a gene sequence does not necessarily indicate its transcriptional activity — there is evidence for sequences being decondensed in *anticipation of activity*. Therefore, sometime may elapse before the gene is actually traversed by RNA polymerase and message is produced. But, as will be discussed in some detail later in this chpater, the correlation between DNase I digestibility and transcriptional activity of a gene sequence is strong. Given this, it is significant to observe that by using this procedure globin genes can be digested away in chromatin derived from erythroid cells but not in chromatin derived from oviduct.

In other words, gene sequences coding for cell-specific proteins are preferentially decondensed, and no doubt also preferentially transcribed, only in the specialized cells and at the times when these genes and their products are characteristic.

Some DNA is never Transcribed in any Cell

Analysis of the types of DNA sequences occurring in eukaryotic cells reveals that some DNA is comprised of tandemly repeated short sequences that are concentrated in heterochromatin such as the centromeres of chromosomes, and the Y chromosome. Present evidence suggests that much of this DNA is never transcribed in any cell. Some spacer sequences occur between genes, for example between multiple copies of genes for ribosomal RNA. Such spacer sequences are often take to be untranscribed some of the abundant small nuclear RNA molecules may be derived from some DNA spacers previously assumed to be untranscribed. But the very large size of the genomes of the higher eukaryotes certainly indicates that much of the DNA is redundant and probably not utilized as either coding or regulatory sequence. The pseudogenes found in many gene families, often presumed to have arises as cDNA copies of reverse transcribed message, not only lack introns but are often liberally supplied with stop signals that will prevent RNA polymerase molecules from transversing very far along them.

Let us now attempt to summarize the conclusions that can be drawn from these various observations. All genes are amenable to some degree of regulation in some cellular situations, even in house-keeping genes, which are normally active in almost all type of cells. But in the main it is reasonable to distinguish between the constitutively expressed house-keeping genes, routinely transcribed in most cells, and the much greater number of genes coding for cell-specific proteins. The products of these genes may appear for only confined periods of a cell cycle, or only in one or two special tissues or cell types. But this is not to say that genes are either on or off. Far from it. Rate control is a highly important aspect of gene regulation, and the impression of transcription being either on or off is almost certainly a superficial impression created by our still very limited understanding. Having established the main pattern of gene expression at a transcriptional level, it is now timely to move on to examine in greater detail the various mechanisms of gene regulation that are known, and to try to determine how they jointly operate to provide the astonishing variation and fine sensitivity of selective gene activity that is so fundamental a trait of cell differentiation and compartmentalization.

Mechanisms of Gene Regulation and Expression

It is perhaps permissible in a book on differentiation to concentrate on protein synthesis in our consideration of gene expression at the phenotypic level, and to allude only briefly to the fact that the post-translational regulation is an important aspect of phenotypic modification and therefore, ultimately, of gene expression. Thus, although a mutation that prevents conversion of proinsulin to insulin will present itself as an organism deficient in insulin, a useful discussion of gene expression must, be set within limits, and the main limit set here will be at the level of protein synthesis. Very briefly, however, we should note that the stability of different proteins in different cells is very variable. Thus, as originally set out by Schimke and Doyle (1970), within rat liver tissue the half-lives of different types of protein molecules range from 1 hour to 5 days; the same protein may also enjoy different life expectancies in different cell types, as demonstrated for the LDH isozymes. Thus LDH 5 in the rat has a half-life of 1.6 days in heart muscle cells, 31 days in skeletal muscle cells, and 16 days in liver cells. So in concentrating on transcriptional and post-transcriptional levels of gene expression we should be mindful that other variables that affect gene expression are operative at higher levels of the cell and the organism. But we do not know of any example in which the restriction of a particular protein to a particular type of differentiated cell is determined by protein stability. The interesting example of the limitation of histone availability to particular times in the cell cycle, which is in part a result of controlled degradation, will be discussed later under post-Transcriptional regulation. This leaves open for consideration both transcriptional regulation at the level of the gene, and post-transcriptional regulation between the RNA and the protein product. It is best to begin with the gene, and consider first the mechanisms that are known to determine and affect gene activity at the level of transcription.

Transcriptional Regulation–Chromatin Conformation

It is now well established that chromatin consists of the more or less linear molecule of DNA wound around beads of histone molecules in complexes called nucleosomes.The molecular structure of the nucleosome is that of a central aggregate of two molecules each of histone H3 and H4 complexed with two dimers each made up one histone H2A and one H2B, on either side. This provides a total aggregate of eight molecules of histone. Almost two complete turns of DNA, approximately 150 bases long, is wound around this complex, and another 50 based of DNA acts as a linker between adjacent nucleosomes. Since only about 200 bases of DNA are involved in each nucleosome, and a gene is on average close to a kilobase in length, it is clear that a nucleosome cannot in any strict sense be a genetic unit of function. Present evidence favours the view that nucleosomes continue to be present on most transcriptionally active DNA sequences, but they are probably reduced in number. Thus, although evidence from some laboratories suggests that the ribosomal genes of the amphibian nucleolus lack nucleosomes (the small beaded structures visible on such genes during transcription are almost certainly the molecules of RNA polymerase I), active gene loci in *Drosophila* give a positive reaction to antibodies against H3 and H4, indicating that at least these, subunits of the nucleosome persist on such DNA. The persistence of nucleosomes in the promoter regions of genes is also open to question, and certainly in some active genes the nucleosomes are displaced or 'phased' in these regions.

Whether nucleosomes remain on all active genes or not, it is clear that the conformation of the chromatin is dramatically altered to allow its decondensation in regions where transcription is occurring. Such chromatin condensation is itself a very important aspect of eukaryotic gene regulation. Although the puffing of confined regions of the dipteran chromosome had long suggested a correlation between chromatin decondensation and gene activity, it was not until the use of specific DNase enzymes became commonplace in the late 1970s that the strict correlation became apparent. DNase I is an endonuclease that introduces single-strand nicks into DNA; when it is used at relatively low concentrations, as in the experiments of Weintraub & Groudine (1976), it is found to have a marked specificity for gene sequences that are active in transcription. Thus the enzyme will preferentially cut the globin gene sequence in erythroblasts but not in oviductal tissue, and it will selectively cut the ovalbumin gene sequence in oviductal tissue but not in erythroblasts. Now it must be stressed that this correlation is not proof of causality — we do not know for sure that the open conformation invariably precedes and permits transcription, since both of these could follow independently from some other event. But most evidence favours the view that decondensation does indeed often precede transcription. In the erythroblast of the chicken, it is possible to detect the decondensation event in the globin gene before transcription has begun. Curiously, genes that are transcribed at very low rates seem to be just as easily digested as those that are maximally transcribed, so the state of decondensation does not itself determine the rate of polymerase activity and movement. Chromatin decondensation seems to be a necessary prerequisite for gene activity, but it alone is not sufficient to ensure gene activity.

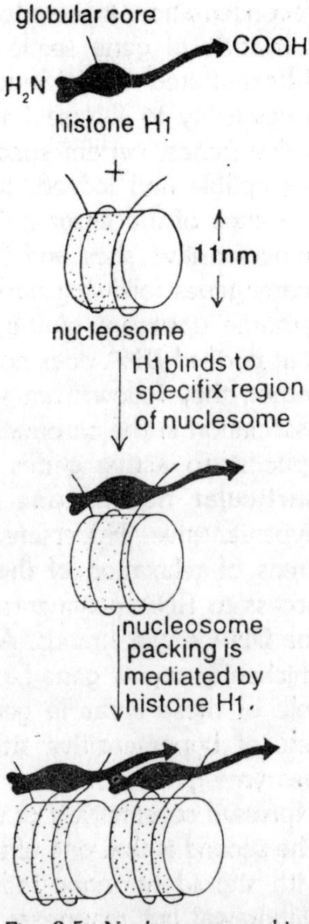

Fig. 3.5. Nucleosome packing to form solenoid.

Two additional points need to be made about chromatin decondensation. The first is the interesting observation that the decondensation event can often involve regions that extend thousands of bases upstream or downstream from the transcribed sequence. This seems to be an observation of particular significance since it has long been known that the puff of the polytene chromosome similarly extends over a region of chromatin much larger than a gene. So, whatever the significance is of this extended decondensation, it seem to be a characteristic aspect of the specific activation of gene sequences coding for cell-specific proteins in differentiated cells. Secondly, in addition to the generalized accessibility to DNase I that characterizes chromatin containing active genes, certain specific sites in such genes are particularly susceptible and indeed, tend to suffer double-strand cuts in the presence of the enzyme. These sites have come to be known as hypersensitive sites and have been located in the sequences of many genes following nuclease attack, especially in regions of the genome upstream of the coding sequences. It must be stressed that purified DNA does not demonstrate these hypersensitive sites; rather, they follow from some aspect of DNA folding or protein association in the chromatin. They are relatively but not absolutely specific to active genes and may result from the binding of particular non-histone proteins to the sites that become hypersensitive. Hypersensitive sites may also represent localized areas of relaxation of the DNA double helix that would permit access to RNA polymerase and, inadvertantly, nucleases that cut the DNA single strands. An analysis of hypersensitive sites in the chicken lysozyme gene has revealed two additional aspects of the role of these areas in gene regulation. The first is that distinct sets of hypersensitive sites have been found upstream of the lysozyme gene promoter, depending on whether the gene is being expressed constitutively or under specific steroid hormone induction. The second is that one of the DNase hypersensitive sites correlated with steroid hormone induction disappears on steroid hormone withdrawal but reappears on secondary induction. This suggests that some sites are present upstream from potentially active genes, while others are present only in association with genes that are truly active.

A further twist to the story of DNase I hypersensitive regions involves the discovery that most of these regions contain specific sites that are sensitive to enzymes such as S1 nuclease (a nuclease from Aspergillus that degrades single stranded DNA), and to bromoacetaldehyde (a chemical that detects non-B-from DNA). There is evidence to suggest that these sites are supercoiled and under torsional strain, at least for part of the time. Some of these sites may contain Z-form DNA or be composed of short regions of supercoiled DNA that is recognized by regulatory molecules.

Supercoiling is known to be important in prokaryotic gene regulation and there is some evidence to implicate it in that of eukaryotes also. Torsional stress could have the capacity to force open the double helix in specific sites, thus permitting improved access to the base sequence by gene regulatory molecules and transcription factors.

Experiments with the virus SV40, which has its genome a chromatin complex of DNA in nucleosomes, suggest that DNase I hypersensitive sites are free from nucleosomes and are located in the SV40 circular genome near the points of initiation of both

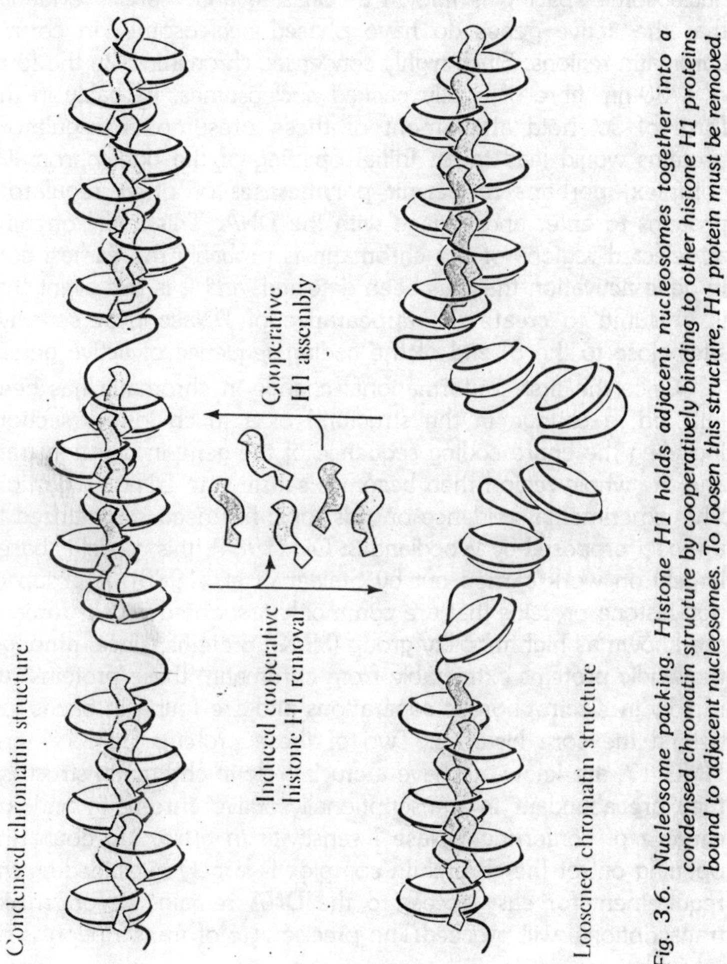

Fig. 3.6. Nucleosome packing. Histone H1 holds adjacent nucleosomes together into a condensed chromatin structure by cooperatively binding to other histone H1 proteins bound to adjacent nucleosomes. To loosen this structure, H1 protein must be removed.

DNA replication and transcription. If this interpretation of DNase I hypersensitivity is applied to the eukaryotic genome, it would be assumed that such sites become apparent by the removal of nucleosomes from defined areas of chromatin close to the promoter regions of genes. What leads to this stripping of the nucleosomes? If we knew this it seems we might be close to understanding the primary signal for gene activation. There is some evidence to suggest that specific regulatory proteins may interact with the chromatin in these regions, thus destabilizing the nucleosomes and leading to loss of normal nucleosome spacing. This change in nucleosome spacing is known as phasing and there is evidence that the active genes do have phased nucleosomes in certain chromatin regions. Since highly condensed chromatin is in the form of a 30-nm fibre of tightly packed nucleosomes, probably in the form of solenoid attachment of these presumptive regulatory proteins would lead to an initial opening of the tight chromatin complex, perhaps to permit polymerases or other regulatory proteins to enter and engage with the DNA. This initial opening at selected regions of the chromatin is probably the earliest step in gene activation that has been detected, and it is this event that is presume to create the appearance of DNase hypersensitive sites close to the 5' end of the coding sequence of active genes.

Once the first conformational change in chromatin has been initiated, a change in the structural of a much longer section, including the entire coding sequence of the gene in question train and the whole region then becomes sensitive to DNase digestion. The experimental evidence on this topic has been summarized in a model proposed by Woodland & Old (1984); this model is based largely on work carried out by Stalder et al. (1980). A group of non-histone proteins that are commonly associated with chromatin are known as high mobility group (HMG) proteins, since, amongst the acidic proteins extractable from chromatin, these proteins run rapidly in electrophoretic separations and are found in bands just behind the core histones. Two of these proteins, HMG 4 and HMG 17, are known to have a crucial role in chromatin structure; they are abundant in transcriptionally active chromatin and are capable of conferring DNase I senstivity *in vitro*. No doubt the opening out of the chromatin complex is largely explained by the requirement for easy access to the DNA remains decondensed, transcriptional will proceed.The precise rate of transcription is no doubt governed by many others factors and will vary from gene to gene and from cell to cell. As was stated above, this decondensation event does not in itself guarantee transcriptional activity since gene sequences have been detected with either DNase I hypersensitive sites, or generalized DNase I sensitivity, or both, but in which three has been no indication or transcriptional activity. There is, however, a reasonably good general correlation between these events and the activity of genes affected by them.

A number of interesting question arise from this brief account of chromatin decondensation events. The first is the nature of the factors that may be responsible for the initial induction of hypersensitive sites. Although still uncertain, the answer to this question will be attempted later in this chapter in sections dealing with gene regulatory molecules. A second question relates to the likely roles HMG 14 and HMG 17 play in rendering nucleosomes amenable to nuclease digestion through decondensation. Weisbrod & Weintraub (1979) have examimed this question by stepwise depletion of chromatin protein, followed by exposure to HMG 14 and HMG 17, and subsequent assay of DNase I susceptibility. These authors found that all proteins except H3 and H4 can be removed without obliterating the endonuclease sensitivity induction effect of HMG 14 and HMG 17. So it may be that the acidic HMG proteins are binding to the central core histones, but whether this truly initiates the nuclease sensitivity is less certain, since clearly some conformational change following the earlier development of hypersensitive sites may itself be necessary to permit the binding of the HMG molecules. The mechanism by which generalized decondensation occurs is still not understood, and the observation that puffing can be infectious between homologous regions suggest that it may have an autocatalytic aspects.

Transcriptional Regulation–DNA Methylation

The genomic DNA of higher eukaryotes is modified following replication so that a large proportion of the cytosine (C) residues are present as 5-methylcytosine (5mC). Curiously such methylation has not been detected in the DNA of lower eukaryotes such as yeast and *dictyostelium*, nor in the fuit fly, *Drosophila*.The percentage of methylated C residues in DNA relative to unmethylated C residues is highly variable, from less than 1% in some insects to over 505 in some higher plants and vertebrates. These generalizations are of only passing interest, however, since

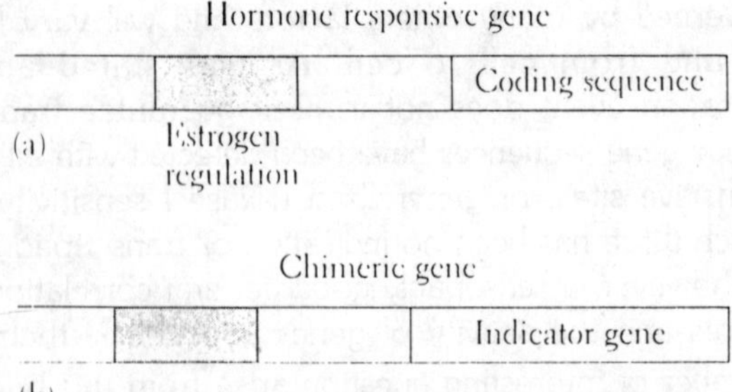

Fig. 3.7. Model demonstration of gene regulatory regions. DNA sequences near the 5′ end of genes appear to control gene expression by binding to regulatory proteins. (a) A short (10–50 bp) region near the gene is sufficient to confer regulation of the hormone estrogen. (b) If the regulatory region is placed in front of another gene, the expression of that gene will respond to estrogen.

it is probable that only the methylation values of specific genes and gene regions are of particular significance in terms of gene regulation. Most, though not all, of the methyl C residues occur in a dinucleotide sequence with guanine (G), 5'-CG-3', and more than 50% of such sequence are modified in most eukaryotic chromatin. Invariably the methyl groups are associated with symmetrically opposed CG pairs so that CCGG on one DNA strand is hydrogen bonded to a GGCC on the other strand, the starred cytosine being methylated in each case. This symmetry of modifications also ensures that a methylation pattern in DNA is faithfully copied by post-replication modification, so daughter strands come to have the same methylation pattern as the mother strand.

Here then is a permanent modification to DNA which, once effected, is passed on to successive generations of molecules.This is certainly an interesting molecular phenomenon in view of the remarkable stability of cell differentiation, and in recent years many biologists have wondered whether genetic shut down by DNA methylation could be one of the factors involved in accomplishing the stable patterns of gene expression that characterize different cell types. DNA methylation is relatively easy to detect since pairs of restriction enzymes are known in which one of the pair is insensitive to methylation in its nuclease action and the other is not. Such a pair of isoschizomers are Hpa II and Msp I, which both cleave the target sequence CCGG, but whereas Msp I will

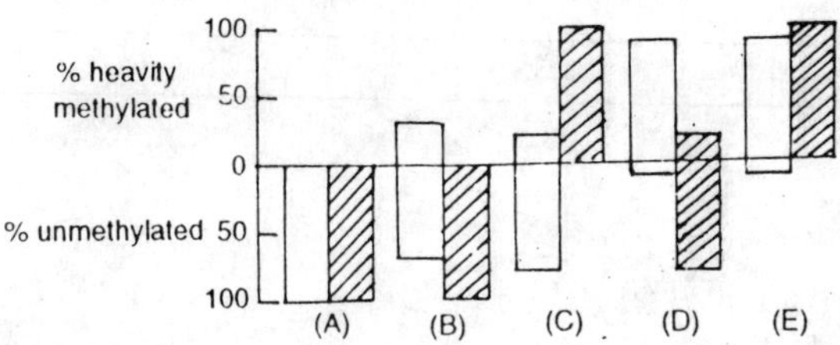

Fig. 3.8. The distribution of total DNA (open bars) and rDNA (hatched bars) between methylated and unmethylated fractions of the genome. (A)–(E) represents one combination that has been observed. At one extreme (A) are many insects in which 100% of rDNA and total DNA is unmethylated (within the sensitivity of the experiments). At the other extreme are some vertebrates and green plants in which most genomic DNA and all rDNA is in the heavily methylated compartment. In cases where the DNA is divided between heavily methylated and unmethylated compartments, the values are not precise.

cut this sequence when the internal C residue is methylated, Hpa II will not cut in these circumstances. Genomic DNA can thus be cleaved with either of these enzymes, and the restriction fragments can be run on an appropriate gel electrophoresis system and subsequently transferred with Southern blotting to nitrocellulose filters. By hybridizing these filters with suitable radiolabelled DNA probes, followed by stringent washing autoradiographs may be prepared that reveal details of methylation within the sequence defined by the probe. Thus, when sites within the probed sequence are methylated, sub-bands that hybridize with the probe will be generated by the Msp I enzyme, but these will be missing from the profile of fragments resulting from the Hpa II digestion. Unfortunately, not all CG sites occur within the sequence CCGG, so only a subset of possible methylation sites can be studied with these enzymes. However, other restriction enzymes are available for a more through localization of modified residues.

It is now clear that the most useful information that can be derived from such studies must include precise localization of the methyl C within a sequence, and indeed probably the most interesting localizations are those that fall within or close to promoter sequences. These comments stem from the fact that there is a broad but not absolute correlation between the frequency of 5mC and the inactivity of the coding sequence. The higher the methylation, the lower the transcriptonal activity is likely to be. But it also seems that methylation of residues within promoter

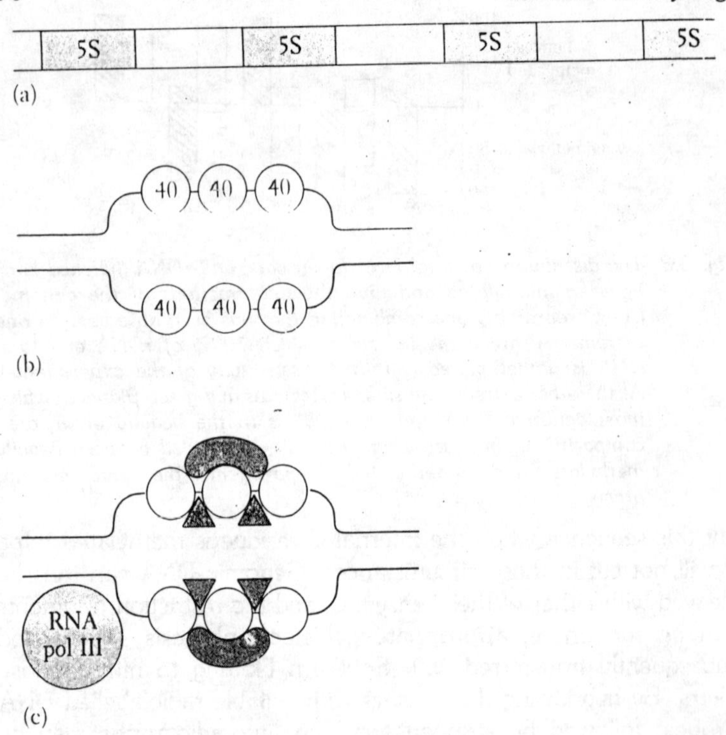

Fig. 3.9. The regulation of 5S ribosomal genes in Xenopus. (a) The 5S genes are arranged in a long tandem array. (b) To initiate transcription, there must be cooperative binding of several molecules of a 40,000 MW protein within the 5S gene sequence to pull the stands of the double helix apart. (c) When this protein binds, other proteins cooperatively bind to make a complex recognized by RNA polymerase III, the enzyme catalyzing the transcription of this type of gene.

regions and other sequences 5' to the coding sequence it most significant. The problem is that although many active genes are found to be under-methylated and thier inactive counterparts in other tissues to be substantially methylated, the correlation is far from absolute and many exceptions exist. A much tighter correlation exists between the methylation or under-methylation of the sequence in the vicinity of gene promoters. DNA of sperm is highly methylated, as is the DNA of the oocyte-specific 5S ribosomal genes in adult tissues, whereas the sites around the coding region of genes such as adult globin, ovalbumin, and immunoglobulin and under-methylated in tissues in which they are expressed but are substantially methylated in other cells in which they are not expressed. However, some genes that are active as determinants of cell-specific molecules in specialized cells have been found to be also rather heavily methylated, and it is premature to draw strict conclusions. The correlation might be much stronger if we were able to compare the degree of methylation in regularity regions upstream of the coding sequences, or even at particular sites in these regions.

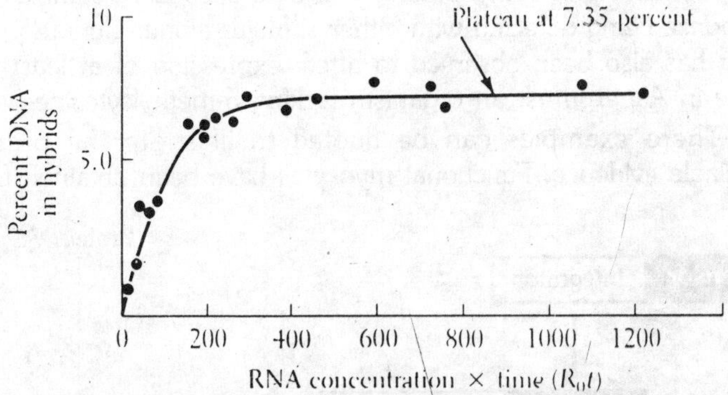

Fig. 3.10. Determination of the number of transcribed genes using saturation hybridization of radioactive DNA with an excess of cellular RNA. R_0t refers to the product of RNA concentration (R_0) multiplied by the length of time (t) of the reaction.

A number of important additional observations about DNA methylation are relevant and are best set out in an ordered arrangement. It will be noted that a number of unresolved questions remains about DNA methylation. Sperm DNA is very highly methylated, yet the paternal genes are widely expressed in the adult and are undermethylated in appropriate tissues. Therefore, some demethylation process must occur. No demethylating enzyme has yet been discovered, and it must be assumed that the reduced methylation occurs by a programmed omission of methylation particular sites following DNA replication. *De novo* methylation of DNA sequences is infrequent but certainly occurs. Unmethylated sequences introduced into vertebrate cells can subsequently become methylated.

The nucleoside analogue 5-azacytidine is believed in inhibit enzmes that methylate cytosine residues following DNA). Thus it might be expected that the addition of this compound might lead to gene activation following a few rounds of DNA synthesis and cell division. Data from experimental used of the drug have been striking although complex. In some situations selective gene activation has indeed been obtained, and even dramatic changes in the differentiated state of cells. One of the confusing aspects of these experiments is that 5-azacytidine is also a mutagen, and some of the affects recorded may be result of mutation. It is also powerfully cytotoxic, and cell selection in treated cultures may be important. There are also several genes where activation requires 5-azacytidine plus some other agent as a secondary stimulus, no response being detected with either stimulus alone. Curiously, the drug has also been observed to affect expression of at least one gene in *Aspergillus*, an organism lacking 5-methylcytosine.

There examples can be quoted to illustrate the present available evidence. Functional myocytes have been obtained from

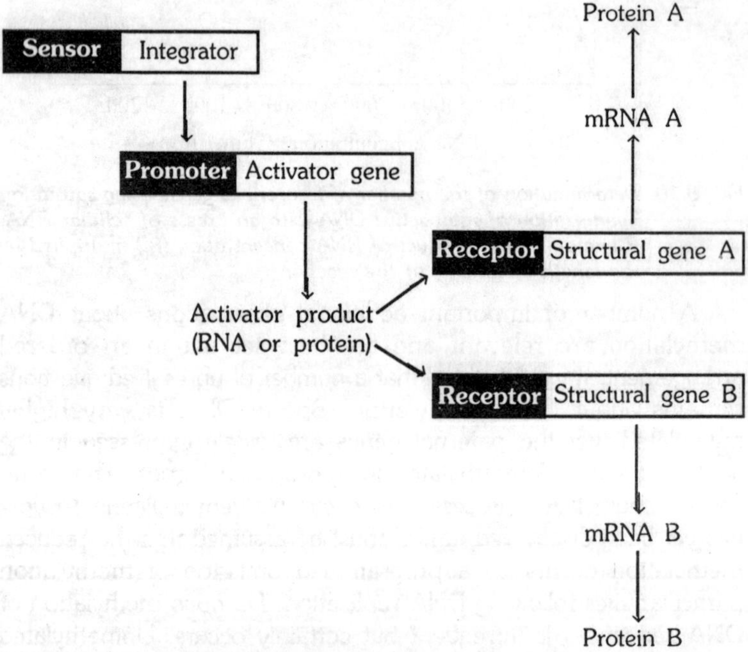

Fig. 3.11. The Britten-Davidson model of gene regulation during development. According to this model, a sensor-integrator gene responds to molecular signals, such as the presence of a hormone in the nucleus, by producing an RNA or protein product called activator. The activator, in turn, binds to the regulatory regions of numerous genes to stimulate their transcription.

cultures of non-muscle mouse embryo cells following a two-week exposure to the drug, and these remain functionally normal in its absence.Thymidine kinase synthesis, although admittedly the product of a house-keeping gene,in increased by up to 10^6-fold following exposure to the analogue. 5-azacytidine has been administered to patients with thallasamemia and sickle cell anaemia, and enhanced synthesis of foetal haemoglobin has been reported. It must be emphasized the specific gene activation in response to this drug is not what would be predicted if general hypomethylation (under-methylation) is occuring. The use of the drug does indicate that DNA methylation is important as a gene regulatory mechanism.

At present the most plausible conclusion from the data is, to quote Jones (1985), collectively, these data argue that 5-azacytidine activated genes by changing their methylation status, but we obviously have much more to learn. The *Xenopus* 5S genes can be highly methylated as well as being transcriptionally active. 5S genes of the oocyte specific series are heavily methylated in somatic cells, yet when these are injected into oocyte nuclei (the genes themselves being within somatic cell nuclei), they become active in RNA synthesis. Assuming that they are not rapidly demethylated by an as yet unknown process, it must be assumed that the regulation of these genes in oocytes is independent of methylation, or at least that the effects of hypermethylation (over-methylation) can be overriden.

There is evidence that the cytosine methylation in DNA alters the structure of the double helix in a fundamental way and favours the transition from B-form to Z-form DNA (Bele & Felsenfeld, 1981). B-form DNA is the most frequent molecular conformation of double-stranded DNA in nature and is right-handed, that is, the helix turns in a clockwise orientation, whilst Z-form DNA, a minor form of molecule of uncertain occurrence in nature, is left-handed. It is conceivable that the B-Z transition is itself involved in gene regulation and that this may be the way in which DNA methylation has its effects on transcription. Although DNA methylation is rare or absent in lower eukaryotes, it is widespread in bacteria and indeed is exploited as a means of protecting the genomic DNA, against endonuclease attack, the cellular restriction endonucleases being directed against invading non-methylated viral DNAs, Methylation in bacteria involves both adenine and cytosine bases, and although both types are commonly present together, some *E. coli* strai ıs possess only 6-methylcytosine. In a somewhat different strategem, foreign DNA injected into eukaryotic cell nuclei is often methylated by the cell, and this may be a means of reducing the likelihood of causallly acquired foreign DNA being expressed to be deteriment of the cell. That this is not always the case is illustrated by the accurate transcription and translation that has followed transfection of some foreign genes into mice and *Drosophila* and the ultimate demonstration of gene-line transfomation of the injected animals and their progeny.

In some recent experiment on mice involving integration of a retrovirous. The Moloney murine leukaemia virus, it was established that not only the genomic proviruses were heavily methylated, but also regions of mouse genomic DNA adjacent to the provirus DNA. When the provious was integrated into a previously hypomethylated DNA region, the virus and adjacent DNA subsequently became hypermethylated. This methylation of genomic DNA was also associated in some mouse strains with an altered chromatin conformation. It is not easy to determine whether the methylation of the adjacent genomic DNA precedes the conformational change in the chromatin or results from it, but it certainly suggest that virus-induced DNA methylation could interfere with appropriate gene activation during embryonic development.

Transcriptional Regulation–Histone Acetylation

Although methylation is the sole modification that DNA normally undergoes in the cell, the histone component of chromatin is subject to three different post-synthetic modifications. Only one of these has been proposed as a major aspect of eukaryotic gene regulation, but all three will be briefly discussed since they do have significant roles to play in cell differentiation. The first, histone methylation, affects only histones H3 and H4, and involves the irreversible methylation of a few lysine residues, all located in the basic amino-terminal regions of these molecules.

Although the methylation of the lysine residues does not affect their net charge, it does alter the hydrophobic nature of the side chain. A second post-synthetic modification process affecting histones is phosphorylation. It affects serines and threonines, changing them from a state of neutral charge to one of negative charge, and is reversible reaction. Histone H1 is the only histone that undergoes major phosphorylation, and this occurs solely in its basic amino-

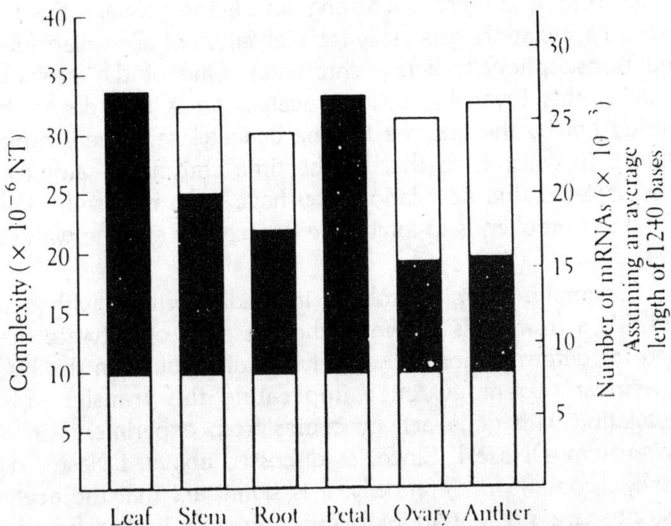

Fig. 3.12. Measurement of the complexity of RNA in different organs of tobacco. The shaded part of each bar represents mRNA types present in all organs. The black zones represent mRNA types shared between leaf and each other organ; note that petals, which are modified leaves, share all transcripts with green leaves. The open part of each bar represents organ-specific mRNAs.

acid carboxyl-terminal regions. The state of phosphorylation of H1 varies through the eukaryotic cell cycle, and after H1 phosphorylation, chromatin becomes much more strongly condensed, as it does in mitotic chromosomes. There is indeed evidence that activation of the histone kinase enzyme that is responsible for H1 phosphorylation may be the first step in the chain of events that leads on to eventual chromatin condensation and mitosis prior to cell division.

The third type of modification undergone by the histones is acetylation, and this is of two kinds. The first is the irreversible acetylation of the amino-terminal serines of histones H1, H2A, and H4. These modifications seem to be associated with histone synthesis. But a second type of acetylation, which is readily reversible, involves lysine residues in the amino-terminal regions of H2A, H2B, H3 and H4. This acetylation converts the normally basic lysine side chain to a neutral acetyl lysine, and thus reduces the net basic charge of the amino-termminal ends of the affected histones. Both H3 and H4 can have up to four lysines in the acetyl form, and there is a strong correlation between this type of histone acetylation, especially tetracetylation of all available lysines, and transcriptionally active chromatin. One of the problems of studying this reversible histone acetylation is that deacetylation, carried out by the enzyme histone deacetylase, is very rapid and difficult to control, so that by the time chromatin samples have been isolated, the acetylation may have been reversed. One way round the problem is to inhibit the deacetylase enzyme with butyric acid.

Chromatin samples isolated in media fortified with butyrate and taken from cells grown in the presence of butyrate may be used for determination of acetyl lysine distribution in the hisones. The most cogent evidence implicating this transient histone acetylation with gene activity comes from experiments involving the enzyme DNase I. Since, as discussed above. DNase I digests active chromatin preferentially, it is significant that the acetylated histones are present in high concentration in the fractions of histones first released from initial digestion of chromatin with this enzyme. Of course, this does not by any means establish a causal relationship. Active chromatin, because of its more open conformation, may have a greater chance of having its H3 and H4 histones acetylated, perhaps because the nucleosomes themselves are in some way relaxed or structurally altered. But there is not doubt that acetylation of the core histone lysines would tend to loosen the nucleosomal structure. So it may be that this transient hisone acetylation is indeed one of the mechanisms employed, along with others, to bring about the transition from a silent condensed gene to a transcriptionally active and extended one.

Transcriptional Regulation–Protein A24

A 24 is an unusual hybrid protein, being a complex of histone H2A and the non-basic protein ubiquitin. The ubiquitin is covalently bound via the side chain amino group of lysine 119 of the histone. Some 10% of H2A molecules are in the form of A24, and these specialized histones seem to be confined to interphase chromatin, disappearing as the chromosomes condense. A24 is highly abundant in the chromatin of active genes, but once more it is uncertain whether this A 24 enrichment of nucleosomes is a cause or an effect of transcriptional activity. The molecule of ubiquitin itself also has a role in moderating protein breakdown.

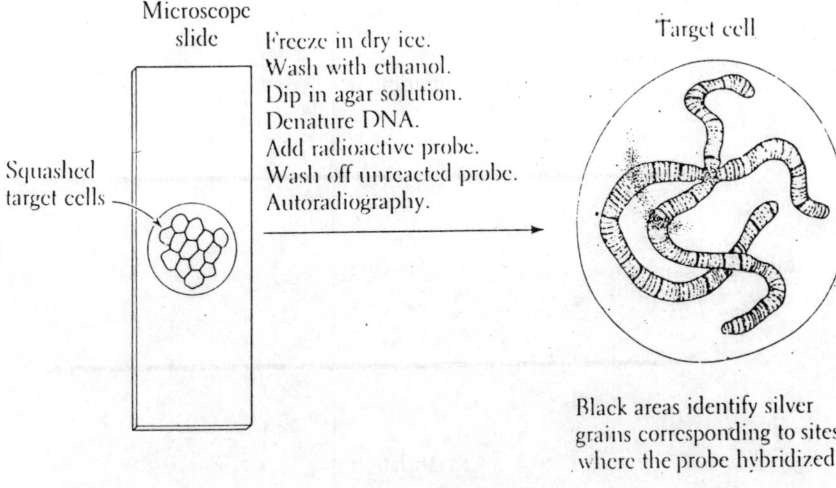

Fig. 3.13. Model of polytene chromosome puffing. The tightly coiled chromatin loops within a band unwind.s

Transcriptional Regulation–Gene Regulatory Molecues

It has now been assumed for many years that the eukaryotic genome is regulated, at least at one level, by specific gene regulatory molecules. Such molecules either are the products of specific regulatory genes or carry information from the cytoplasm or the cell surface and so permit transcriptional response to change outside the cell. Belief in such molecules was greatly strengthened by the isolation of the regulatory protein of the *lac operon* in *E. coli*, although the absence of any discrete operons from eukaryotic genomes has prevented a strict application of our knowledge of that regulatory system to eukaryotes. Not the least of the problems in understanding eukaryotic gene regulation has been the difficulty of devising experiments that would permit clear identification of regulatory molecules. Since there are a multitude of factors in cell cytoplasm that will affect transcription, it is necessary to define rather strict criteria for the identification of such gene regulation. As we shall see, it is easient to begin by stating what gene regulatory molecules are not and then to proceed to define what they are. Let us briefly review the various classes of cell factors that are likely to affect transcription, either by inducing or

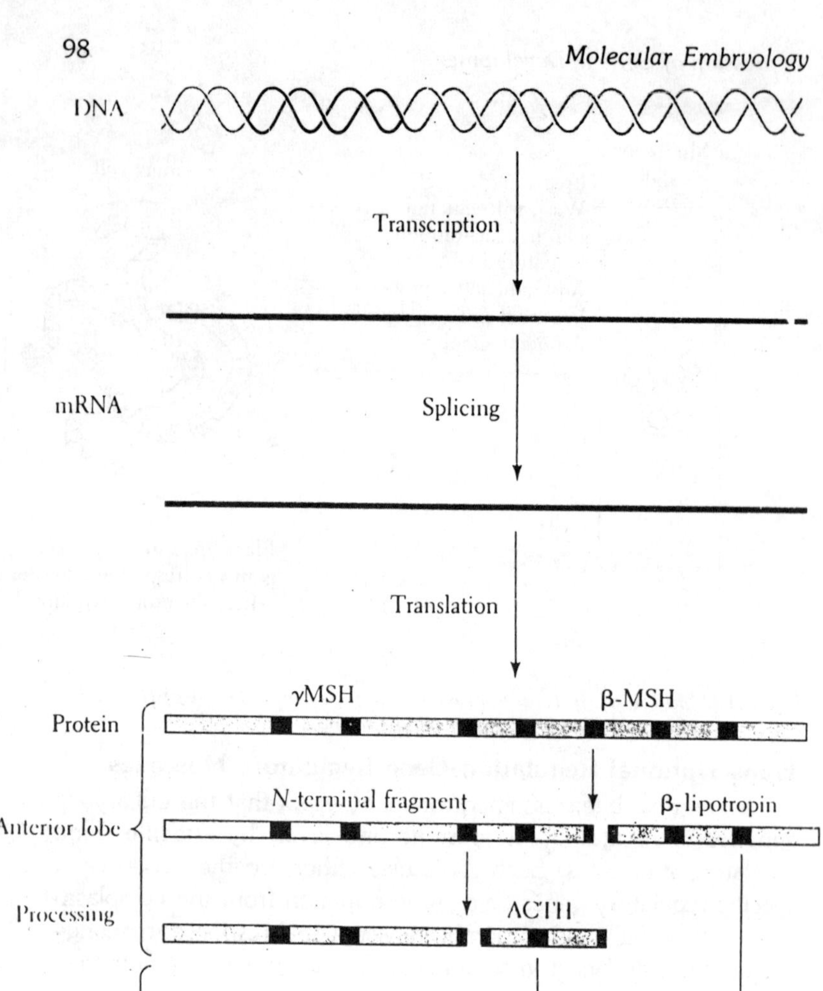

Fig. 3.14. Differential proteolytic cleavage yields multiple protein products from some genes. The processing pathway of the preopiomelanocortin (POMC) gene is shown.

preventing transcription on specific sequences, or by generally increasing or decreasing transcriptional rates in the cell. Most of these factors are clearly not what is normally meant by gene regulatory molecules, but these factors might easily be confused with actual gene regulatory molecules in loosely defined experiments.

RNA Polymerases

These enzymes are necessary for transcription and if in short supply, would sharply affect it. They are of three basic types in eukaryotes; type I, which transcribes the large ribosomal RNA genes; type II, which transcribed protein-coding genes into heterogeneous nuclear (hn) RNA and mRNA: and type III, which transcribes sequences coding for the small 5S ribosomal RNA and the tRNA genes. Presumably there is competition for type II polymerase by the various promoter sequences that lie upstream of the protein-coding genes.

Endonucleases

Such enzymes are likely to affect transcription, especially with *in vitro* cell-free systems, by introducing into DNA nicks that may serve as initiation sites for some polymerases. At low concentration some endonucleases are likely to enhance transcription in cell-free experiments, but at high concentrations the endonucleases will sharply reduce it.

Topoisomerases, helicases, and other helix-destabilizing proteins

A number of different proteins are known that alter the three-dimensional structure of DNA and render it more available for processing. Most of the molecules so far recognized have specific roles in DNA synthesis, but other may affect transcription.

DNA methylase

As mentioned earlier, this enzyme is likely to render DNA less available for transcription, and factors that antagonize methylation would, in the long term, enhance transciption.

Histone acetylases and deacetylases

Such enzymes, by modulating histone acetylation as already discussed, may have profound effects on transcriptional rates.

Factors such as ATP

These molecules may not show any strict affinity for either nucleic acid or chromatin, but by changing the available energy or, in the case of cyclic AMP, by affecting the catabolite activator protein on the CAP site of the *lac operon*, they may influence transcriptional rates.

Other Transcriptional Factors

Many other molecules besides polymerases are necessary for and involved in the process of transcription, either by making nucleotides more or less available to the polymerization process or by affecting the polymerase itself.

Ions and other small molecules

The conformational state of chromatin has been seen to be a major factor in modulating gene activity, and it is known that many ions, especially those of calcium, magnesium, and maganese, directly affect chromatin conformation. Many also affect the RNA polymerases. Ionic effects may be difficult to control in cell-free transcription system in that optimal calcium levels for the polymerase may not be optimal for chromatin conformation. Within the intact cell, localized microclimates of ionic concentration are no doubt part of the everyday mechanisms of cellular homeostasis. Having underlined what is not meant by gene regulatory molecules, we are now in a position to address more positively the subject of these factors. They are molecules that are most likely to be either RNA or protein in nature to provide the necessary structural complexity. They have as their main function the regulation of specific gene sequences.

One of the earliest efforts to define such factors functionally was made by Davidson & Britten in their attempt to model eukaryotic transcription on an extrapolation from *E. coli* lac operon regulation. These authors suggested that such molecules would most probably the RNA in nature; they would act as positive signals to a variety of different genes, themselves being the products of one regulatory gene; and they would thus allow the more or less co-ordinate activation of a series of cellular differentiations. Some hypotheses, such as that of Holliday & Pugh (1975), emphasize a non-reversible repression by DNA modification within cellular compartments. But rather than continue with the doubtful profitability of further hypothesis, let us turn to a brief review of the positive evidence for gene regulatory proteins and then proceed to draw general conclusions from the data.

Evidence for Specific Gene Regulatory Molecules

The *lac repressor* has been isolated. It is a tetrameric protein of four identical subunits with a combined molecular weight of 150000. It is the translational product of the regulatory gene 'i' and acts as an allosteric protein, with separate binding sites for the operator gene and the induter molecule. It acts negatively, so that gene activation requires de-repression, i.e. the removal of the repressor protein from the operator gene by the inducer. Other protein regulators have been identified in other bacterial operons, most being gene repressors, but some are capable of inducing gene activity positively. Most seem to act by facilitating the attachment of the RNA polymerase to the promoter sequence, while some can act positively at one promoter and negatively at another, depending on the specific binding site for the protein within the promoter. When an E. coli cell is invaded by the bacteriophage lambda (λ), one of the two possible pathways of interaction ensue. The virus may multiply within the host cell and ultimately kill it by lysis (the lytic pathway), or the virus may be incorporated into the host genome and be replicated as part of the normal bacterial chromosome (the lysogenic pathway).

The results of this decision are important not only for the host cell; they also involve very different strategies of transcription for the virus. In the lysogenic state it is crucial that the viral genes are kept silent. The products of these viral genes could be organized to produce new viruses. A lambda gene regulatory system is responsible for controlling the decision. It is based on two virally coded regulatory proteins, the cro protein and the lambda repression protein each of these proteins blocks the synthesis of the other gene sequence. In the lytic state, the crop protein is expressed and the lambda repressor protein synthesis is blocked, while in the lysogenic state the opposite situation occurs. Many factors, such as the conditions of the bacterial cell determine the initial choice of pathway, and the lysogenic pathway can convert to the lytic pathway if the cellular environment changes. The expression of each of these proteins regulates another set of viral genes, ensuring widespread transcription in the lytic choice, and comparative transcriptional quiescence in the lysogenic pathway.

The particular relevance of the phage lambda story here is that it indicates not only how one gene regulatory mechanism works in prokaryotes, but also the possible ways in which eukaryotic ones might also operate. Turning now to eukaryotes, the first example involves the induction of puffing in the *Drosophila* polytene chromosome in response to the hormone ecdysone. Ecdysone binds within target cells to a specific receptor protein which then associates with specific chromatin sites and activates puffing, itself a reflection of localized transcriptional activity. These activated genes, the so-called primary response genes, produce various products, some of which shut off the primary response by a negative feedback mechanism, other of which proceed to turn on a larger set of genes, which constitute the secondary response genes.So here is a rather simple two-tiered system of gene regulation by a hormone a specific hormone-receptor protein, and a family of gene products that result from the activity of the ecdysone primary response puffs.

The next example also involves puffing in Drosophila polytene chromosomes, this time in response to heat-shock stimulus, and it provides perhaps the best evidence to date of a specific gene regulatory protein in eukaryotic cells. When Drosophila larvae are exposed to elevated temperatures of about 40°C, certain specific puffs appear, known as heat-shock puffs. The products of these puffed genes are assumed to be heat-shock proteins that help to protect the cell against the damaging effects of increaed temperature. Heat-shock protein production is widespread in nature, being known in organisms ranging from yeast to the human, but what is so remarkable about the *Drosophila* system is the way in which the heat-shock signal activates the heat-shock genes. Compton & McCarthy (1978)were able to demonstrate that cytoplasm from heat-shocked *Drosophila* tissue culture cells would initiate the production of heat-shock puffs in isolated nuclei from salivary gland, following exposure of the nuclei to such cytoplasm. Capitalizing on this elegant *in vitro* assay.

Craine & Kornberg (1981) have attempted to analyze the nature of the cytoplasmic factor and found them to be both protease-sensitive and heat labile, that is, almost certainly proteins. They further demonstrated that, following such exposure to heat-shock cytoplasm, the nuclei are more than 100-fold enriched for heat-shock-gene-specific mRNA. More recently, further work on *Drosophila* heat shock response has identified a protein, termed heat-shock transcription factor B, that binds to a 55-bp region upstream of a heat-shock gene TATA box and apparently functions by facilitating RNA polymerase II attachment. This factor seems to have all the characteristics expected of a classical gene regulatory protein.

A regulatory protein termed GAL4 has been identified in yeast. It binds to four separate sites, each of approximately 17 base pairs with a promoter region that is common to two separate coding sequences, one on one strand and one on the other. Although each of these separate sites acts as a recognition sequence, there is evidence that the sites act synergistically when occupied and that this leads to rapid transcription of the GAL1 and GAL10 genes. Also each recognition sequences shows dyad symmetry and so may be recognized on the basis of its secondary structure rather than its base sequence. The GAL4 protein may be a dimer or tetramer, allowing easy stoichiometric interaction between the protein and the three-dimensional symmetry of the DNA dyad.

There is a protein that controls mating type in yeast. Two distinct mating types, *a* and α, are known in the yeast *Saccharomyces*. The mating type is controlled by MAT, a single locus with two alternative alleles, the *a* allele and the α allele. The gene products of these alleles are regulatory proteins. One of these proteins, termed α2, is one of two products of the α allele; it binds to a specific sequence upstream of the a-specific genes and thus brings about their repression. The other protein, α1, activates the specific gene relevant to the a mating type. In the mating-type cells, the α-specific genes lack α1 and are therefore not active, and the a-specific genes are activated due to absence of α2. What happen in diploid a/α cells? Apparently a completely new pattern of gene expression is initiated by the combinatorial effect of regulatory proteins from both MAT alleles. The sporulation genes are activated in the diploid, quite probably because an α2-a1 complex (protein a1 is the product of the a allele) binds to a regulatory site close to a gene *RME1*, which has to be repressed before sporulation can occur.

An interesting protein has been identified in connection with transcriptional of the *Xenopus* somatic 5S genes. A repeating unit of 5S DNA has been cloned, and regulation of its transcription has been studied by deletion of parts of the sequence followed by its exposure to RNA polymerase III enzyme in a transcribing medium. Somewhat surprisingly, it was found that deletion of all or part of a 30-nucleotide stretch in the centre of the 120 nucleotides of coding sequence abolished transcription. It was thus clear that a control region, recognized by the polymerase, was placed in the middle of the gene. In addition to this astonishing discovery, it was found that a protein transcription factor other than the polymerase itself was necessary for accurate transcription; it also bound to this central control region, and it had a molecular weight of about 40000. This protein, now often referred to as TF IIIA was also found to be similar or identical to a protein that could be isolated from 7S cytoplasmic particles. Each of these 7S particles consist of one moecule of 5S RNA plus one protein molecule; the 7S particle seems to be a storage form of 5S RNA in the cytoplasm. The location of the same protein on both the 7S particle and the 5S coding sequence neatly explains another observation about the activity of the 5S gene, namely that in a cell-free system, addition of free 5S RNA will slowly inhibit the transcription of the 5S genes present in the system.

It seems that their is competition between the gene and its product for the same protein and so a form of end-product inhibition is built into regulation of this gene sequence. By inference, it seems that this transcription factor is active in somatic cells of *Xenopus* tissue providing regulation of the amount of somatic-type 5S RNA that is produced. In summary, it can be concluded that evidence exists for certain proteins having a role as regulatory molecules in the control of eukaryotic gene expression. It should be stressed that there is no reason why RNA should not also fulfil this function, and indeed this was the supposition in the original model of eukaryotic gene regulation proposed by Davidson & Britten (1973). All that can be said with certainly is that there is at present no good evidence for RNA is of uncertain function and there might easily be regulatory RNA amongst the heterogeneous nuclear (hn) RNA or small nuclear (sn) RNA, although much of the former seems to be message precursor and at least some of the latter seems to have a role in splicing out introns from hnRNA.

Knowledge of protein gene regulatory molecules is meagre largely because of the difficulty of designing experiments that will unequivocally reveal such molecules. One of us (N.M.) was involved for some years in experiments in which transcriptionally inert nuclei from *Xenopus* erythrocytes were isolated, exposed to fractions of cytoplasm from *Xenopus* erythrobasts that were transcriptionally active, then washed free of cytoplasm, prior to incubation in a transcribing medium. Following transcription, which was indeed evident after exposure to such cytoplasm, and in the presence of additional Xenopus endogenous polymerase, radiolabelled transcripts were isolated from the nuclei and identified by hybridization to specific DNA probes. These experiments suffer from the following difficulties of design and interpretation. Firstly, the transcripts are few and, after isolation, are at or near the limits of detection by specific probes. Secondly, the question of whether a gene is being activated or merely induced to transcribe faster is difficult to determine and, again, dependent upon the sensitivity of the assays. Also in attempts to isolate specific protein gene regulators from the cytoplasm, proteins are separated by column chromatography before being used for nuclear activation.

The difficulty here, as referred to in the previous section, is in being certain that the proteins that are being isolated are not endonucleases, polymerases, methylases, or one of a host of proteins that might alter rates of transcription in relatively unspecific ways. Some possible avenues of experimentation have now emerged that can be expected to overcome many of these shortcomings. One of the most obvious is the use of cloned sequences of specific genes, including their use in either cell-free systems or the amphibian oocyte 'test-tube'. The recognition of the 5S regulatory proteins stems from such an approach. It is also possible to use such cloned sequences to 'fish out' regulatory molecules by recovering the cloned sequences after incubation and attempting to recover regulatory molecules already engaged with them.

So both *cis*- and *trans*-acting factors as well as mechanisms of DNA methylation, the role of enhancer and promoter sequences, and RNA splicing events can all be studied in this way, and many such experiments are currently under way in laboratories around the world. To name but two, Chada, Magram & Constantini (1986) have examined the regulation of human foetal globin genes by injecting cloned copies of a human gamma (γ) globin gene into fertilized mouse eggs and looking at its regulation in resulting transgenic mice. They found that the foetal gene is expressed embryonically in the mouse, indicating that conditiond in the mouse embryonic erythroid cells favour its activation, just as do those in the mammalian foetal (but not embryonic) erythroid cells, a *trans*-acting factor is thus deemed to be commonly involved, since the site of integration is probably random.

A second line of experimentation is being followed in the laboratory of Dr. Laurence Etkin (personal communication), in this case exploiting the *Xenopus* oocyte. Two variatnt H2B genes of the sea urchin are used, largely because these two genes are known to show differential regulation during sea urchin development. If cloned copies of the late H2B gene from sea urchin are injected into the oocyte germinal vesicle, no transcıiption occurs. But if such injection is preceded by injection of sea urchin cell protein extracts into the oocyte cytoplasm some 2 hours previously, then the sea urchin H2B genes are activated. It is known that many proteins will migrate from the cytoplasm to the nucleus of the Xenopus oocyte, and these are presumed to be proteins that have a normal nuclear function. by making salt extractions of sea urchin cytoplasmic extracts. Etkin has identified a protein factor that will migrate into the oocyte nucleus following its injection into the cytoplasm. It will then preferentially activate the H2B gene inthe nucleus and, interestingly, is known to bind to the 5' end of the H2B gene sequence. So here the oocyte has been elegantly and effectively used to identify a protein with specific gene regulatory properties.

It is also useful to ask what precise roles we might expect gene regulatory proteins in eukaryotes to fulfil. They may interact with RNA polymerases, either ensuring engagement of polymerase or, in the case of negative control, block engagement of polymease. They might also be involved in opening up the chromatin conformation, since this seems to be such a fundamental aspect of transcriptional activation in eukaryotes. In line with the appearance of hypersensitive sites around the initiation and promoter sequences of recently activated genes, it may be that nucleosomes are actually disoldged from this area and that this facilitates entry of the RNA polymerases. Alternatively, nucleosome structure may be modified in this region, and it has been suggested that HMG proteins 14 and 17 may play some role in such destabilization since they are considerably abundant inactive chromatin. Actually it seems unlikely that repression of gene activity in eukaryotes relies chiefly on specific regulatory proteins. Most of the genome is transcriptionally repressed in all cells at all times (perhaps the only exception being the widespread transcription apparently occurring in the lampbrush chromosome stage of the vertebrate meiotic diplotene). Almost certainly this is achieved by altering the conformational state of chromatin in restricted areas of the genome, with or without parallel DNA methylation. More probably, most eukaryotic gene regulators act by positive control and are only partially responsible for gene activation in particular differentiated cells. Specific interaction between regulatory molecules and enhancer sequences as discussed in the next section may explain the tissue-specific role of some of these sequences in eukaryotes. It should also be remembered that interaction between protein and DNA need not be in terms of base sequence recognition: indeed, on the GAL promoter region of yeast the sites recognised by the GAL4 protein appear likely to interact on the basis of dyad symmetry.

But whatever the particular roles are of specific gene regulators, at least in eukaryotes they take their place with other mechanisms such as chromatin decondensation, histone acetylation, and DNA methylation. All of these mechanisms help to regulate in a very precise way the enormous number of genes that constitute the genetic potential of a eukaryotic cell. At the moment it looks as if demethylation of regions upstream from promoters, perhaps followed by their decondensation and conversion to Z-form DNA, may be a common feature to transcriptional activation of eukaryotic genes.

Transcriptional Regulation–Promoter and Enhancer Sequences

The *lac* operon of the *E. coli* is regulated by various sequences upstream from the coding sequence, including a regulatory gene *i*, a promoter sequence P, and the operator sequence O. In eukaryotes the DNA upstream from a coding sequence may affect its transcription. This DNA sequence that affects transcription may be very long, in some cases many thousands of bases. With the advent of assaying transcription of cloned DNA sequences by transfecting cells with such DNA, it has become relatively easy to examine the effects of adding or substracting various regions of upstream DNA on the efficiency of transcription of coding sequence itself.

Comparison of 5' upstream sequence of a large number of polymerase II transcribed genes reveals two conserved regions. One of these, located some 30-base pairs, upstream from the initiation site of the gene itself, is an (A+R)-rich region with a consensus sequence reading TATA. This is the Goldberg Hognes of TATA box. A second conserved sequence lies some 80 base pairs upstream and is the CCAAT box. There are strong grounds for believing that these sites are recognized by the polymerase II molecule, and thus any modification or conformational change in this area of the chromatin would be expected to have direct effects on the act of transcription. Genes transcribed by the polymerases I and III have polymerase-binding sites also, and as discussed elsewhere in this chapter, a promoter site of the ribosomal 5S gene is actually in the middle of the coding sequence. In some promoters there are also specific regions that are positively regulated by molecules such as steroids or heavy metals. For example, the metallothionein genes have separate sequences within the prmoter region that specifically bind steroids, on one hand, and heavy metals such as cadmium and zinc, on the other; when these sites are occupied, transcription is induced.

A recent development in our knowledge of transcriptional regulation in eukaryotes is the discovery of enhancer sequences. Most present knowledge of enhancer sequences relates to viral enhancers, for example those in the genome of SV40, but enhancers have also been discovered in relation to genes for immunoglobulin, insulin, and alpha amylase. In the spacer regions between Xenopus larger ribosomal genes there are multiple regions 60-80 base pairs long that confer a 20-fold increase in transcription rate compared with genes lacking them. These enhancer elements can work in either orientation, or when located 4-5 kb away from the promoter, or even when located in the gene coding sequence. Although the mode of action of enhancer sequences remains unclare, the following properties have been noted: (1) they enhance transcription from the cap site of a linked gene, (2) they can work in either orientation and may be upstream or downstream, (3) they work over long distances, often many kilobases, and (4) they are relatively short, usually between 50 and 300 base pairs. Perhaps the most relevant and exciting observation about enhancer sequences is that many are tissue-specific and presumably acquire enhancing activity through interaction with other molecules, perhaps the much sought specific gene regulators.

We should certainly bear in mind, when considering the effects of upstream sequences on transcription, that there is good evidence that proteins other than the polymerases specifically attach to these regions. Some of these proteins may facilitate polymerase binding, as does the 5S transcription factor IIIa, and others may impede the attachment of polymerase. Many of these factors form very stable complexes with specific DNA sequences, and at least in part, may explain the stability of gene expression in differentiation.

The *Xenopus* oocyte has proved to be a fruitful experimental 'test-tube' for the study of enhancer sequences, since cloned genes can be engineered to possess or not possess these sequences in their flanking regions. By having a heterologous gene both with and without an enhancer sequence, the transcriptional performance of the gene can be assayed in the oocyte. Alternatively, gene can be transfected into tissue culture cells of different types and their origins and expression can be monitored. Many recent publications report tissue specific expression of artificially introduced genes, but only in cells of appropriate types. Thus Chada et al. (1986) have demonstrated the tissue-specific and developmental-time specific regulation of a foetal globin gene.Tilghman (1985) provides good evidence for the dependence of tissue-specific expression of albumin and a fetoprotein genes on 5' flanking sequences.

Post-transctiptional Regulation

Regulation of gene action is not confined to the level of transcriptional and very many steps, allowing possible control, intervene between a gene and its final proudct in the phenotype. Steps that come between transcription and translation are described as post-transcriptional and steps following initial protein synthesis are described as post-translational. We will now examine some of these processes of post-transcriptional regulation.

Some RNA is Capped and Tailed

All polymerase II transcripts in the nucleus are initially in the form of heterogeneous nuclear RNA (hnRNA). Some especially those transcripts that are destined to become message, are covalently modified at both ends. First, the 5' and, which is the end first synthesized during transcription, is capped by the addition of a 7-methylguanosine residue, even before the completeion of transcription, of the rest of the molecule. The 3 terminus is modified after completion of the transcript by the addition of between 100 and 200 residues of adenylic acid to form a poly A tail, the addition being carried out by a unique polymerase enzyme. The precise functions of capping and tailing are not known, but they seem to serve to identify a message or potential message and tailing may also help in the final export of this message from the nucleus. Although hnRNS account for only 7% of and 3%

respectively, of the total steady-state amount of RNA in a cell (71% is cytoplasmic ribosomal RNA) the hnRNA accounts for 58% of the total synth :tic RNA product (ribosomal RNA being 39%). This reflects the fact that the hnRNAs and mRNAs are relatively short lived and turn over rapidly in the nucleus of cytoplasm. But this should to be interpreted to mean that eukaryotic message is only used once. Far from it—we know that some message are used many hundreds of times, and experiments with anucleate *Artabularia*, reveal that some eukaryotic mRNAs are very song lived with a half-life of months. Bacterial message, on the other hand, has a half-life of minutes and turns over very rapidly.

RNA is Processed to Remove Intron Sequences

Most, but not all, hnRNA contains intron sequences, representing the non-protein determining parts of a gene sequence. Some genes, such as those coding for histone, have no intron sequences, but most have two or more, and some have more than 50, as in the extreme case of the procollagen alpha gene. We will not expand here on the interesting but puzzling problem of the origins and evolutions of interrupted genes, except to say that it now seems that the presence of introns is primitive to cells, and that they have been secondarily lost by prokaryotes. Also, they serve a function in genomic evolution by facilitating rapid evolution of new proteins by the novel splicing together of exons coding for existing protein domains. Let us discuss briefly, however, what is known about how these introns are handled by the cell and what their possible significance may be in terms of gene expression and differentiation.

Introns removal and the splicing together of the exons remaining must be absolutely precise. This is in part engineered by a district group of nuclear particles, the small ribonucleoprotein particles, consisting of specific proteins conjugated with a molecule of small nuclear RNA (snRNA). In the eukaryotic genome there are many repetitious regions that code for this class of RNA, and there is evidence to suggest that part of the sequences of snRNA is complementary to sequences at the boundaries of the introns of hnRNA.

What is perhaps more relevant to the subject to this book is the fact that differential splicing is used in different lymphocyte cells to produce differeing proteins from the same hnRNA molecule. As originally discovered by Early et al. (1980), the two

Table 3.1. Sizes of MInimum primary Transcripts as compared with the sizes of Messenger RNA for various class II Transcribed Genes

Gene	*mRNA size (nucleotides)*	*Minimum transcript size (neucleotides)*	*Ratio*
β-globin (rabbit	589	1295	2.2
β-blogin major (mouse	620	1382	2.2
Rat preproinsulin I	443	562	1.3
Rat preproinsulin II	443	1061	2.4
Chick ovalbumin	1859	7500	4.0
Chick ovomucoid	883	5600	6.3
Chick lysozyme	620	3700	6.0
Chick pro-α 2 collagen	5000	38000	7.6
Xenopus vitellogenin A1	6300	21000	3.3
Xenopus vitellogenin A2	6300	16000	2.5
Mouse dihydrofolate reductase	1600	42000	2.6

differing mRNA molecules are produced by part of an intron being omitted from one form of the mRNA but included in the exons splice used to produce the other mRNA. This allow production of two distinct proteins, both immunoglobulins, but one with a long strand of hydrophobic amino acids at its carboxyl terminus, and the other with only a short length of relatively hydrophilic amino acids. The Ig molecule with the long hydrophobic peptide is membrane bound within a lymphocyte whilst the molecule with the terminal hydrophilic peptide is membrane bound within a lymphocyte, whilst the molecule with the terminal hydrophilic peptide is secreted from the cell. This change in splicing takes place within the life of a single lymphocyte cell and neatly explains the following observations. Immature lymphocytes retain antibody and simply insert the ig molecules into their plasma membranes, whereas following stimulation with antigen, the same lymphocyte becomes secretory, releasing antibody molecules into circulation. This example not only illustrates the advantage that introns may confer on cellular synthetic processes, but it not doubt illustrates a process that is not confined to immunoglobulins. It also seems likely that differential splicing of introns provides cell with a rapid means of evolving novel proteins composed of peptide domains not previously bonded together. These may be arranged in new and possibly advantageous ways. Intron/exon splice junctions code for regions on the surface of the peptide produced, thus suggesting evolution of proteins as complexes of peptide building blocks.

Most RNA in Never Exported From the Nucleus

Only about 5% of the RNA transcribed is ever permitted to leave the nucleus. This is explained partly by removal of intron RNA, but also by the fact that many entire RNA molecules simply turn over within the nucleus.The implications of this remain largely obscure, but some clues about the identification of RNA for export are coming to light. Although no tall genes contain introns, most do, and it seems that the presence of some of these introns is essential for RNA export. In other words, introns are used a means of identifying or ticketing the molecules that are to be passed out of the nucleus. This follows from experimental removal of intron sequences from cloned genes and evidence that of mRNA synthesized from them is retained in the nucleus. It is possible that nuclear export is actually a highly important way of screening the RNA to be expressed at any one time, but for the present there is insufficient evidence to allow a firm conclusion along these lines. Certainly there is no indication that post-transcriptional control, including nuclear export, is nearly as important as transcriptional regulation itself in determining which genes are expressed at any particular time and in which cell compartment they are expressed.

Message Degradation Rates are Significant

The rate in which eukaryotic mRNA is degraded in the cytoplasm is highly variable. This implies that differential message breakdown is, in some circumstances, an important method of regulating not only the rate of gene expression, but also the lag between transcriptional shut-down and the cessation of specific translation. One situation will serve to illustrate this point very well, namelythe survival of histone mRNA during the cell cycle. New histone is required in massive amounts immediately at the start of the S period of DNA synthesis to provide the new DNA with nucleosomes. We now know that the restricted availability of histone message is not achieved as a result of transcriptional control alone, but by differential breakdown rates for histone message. Indeed in some organisms the differential degradation of histone message seems to be the chief regulatory mechanism to achieve the marked differences in rates of histone synthesis during the cell cycle. Histone synthesis has been the most thoroughly investigated example, but it is likely that similar mechanisms operate with other specific messages, and therefore possible that the expression of other genes is powerfully influenced by regulation at this level.

There is no Evidence for special Regulation During Translation

Quite contrary to assumptions in the early days of cell biology before the discovery of messenger RNA, ribosomes do not show specificity for particular messages, not do they themselves determine the nature of the protein population being synthesized in the cell. There is a long standing assumption that protein destined for export from the cell is chiefly synthesized on membrane-bound ribosomes and that protein destined for retention is synthesized on free ribosomes. This is partially, but certainly not absolutely, true.

Post-translational Modifications to Proteins

Some proteins are altered after synthesis, usually by partial degradation or trimming, as, for example, by the enzymatic removal of the central section of the proinsulin molecule to yield the active protein, insulin. For their activity many proteins also depend on being complexed into compound proteins together with other subunits, either the same or different in nature. Such post-translational control mechanisms do play a significant role in determining the activities of differentiated cells.Thus haemoglobin production is highly dependent on the availability of them to complex with globin subunits and may be deficient in cases of iron-dependent anaemia. The tetrameric protein lactate dehydrogenase comprises subunits of only two kinds, but the tetrameric arrangement varies with the cell type and this in turn results from, amongst other factors, the differential rate of breakdown of different tetramers in the various cell types.

Gene Expression in Mitochondria and Chloroplasts

All non-prokaryotic cells contain mitochondria, which are autonomous, self-replicating organelles; plant cells, in addition. possess chloroplasts. No copy of the mitochondrial or the chloroplast genome exists in the main cellular genome in the nucleus, and in all cases so far studied, mitochondria are exclusively maternally derived. Although a mitochondrion is indeed present in the sperm, it is destroyed at or after fertilization, and only the egg mitochondria populate the new organism. Although there is no evidence to suggest it, it is possible that more than one clone of mitochondria or chloroplasts exists in any one organism and that these are unequally segregated into certain types of differentiated cells. This would implicate mitochondria and chloroplasts in the process of differentiation, but to date the suggestion remains entirely specultative. All that can be observed is that some cells have mitochondria of strikingly different morphology from those found in other cell types, and that some cells possess mitochondria with more than on morphological form.

A few observations on the DNA of these organelles are appropriate in a book of this sort. The first is that the genetic code utilized in mitochondria is slighly different from the normal universal code. The codons for the stop signal and for a few amino acids differ in mitochondria, and even between mitochondria of different organisms. Secondly, these organelles are clearly analogous to prokaryotes in a number of ways, not least in the fact that they lack chromatin and have a circular DNA chromosome. They do, however, have some peculiar non-prokaryotic features, most importantly the presence of introns sequences in some years mitochondrial genes and the tailing of some mitochondrial messenger RNA. Mitochondria and chloroplasts and therefore very specialized organelles, but whether they play a definitive role in differentiation is hard to say. Since they possess some cytoplasmic proteins and synthesize parts of others, they could well play an important but subsidiary role in determining the pattern of gene expression in a cell.

A Note on Gene Amplification

In out earlier discussion of gene activity and regulation and the different categories of nuclear genes, no mention was made of a remarkable cellular phenomenon. This phenomenon is highly specialized and not known to be of general application, but is of considerable importance.This is the temporary increase in gene copy number termed gene amplification.

Three situations are known in nature in which gene amplification occurs. The best known situation applies exclusively to the large ribosomal RNA genes and occurs in the oocytes of some amphibians, including Xenopus. In this process, the few hundred copies of the 18S and 28S genes in the main genome are selectively amplified by replication so that between 1 and 2 million copies exist. This staggering number of gene copies is used to provide a large store of ribosomes for the relatively large amphibian egg cell. Curiously the ribosomal 5S genes are not amplified: rather the amphibian genome is permanently endowed with a large number of copies of oocyte-specific 5S genes, since one molecule of 5S RNA is needed for every molecule of 28S RNA. A second natural situation is in connection with the synthesis of chorion proteins in the ovarian cells of Drosophila, in which there is specific and temporary amplification of a number of genes coding for the chorion (egg coat) protein. The third is the amplification of the metallothionein gene in livers of mice treated with excess amounts of cadmium.

There are a few related situations in which temporary of permanent increases in specific gene copy number occur. One is the permanent increase in the copy number of the dihydrofolate reductase gene in vertebrate cells exposed to increasing doses of the drug methotrexate. This 'forced amplification' is really a type of very rapid evolutionary selection and results in tissue culture cells with a visible cytogenetic aberration. In these highly selected cells a larger block of chromatin becomes visible: this represents the massive repeitious block of sequences coding for dihydrofolate reductase. Another increase in gene copy number takes place in *Drosophila* in relation to the ribosomal RNA genes. The *Drosophila* mutant *bobbed* has a reduced number of cistrons for rRNA gene number, indicating that some correction mechanism is occurring. This has come to be termed gene modification.

Although gene amplification and other processes that rapidly alter gene copy number are certainly fascinating, there is no indication that they are widely used in nature to accomplish selective gent expression in differentiated cells. In situations such as that in which massive production of fibroin by the silk glands of *Bombyx* is observed, no specific gene amplification occurs. Nor does it occur with other massive output genes coding for cell-specific products, such as those specifying globins, collagens, casein, cocoonase, or keratin.

Gene Regulation and Differentiation

Having now briefly reviewed what is known about gene expression and its regulation. It is opportune to put it in perspective in relation to our present quest — an understanding of differentiation. Does our understanding of gene regulation and selective gene expression explain differentiation? Of course, it does not. It explains an important process of differentiation, but it provides little clue as to how initial choices are made regarding cell destiny. Are there any tiny windows of light to illuminate this puzzling question in our scan of gene regulation? What we should perhaps be concentrating on is genetic or cellular evidence for genes the make cellular decisions that control other batteries of gene. In some sort of cascade of genetic response. It so happens that some sort of cascade of genetic response. It so happens that some genes with just such properties have recently come to light. They are the genes involved in the early development of *Drosophila* and are collectively termed homeotic genes.

As we discuss later, it may be that some cell commitment has a purely genetic basic and is accomplished by a mitotic clock or some other special genetic mechanism. But for the moment it must be concluded that studies of gene regulation go some way to explain how cell differentiation is engineered, but not how the choices of design and cell specificity are actually made.

Translational Regulation of Gene Expression

Almost all embryos make use of a two-past strategy to meet the demands of cleavage and early development. Pools of structural proteins, enzymes, ribosomes and mRNAs are synthesized during oogenesis and stored in the mature oocyte. The activated oocyte and early embryo draw on these pools in an orderly temporal and spatial way to produce cascades of metabolic changes and specific sets of poreteins and organelles. Almost all of these early changes are independent of any ongoing nuclear transcription from the zygotic and embryonic genome. As cleavage proceeds, the maternal mRNA and protein pools are gradually depleted. At the same time, the numbers of embryonic nuclei increase exponentially and their transcriptional contributions become more significant. Finally, by the late blastula or equivalent stage most of the original maternal contributions are exhausted and further development is dependent on transcripts produced by the embryo's genome. The exact details of the extend to which the embryo relies on maternal vs. embryonic contributions and the timing of this transition from one source to the other very both among individual organisms and according to the individual component. However, in virtually all systems studied the immediate changes in protein synthesis what accompany oocyte activation and fertilization are due to changes in the translatability of the maternal mRNA pool.

From the few organisms that are accessible to studies of oogenesis, we know that a large fraction of the oocyte's store of maternal mRNA is synthesized very early in oogenesis. For example, pre-vitellogenic *Xenopus* oocytes contain nearly their full complement of ply(A)$^+$ RNA and mRNA sequence complexity. Studies of histone mRNAs and poly(A)$^+$ sequences complementary to cloned cDNA probes show that these individual mRNAs follow the same pattern of accumulation early in oogenesis. In sea urchins, about half of the maternal RNA sequence complexity is synthesized prior, to the onset of vitellogenesis. The remaining sequences accumulate during the latter part of oogenesis. The bulk of the maternal histone mRNA pool is made very late in oogenesis, between the time of meiotic maturation and fertilization.

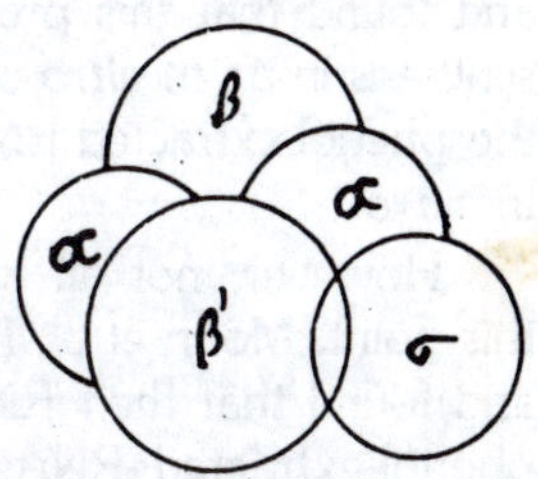

Fig. 3.15. A model of the structure of prokaryotic RNA polymerase showing association of five polypeptides ($\alpha_2\beta\beta\sigma$)

The maternal mRNA pool is phenomenologically a very special class of mRNA. Some of these molecules (generally less than 1%), are actively translated on polysomes in the oocyte, bu the vast majority of maternal mRNA sequences in the mature oocyte are translationally inert. They are in some way "stored" for future use. Fertilization (or meiotic activation) triggers a rapid mobilization of some of these stored mRNAs onto polysomes, and, in some organisms, the reduction or cessation of the translation of certain other previously active mRNAs.

Despite much work in this area over the past 10 year, we understand very little about the processes that control these translational changes. The early experiments of gross et al. (1973) demonstrated that the sea urchin egg contains maternal mRNAs (including histone mRNAs) that can be recovered after phenol extraction and are perfectly active in heterologous cell-free protein synthesizing systems. This result firmly established the idea that the mRNA itself was perfectly OK and that someother component in the egg was responsible for maternal mRNA unavailability. Several experiments supported the idea that maternal mRNAs are associated with proteins in messenger ribonucleoproteins (mRNPs), that these proteins "mask" the translational activity of most maternal mRNAs in the egg and that fertilization leads to an unmasing of mRNAs which results in an increase in protein synthesis. For example, Kaumeyer et al.,(1978) isolated egg RNPs and found that this preparation was inactive in directing protein synthesis in an *in vitro* cell-free protein synthesizing system, whereas the phenol-extracted RNAs from the RNP preparation were active *in vitro*.

However, not all workers in this field are in agreement on this point. Moon et al. (1982) carried out very similar experiments and found that their RNP preparations were just as active as the phenol-extracted RNAs isolated from the RNPs. Difficulties in precisely quantifying the mRNA content of various mRNP preparations and the temperate-activity of such diverse mRNA populations may contribute to the disparity of the two kinds of results. The use of clones probes complementary to individual mRNAs should help to resolve this controversy. In contrast to this idea of masked mRNA, some maternal mRNA sequences may be stored as larger, incompletely processed precursor RNAs. Constantini et al. (1980) found that 70% of the poly $(A)^+$ RNA complexity in the unfertilized sea urchin egg is present in very large RNA molecules that have several features of nuclear RNA. These RNAs appear to consist of interspresed, covalently linked transcripts of unique and repetitive sequence. Such molecules are not detected on polysomes in the embryo. One interpretation of this finding is that these sequences could represent translationally inactive mRNA precursors that are processed after fertilization to produce active mRNA sequences.

So far, there is only one example of bona fide mRNA sequence of known coding assignment that fits this scheme. a-tubulin RNA sequences in the mature sea urchin egg are considerably larger than the usual (~2000 nucleotide) length found in embryos and adult tissues. These egg sequences are reduced to the usual size within 30 min of fertilization. However, this phenomenon cannot explain the translational block on all stored mRNAs, since many abundant sea urchin maternal mRNAs are active in cell-free systems. Also, at least one class of maternal mRNAs, those encoding the histones, do not undergo any detectable size change after fertilization.

Various other components of translational machinery have been implicated in these translational, switches. In urchins there is a 2-3 fold increase in the elongation rate, but this is too low to account for the overall 10-30 fold increase in the rate of protein synthesis. Ballinger and Hunt (1981) found that fertilization causes a rapid and complete phosphorylation of one of the 40s ribosomal subunit proteins. However, careful comparisons of the kinetics of this phosphorylation with those of the rate increase, and of the representation of this protein in polysomal ribosomes vs. unrecruited ribosomes appear to rule out a direct and causal relationship. Finally, the experiments of a Laskey et al. (1977) and of Richter and Smith (1981) show that globin mRNA injected into *Xenopus* oocytes competes on a 1:1 basis with endogenous polysomal mRNA and not a 1:100 basis with the entire pool of maternal mRNA. Their results, as well as those of Lingrel and Woodland show that there is no spare translational capacity in the *Xenopus*

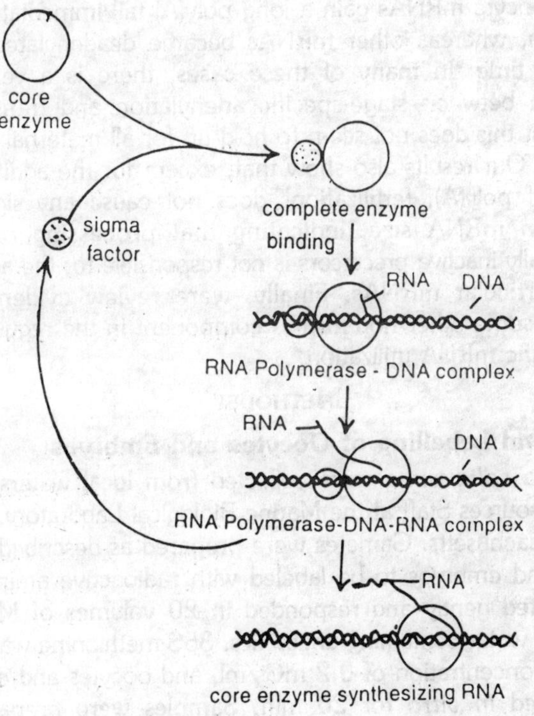

Fig. 3.16. Role of sigma factor and core enzyme of RNA polymerase during transcription.

oocyte and indicate that mRNA availability cannot be the sole factor regulating the rate of protein synthesis. We have been investigating the molecular mechanisms of translational regulation in the marine mollusc *Spisula*. Oocytes and early embryos of this common claim contain mRNA populations that are indistinguishable by cell-free translation of their phenol-extracted mRNAs.

When the oocytes are fertilized, there is a rapid and easily observed change in the pattern of protein synthesis. In this paper, we briefly review earlier results demonstrating that this change is regulated entirely at the translational level, and then present some recent experiments designed to get at the underlying mechanisms of this change. cDNA clones containing sequences complementary to several translationally regulated maternal mRNAs have been constructed and used to directly investigate changes in mRNA structure and activity during early development.

Almost all of the *Spisula* maternal mRNAs examined so far show a change in polyadenylation at fertilization: some poly(A) deficient oocyte mRNAs gain a long poly(A) tail immediately after fertilization, whereas other mRNAs became deadenylated during this same time. In many of these cases, there is a very close correlation between stage-specific adenylation and translational activity, but this does not seem to hold up for all maternal mRNAs examined. Our results also show that, except for the addition and removal of poly(A), fertilizations does not cause any significant changes in mRNA size, indicating that processing of larger translationally inactive precursors is not responsible for the activation of the particular mRNAs. Finally, were review evidence that implicates some other non-mRNA component in the regulation of stage-specific mRNA utilization.

Methods

Culture and Labelling of Oocytes and Embryos

Spisula solidissima were collected from local waters by the Marine Resources Staff at the Marine Biological Laboratory, Woods Hole, Massachusetts. Gametes were prepared as described earlier. Oocytes and embryos to be labeled with radioactive amino acids were pelleted gently and responded in 20 volumes of Millipore-filtered sea water containing antibiotics. 35S-methionine was added to a final concentration of 0.2 mCi/ml, and oocytes and embryos were labeled *in vitro* for 20 min. Samples were prepared for electrophoresis on polyacrylamide gels containing SDS.

Preparation of 12K Supernatants

Oocytes or embryos were washed twice with cold calcium and magnesium-free sea water and once with homogenization buffer. Cells were resuspended in 3 vol of homogenization buffer and broken in a Dounce homogenizer. Homogenates were spun at 12,000 × g for 20 min. Aliquots of these "12K" supenatants were frozen in liquid nitrogen and stored at –80°C.

Polysome Gradients

12K supernatants were diluted with 45 vol of gradient buffer. Half of each sample was treated with 30 mM EDTA to release mRNA from polysomes. 1 ml of 12K supernatant was layered onto a 15-40% sucrose (w/v) 11 ml gradient made in gradient buffer, and gradients were centrifuged in an SW41 rotor at 40,000 rpm for 70 min. Fractions were collected and RNA was extracted. RNA from each gradient fraction was dissolved in an equal volume of water, usually 30 µl, and stored at –80°C.

Cell-free Translation

RNA samples were translated in a mRNA-dependent rabbit reticulocyte lysate. Total RNA was translated at a final concentration of 600 µg/ml. When polysomal gradient RNA fractions were translated, 2 µl each RNA sample was added to 8 µl reticulocyte lysate. When the translational specificites of unextracted Spisula oocyte or embryo 12K supernatant were assayed, 1 µl of unextracted 12K supernatant was diluted with 1µl of water and then added to 8 µl of reticulocyte lysate. All incubations were carried out for 1 hr at 30°C. Labeled protein samples were analysed on SDS-polyacrylamide slab gels.

Isolation of Cloned cDNAs Complementary to Translationally Regulated mRNA

Double-stranded cDNAs complementary to poly(A)* RNA from 2-cell or 18 hr embyos were synthesized and inserted into the Pst I site of the plasmid pBR322 by the GC-tailing method. Bacterial clones transformed with recombinant plasmids were isolated and recombinant plasmids DNAs from selected bacterial colonies were prepared from bacterial lysates. The cDNA sequences carried by individual clones were determined by mRNA hybrid-selected translation of the complementary mRNA from total 2-cell or 18 hr embryo RNA followed by translation in vitro and identification of the translation products on 1-D gel electrophoresis. Details of this aspect of the work are presented elsewhere.

RNA Blots

RNA samples were electrophoresed on 1% agarose gels containing formaledhyde, transferred to nitrocellulose and hybridized with ^{32}p-labelled plasmid DNA probes. Plasmid DNAs were labeled in vitro with 32p by nick-translation. Total RNA was fractionated on olgio (dT)-cellulose and, in some cases, on poly(U)-Sepharose (Pharmacia).

RNAse H removal of poly(A) Tails

Unfractionated 12K supernatant RNAs were hybridized witholigo (Dt) and the double stranded portions of the hybrids were digested with RNAse H. Nuclease-resistant RNAs were then electrophoresed on an agarose gel, blotted to nitrocellulose and hybridized to 32p-labelled cloned cDNA probes as above.

Results

The Pattern of Protein Synthesis Changes within ten minutes of Fertilization

Fertilization of the *Spisula* oocyte triggers breakdown of the germinal vesicle (10 min post-fertilization), first (40 min)and second (50 min) meiotic divisions. First cleavage occurs 70 min after fertilization and is unequal, producing a small AB and a large CD blastomere which have different developmental fates. Subsequent cleavages occur without any significnat change in the mass of the embryo. About 5 hr after fertilization, the 60-cell embryo gastrulates and develops specialized ciliated regions. The pyramidal trochophore larva takes shape between 9 and 12 hr. At 12hr the viliger larva beings to form: shell is secreted, muscle and nerve cells are produced and the gut becomes regionalized. At 20 hr, the clam-shaped veliger larva contains many differentiated organs, including a highly ciliated velum. This larval form persist for several days until metamorphosis occurs.

During the first 24 hrs of development, there are many changes in the pattern of protein synthesis. The earliest of these occurs between 2 and 10 min of fertilization and is regulated entirely at the translational level. As development continues and the embryo's store of maternal mRNA declines, new transcription makes increasingly important contributions to the total mRNA pool and to the pattern of protein synthesis. In this paper, only the early changes is protein synthesis will be considered.

The Early Change in the Pattern of Protein Synthesis is Controlled Entirely at the Translational Level

Within 10 min of fertilization of artificial activation, synthesis of many oocyte-specific proteins ceases and the zygote begins to make new sets of proteins. For example, the proteins marked X, Y and Z are synthesized by the oocyte, but their synthesis is not detectable in zygotes. In contrast, proteins A, B and C are synthesized at a very low levels, or not at all, in oocytes whereas they are prominently labeled in 10-min embryos. Comparisons of the *in vivo* patterns with the *in vitro* translation products programmed by oocyte and zygote RNAs in a reticulocyte lysate reveal that this abrupt change in the kinds of proteins being made occurs in the absence of any detectable changes in the total maternal mRNA pool. Phenol-extracted oocyte and zygote RNAs appear to contain the same mRNA sets including the sequences for proteins A,B,C,X,Y and Z.

The experiment demonstrates that different subsets of this maternal mRNA pool are found on polysomes before and after fertilization. 12,000 × g supernatants of oocyte and zygote homogenates were fractionated on sucrose gradients into polysomal and post-ribosomal supernatant fractions. RNA was extraced from the these fractions and translated in vitro. There is a very specific subset of oocyte mRNAs are engaged on polysomes and that fertilization causes a dramatic redistribution of sequences between the translated, polysomal compartment and the non-translated, post-ribosomal supernatant compartment. Analyses of the translational status of individual mRNAs by this method are hindered by two problems.

First, when the polypeptides encoded in vitro by two or more mRNAs comigrate, it is impossible to conclude anything about the behaviour of their respective mRNAs. Secondly, since the translation of maternal mRNA is quantitatively and qualitatively regulated *in vivo* and we do not understand the molecular mechanisms responsible for this, seemingly straightforward interpretations of *in vitro* translation assays could be wrong. Therefore, we have used a second mRNA assay method, hybridization to cloned cDNA probes that does not depend on *in vitro* template activity. Recombinant DNA clones consisting of pBR322 and cDNAs complementary to translationally regulated maternal mRNAs were isolated from there cDNA libraries, Rosenthal and Ruderman, unpublished. Among the clones used in the following experiments were those carrying sequences for protein A (clones 1T55)$_2$ protein C (clone 1T43), and a-tublin (clone 3V4). Several others clones that showed hybridizations to RNA on Northern blots of 2-cell RNA, but did not selectively hybridize detectable amounts of template-active RNA, were also used in this study.

These clones were first used to monitor the changes in the translational utilization of several individual mRNAs at fertilization. Oocyte and embryo homogenates were fractionated on polysome gradients. RNAs were extracted from each of the other gradient fractions, electrophoresed on denaturing agarose gels, blotted to nitrocellulose and hybridized with individual 32p-labeled cloned cDNAs. The hybridization pattern obtained with clone 1T55 (protein A) reveals that none of mRNA A is present on polysomes in the oocyte and that all of the mRNA moves onto polysomes right after fertilization. The mRNA for protein C, as well as several other, behaves similarly. Another pattern of maternal mRNA.

Top panels

Translation products encoded by total oocyte RNA. Bottom panels: protein encoded by total embryo RNA. Translation products directed by RNAs isolated from sucrose gradient fractionations of 12K supernatants from oocytes (O) and embryos (E), and of EDTA-treated 12K supernatants from oocytes (^{o}EDTA) and embryos (^{EDTA}E), arrow indicates direction of sedimentation 60S subunits are found in fractions 7 and 8; 40S subunits are found in fractions 8 and 9. Recruitment is expemplified by 6T21mRNA: these mRNAs (liked A and C) are scarcely translated, if at all, in the oocyte but (unlike A and C) fertilization causes only a partial recruitment of these sequences. A third kind of translational change is shown by a-tubulin mRNA (clone 3V4, and several other sequences of unknown coding assignment: a portion of these mRNAs are engaged on polysomes in the oocyte and fertilization results in their abrupt release from polysomes. Finally, a few of the cloned maternal mRNA sequences show no changes in the extent of translation after fertilization.

Sequence-specific Changes in Adenylation are the only Significant Changes in Maternal mRNA Sites after Fertilization

As one way of testing for mRNA precursor-product processing at fertilization, the sizes of the oocyte and zygote versions of several translationally regulated mRNAs were compared on "Northern blots." Equal amounts of oocyte and zygote total cellular RNA were electrophoresed in adjacent lanes of a denaturing formaldehyde-agarose gel, transferred to nitrocellulose by blotting, and hybridizing with ^{32}p-labeled cloned cDNAs. The results obtained for two different sequences that become active after fertilization, mRNAs A (clone 1T55) and C (clone 1T43). Neither RNA sequence shows any evidence of being processed to a smaller size after fertilization. In fact, the zygote mRNAs are slightly larger than their oocyte counterparts, indicating that something has been added to these mRNAs after fertilization. Several other maternal mRNAs also show this slight (50-100 nucleotide) increase in length following fertilization. In contrast, a-tublin mRNA (which is active in the oocyte and inactive in the zygote) becomes slightly shorter after fertilization.

Two kinds of experiments show that these length changes are due to the addition or removal of poly(A) tails at fertilization. First total oocyte and zygote RNAs were fractionated by oligo (dT)-cellulose chromatography into unbound, poly(A)-deficient RNA and bound poly(A) RNA. Whereas the *in vitro* translation products of total oocyte and zygote RNA appear identical, a very different picture is seen when the poly(A)-and poly(A) products are compared. Fertilization causes changes in the abilities of many maternal mRNAs to bind to olgio (dt)-cellulose. The most dramatic examples include the mRNA for proteins A, B and C. The oocyte versions of these three mRNAs are poly(A)-deficient, whereas the embryo sequences are poly(A). In contrast, other mRNAs (such as the one encoding the cell-free translation product marked by an asterisk) is poly(A)$^+$ in the oocyte and poly(A)-in the zygote. These changes occur right after fertilization, just at the time when these mRNAs move onto polysomes. This result shows that the mRNA for proteins A, B and C are poly(A)-deficient in the oocyte and become adenylated after fertilization, whereas other mRNAs become deadenylated during the same interval. This finding is supported by the experiment. Northern blots of total, poly(A)$^-$ and poly(A)$^+$ RNA fractions of oocytes and embryos were hybridized with several different ^{32}p-labeled cloned cDNAs. The hybridization pattern obtained with 1T55 (protein A) shows that essentially all of this mRNA is poly(A)-deficient in the oocyte and becomes poly(A)$^+$ after fertilization. Several other mRNAs, including mRNA

C, show the sam? switch. In contrast, hybridization with the a-tubulin probe (3V I) show that a-tubulin mRNA is poly(A)$^+$ before fertilization and poly(A)$^-$ after.

A second kind of experiment demonstrates that these additions and removals of poly(A) account for most or all of the mRNA size changes that happen at fertilization. When the sizes of the oocyte and embryo versions of individual mRNAs are compared after their poly(A) tails have been removed by RNAse treatment, they are indistinguishable. These and other results demonstrate that, except for the addition or removal of poly(A), there are no significant changes in mRNA sizes that would indicate extensive processing or larger precursors after fertilization. Of course, more subtle alterations such as the removal of just as few nucleotides or the modification of 5' cap structures cannot be ruled out in thes experiments.

The Relationship Between Adenylation and Translational Activity of Spisula Maternal mRNA is Good but not Perfect

In many of the cases examined so far, the correlation between the adenylation and translation status of individual mRNA *in vivo* is intriguingly close. For example, several mRNAs such as A, B, and C are poly(A)$^-$ in the oocyte where they are translationally repressed, and these same mRNAs become polyadenylated at the times as they become translationally active in the zygote. Other mRNAs, such as a-tubulin, are both poly(A)$^+$ and active in the oocyte and become deadenylated and inactivated after fertilization. However, not all mRNAs fit this pattern. A few mRNAs are found in both the poly(A)$^+$ and poly(A)$^-$ fractions in the oocyte but show translational activity at this stage.

Stage-specific mRNA Translation is Mimicked in a Mixed Cell-free Systems

Oocytes and embryos contain the same sets of mRNAs and these mRNAs are equally translatable *in vitro* when the usual reticulocyte cell-free lysate is used. Except for adenylation and deadenylations, no obvious changes in mRNA size occur at fertilization, suggesting that major processing events are not responsible for the translational activation and repression of specific maternal mRNAs. If this is true, then some other component must play an important role in this translational switch. Earlier experiments attempted to reproduce the stage-specific translation of individual mRNAs in a cell-free system. Crude 12K supernatants were preapred from *Spisula* oocytes or embryos homoegenized in a buffer (0.3 M glycine, 70 mM potassium gluconate, 45 mM KC1, 2.3 mM MgC12, 1 mMDTT, 40 mM HEPES, pH 6.9) developed by Winkler and Steinhardt (1981) for a sea urchin egg cell-free system. These 12K supernantants by themselves do not initiate protein synthesis very efficiently. However, when the crude 12K supernatants are mixed with "helper" mRNA-dependent reticulocyte lysate, this mixed cell-free system now initiates and synthesizes protein rather actively. Furthermore, the translation products encoded by the oocyte-reticulocyte mix closely resemble those made by the oocyte in vitro, whereas the embryo-reticulocyte mix yields a pattern of protein synthesis that is like the embryo pattern in vivo.Thus, the initiations that occur in these mixed cell-free systems retain the stage-specificity. Clearly, some translational cues are present in these crude unextracted 12K supernatants and are lost during the phenol extraction of RNA from them. Little is known about the nature of the "information" that is responsible for this stage-specific, sequence-specific translational discrimination. When oocyte and embryo 12K supernatants are mixed in varying proportions prior to adding them to the reticulocyte lysate, the pattern of protein synthesis simply resembles the sum of the input oocyte and embryo patterns. This kind of result suggests that there is no excess of positive or negative diffusible regulatory molecules. One interpretation is that these mRNAs *in vivo* are complexed with proteins as RNP particles *in vivo* and that certain of these proteins can modulate translational activity in different ways under different physiological circumstances. Another possibility is that large pH or ionic differences in oocyte and zygote cytoplasm promote differences in mRNA secondary structure that determine whether or not an mRNA is available for ribosome binding.

Discussion

Fertilization of *Spisula* oocytes causes a rapid change in the pattern of protein synthesis that is regulated entirely at the translation level. Within 10 min of fertilization, many of the mRNAs that were actively engaged in protein synthesis in oocytes are released from polysomes and become translationally inactive in the zygote.Examples include mRNAs, X, Y, Z and α-tubulin. During this same interval other sets of previously stored maternal mRNAs enter polysomes and become translationally active. Some of these mRNAs, such as A and C, are completely mobilized onto polysomes, whereas in other cases only a fraction of the stored sequence is recruited.

The molecular mechanisms that control these switches are unknown. It seems unlikely that large changes in the primary structure of these mRNAs play a role in this phenomenon in *Spisula*. First, except for various adenylations and deadenylations, the mRNA sizes do not change detectably at fertilization. However, small change in mRNA sizes, such as the removal of a few nucleotides or the modification of 5' cap, would not have been seen in our experiments. Second, the oocyte and zygote versions of these mRNAs are, after phenol extraction, equally translatable in the reticulocyte cell-free system.

An important caveat to this second line of reasoning is that at least one change in mRNA structure-namely adenylation—does occur in response to fertilization but this change has no detectable effect on the behaviour of these mRNAs in vitro.They poly(A)$^+$ and poly(A)$^-$ versions of several mRNAs show very different translational activities *in vivo* but are translated equally well in the reticulocyte lysate. Thus, it seems quite possible that other modifications of mRNA primary structure may be occurring and that such modifications could be recognized and obeyed in the oocyte/zygote milieu, but that these differences are simply ignored by a less discriminatory, heterologous *in vitro* system. Only direct sequence comparisons of maternal mRNAs from oocytes and zygotes will properly resolve this tissue.

The role of polyadenylation in translational activation and expression is similarly murky. Several translationally regulated mRNAs undergo a change in polyadenylation soon after fertilization. In some cases, such as mRNas A, B, and C, translational activation of these sequences occurs during the same 10-min interval that they become polyadenylated. In other cases, such as α-tubulin and certain other sequence of unknown coding identity, loss of their translational activity after fertilization is accompanies by deadenylation. However, other mRNAs do not show such a close coupling between adenylation and activity. Most of these comparisons have been carried out using oligo(dt)-cellulose, which fractionates the RNAs into those sequences containing fewer than about 6-10. A residues and those containing longer a tracts. Perhaps some other length differential is used for these translational decisions. If that were so, then the oligo(dT)-cellulose fractionation test would sometimes, but not always, show a correlation between possesion of a poly(A) tail and translational activity *in vitro*.

Changes in adenylation of maternal mRNas have been seen in other species as well. In the starfish, for example, hormonal activation of the oocyte sets off a translational change that closely resembles that in *Spisula*; several of the activated mRNAs are converted form poly(A)$^-$ to poly(A)$^+$ forms. In Xenopus, actin mRNAs become deadenylated after oocyte maturation and cease translation roughly at the same time, but no direct correlation has been made yet. In contrast, deadenylation of the Xenopus oocyte's poly(A)$^+$ histone mRNAs pool at maturation occurs about the same time as the selective recruitment of histone mRNA onto polysomes. In this last case, the relationship between adenylation and translation is the opposite of the one usually seen. However, the failure of all mRNA to follow a single rule may simply mean that there is more than one rule.

In *Dictyostelium*, the transition from a vegetative to an aggregation ("developmental") programme is accompanied by the deadenylation of actin mRNA and the loss of that mRNA from the polysomes. In one of the few relatively direct tests for an involvement of poly(A) in translation, Jacobson and Favreau (1983) found that free poly(A) tracts inhibit the rate of translation *in vitro* of many poly(A)$^+$ RNA whereas the translation of poly(A)$^+$ mRNAs is not affected by the poly(A). These authors suggest that the binding of an mRNA's poly(A) tail to the poly(A) binding protein in some way causes an enhancement of that mRNA's initiation rate. Indeed, early experiments by Jeffery and Brawerman (1975) indicated that the poly(A) tails of mouse sarcoma mRNas interact closely with other parts of the mRNAs. It should be possible to test for such associations in Spisula mRNAs now that cloned probes have been made.

In certain systems, a case can be made that when the overall rate of protein synthesis changes, intrinsic differential efficiences of mRNA initiation will lead to changes in the extent of mRNA competition and therefore to different patterns of protein synthesis. It seems very unlikely that this is the explanation for the changes in maternal mRNAs translation. In *Spisula*, the overall rate of protein synthesis increases only 2-4X after fertilization, yet the pattern of protein synthesis changes very dramatically. In fact, certain mRNAs are to translated at all the oocyte and are completely recruited onto polysomes after fertilization. These same general features appear to be true for starfish and possibly for *Xenopus*. In sea urchins, the rate of protein synthesis goes up sharply, 10-50 fold depending on the species, but the pattern of proteins synthesized by the egg and the zygote are for the most part indistinguishable. Thus, the most dramatic, sequence-specific changes in mRNA translation happen in the face of a very insignificant rate change (Spisula), and the fewest changes occur despite extremely large rate increases after fertilization (sea urchin).

Finally, the role of sequence-specific sequestration and stage-specific release of maternal mRNAs should be considered. This sequestration could be gross, with certain mRNAs being held within organelles, or could be executed at the molecular level by masking proteins that bind to an mRNA and prevent its translation. Both cytological fractionation and *in situ* hybridization experiments show that maternal histone mRNAs in sea urchin eggs are restricted to the egg pronucleus and remain there until just before first cleavage, when they then exist to the cytoplasm. This explanation, however, probably does not hold for the translatinally regulated mRNAs in starfish and *Spisula*. Cytoplasmic fractionation experiments reveal no enrichment for any of these RNAs in the oocyte germinal vesicle.

Molecular sequestration still remains an open possibility. Whereas phenol-extracted mRNAs from oocytes and zygotes are equally active *in vitro* in the reticulocyte lysate, crude unextracted oocyte and zygote homogenates added to reticulocyte lysate programme the oocyte⁻ and zygote-specific patterns of protein syntheiss, respectively. Homogenate mixing experiments suggest that there are no diffusible regulators, a finding also consistent with the idea of masing proteins. However, this is not the only interpretation. Perhaps other RNA-binding molecules arbitrate these translational decisions. New insights will require new experimental approaches.

4

Molecular Regulation of Development

After a messenger RNA has been transcribed, processed, and exported from the nucleus, it still needs to be translated in order to form the protein encoded in the genome. In this chapter, we shall see that regulation at the level of translation is an extremely important mechanism in the control of gene expression. In such cases, the message is already present but may or may not be translated, depending on certain cellular conditions. Thus, translational control of gene expression can be used when a burst of protein synthesis is needed immediately (as we shall find to be the case in the newly fertilized egg), or it can be used a fine-tuning mechanism to ensure that a very precise amount of protein is made from the available supply of messages (as is the case in haemoglobin synthesis). We shall also see that there are several ways to effect translational control and that different cells have evolved different means to do so.

Translation in Eukaryotes

Translation is the process by which the information contained in the mRNA instructs the synthesis of a particular polypeptide. This process has been divided into three components: initiation, elongation and termination. Initiation consists of the reactions wherein the first aminoacyl-transfer RNA and mRNA are bound to the ribosome. The only transfer RNA (tRNA) capable of initiating translation is a special tRNA (tRNAi), which carries the amino acid methionine. As the first reactions involve the formation of an "initiation complex" consisting of methionyl-initiator tRNA bound to a 40 S ("small") ribosomal subunit. This Met-tRNA, is recognized by *Eukaryotic Initiation Factor* 2 (eIF2), which binds it to the 40S ribosomal subunit.

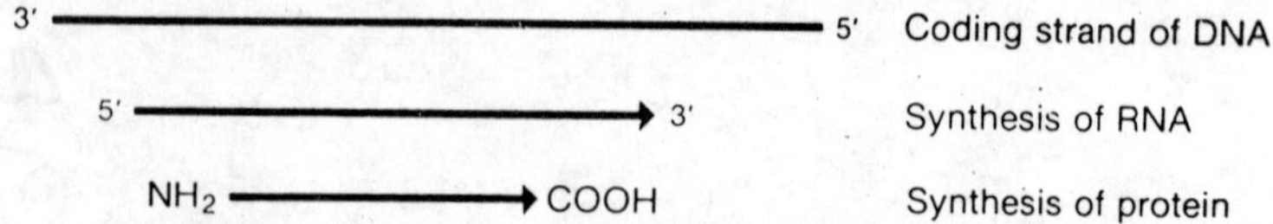

Fig. 4.1. Directions of synthesis of RNA and protein with respect to the coding strand of DNA.

Note that the binding occurs in the absence of mRNA. The mRNA is added next. It first binds its 5' end to the ribosome and then aligns itself so that the methionine-tRNA, is positioned on an AUG nucleotide triplet (codon). This step involves initiation factors 2, 4A, and 4B as well as the 7-methylguanosine "cap" on the mRNA. Without the cap, the binding of mRNA to the ribosomal subunit is not completed. Only after the mRNA has been complexed to the small ribosomal subunit can the 60S ("large") ribosomal subunit bind. This completes the initiation reaction. Elongation involves the sequential binding of aminoacyl-tRNAs to the ribosome and the formation of peptide bonds between the amino acids as they sequentially relinquish their tRNA carriers.

As the amino acids are joined together, the ribosome travels down the message, thereby exposing new codons for tRNA binding. This allows another ribosome to initiate on the 5 end of the message and begin its travelling. Thus, any mRNA usually will have several ribosomes attached to it. This structure is then called a polyribosome-or, more commonly, a *Polysome*. The termination of protein synthesis takes place when the mRNA codons UAG, UAA, or UGA are exposed on the ribosome. These nucleotide triplets are not recognized by tRNAs and hence do not code for any amino acids. Rather, they are recognized by release factors, which hydrolyze the peptide from the last tRNA, freeing it from the ribosome.

Ribosomes and Haemoglobin

One of the major problems in genetic regulation is the co-ordinated production of several products from different regions of the genome. For example, the ribosome of the bacterium *Escherichia coli* consists of 55 protein and 3 different ribosomal

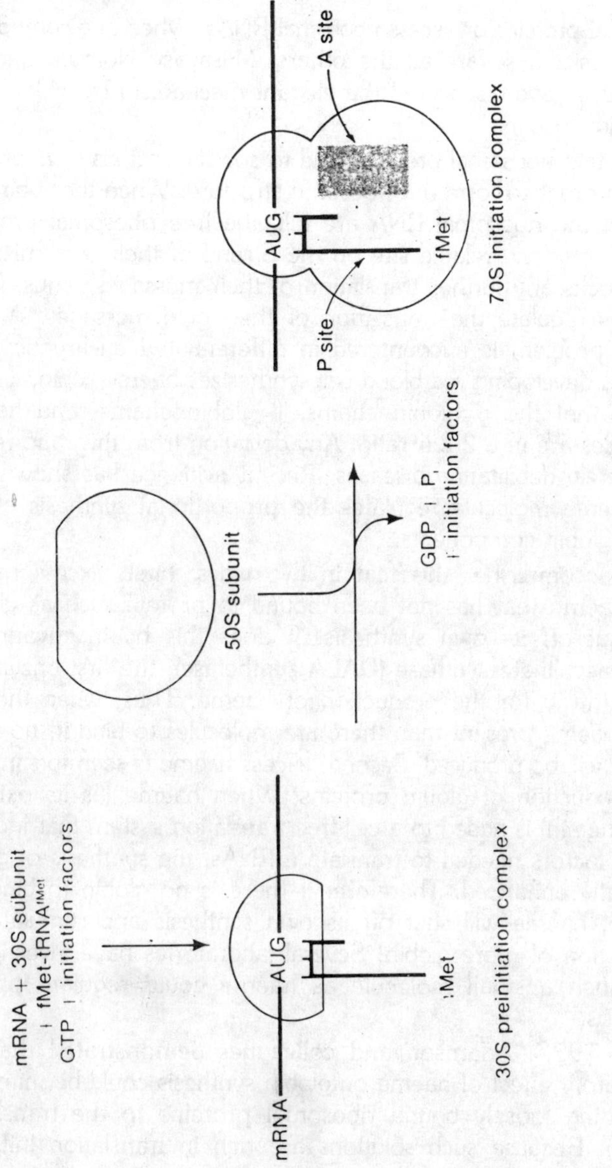

Fig. 4.2. Early steps in protein synthesis in prokaryotes: formation of the 30S preinitiation complex and of the 70S initiation complex.

RNAs. Rather than having so many genes transcribed from the same operon, the ribosomal RNA and ribosomal protein genes are clustered in several spots throughout the *E. coli* chromosome. Still, co-ordinated regulation occurs. One never sees excess ribosomal proteins or excess ribosomal RNAs. When one component is synthesized, so are all the others. Masayasu Nomura and his colleagues have discovered the elegant mechanism by which such regulation occurs.

Certain ribosomal proteins bind to specific regions of ribosomal RNA in order to form the ribosome structure. When these binding sites on the ribosomal RNA are full, the free ribosomal proteins bind to a closely related site on the 5' end of their own mRNAs. This blocks any further translation of their messages. Thus, these proteins regulate the translation of their own messages. A very similar problem is encountered in differentiated eukaryotic cells. When a developing red blood cell synthesizes haemoglobin, it must ensure that the α globin chains, β globin chains, and haeme molecules are in a 2:2:4 ratio. Any deviation from this ratio results in severely debilitating diseases. Recent evidence has shown that the haeme molecule regulates the proportional synthesis of the haemoglobin components.

It accomplishes this feat in two ways. First, excess haeme (i.e., haeme that has not been bound to protein such as globin) will shut off its own synthesis. It does this by inactivating δ-aminolaevulinate synthase (DALA synthetase), the first enzyme in the pathway for the production of haeme. Thus, when there is more haeme present than there are molecules to bind it, no more haeme will be produced. Second, excess haeme is seen to stimulate the production of globin proteins. When haeme (as its oxidized form, hemin) is added to a cell-free translation system that includes all the factors needed to translate mRNAs, the synthesis of globin is greatly enhanced. Therefore, if there is no globin to bind the haeme, haeme will shut off its own synthesis and stimulate the production of more globin! Several laboratories have investigated how such a small molecule as haeme could regulate protein synthesis.

In 1972, Adamson and colleagues demonstrated that the stimulatory effect of haeme on globin synthesis could be mimicked by adding loosely bound ribosomal proteins to the translation system. Because such solutions are rich in translation initiation factors, each factor was tested separately. It was found that eukaryotic initiation factor 2 (eIF2) restored protein synthesis to haeme-deficient lysates in the translation system. This initiation factor is responsible for combining with the initiator tRNA and complexing it to the 40 S ribosomal subunit.

Table 4.1. Components of the in Vitro Translation System Containing Rabbit Reticulocyte Lysate

Concentration	*Concentration*		
Component	*(in 100 µL)*	*Component*	*(in 100 µL)*
Reticulocyte lysate (1: 1)	50 µL	KCL	76 mM
Tris-HCL (pH 7.6) buffer	10mM	Proportional amino acid mixture	6-170 µM
ATP	1 mM	[$_{14}$C] Leucine	0.8 µ Ci
GTP	0.2 mM	"Cold" leucine	26 µM
Creatine phosphate	5 mM	Hemin	10-30 µM
Creatine phosphokinase	10 µg	H_2O to bring total reaction	
Magnesium acetate	2mM	to 100 µL volume	

What, then, was the relationship between haeme and eIF2? To answer this, London and his co-workers added haeme-deficient lysates to haeme-supplemented translation systems. They found that a portion of the haeme-deficient lysate could actually depress the synthesis of globin in the translation system to which it was-added. This indicated that an inhibitor was present. Furthermore, this inhibiting fraction had enzymatic activity. It was a kinase capable of phosphorylating eIF2. To summarize, haeme appears to regulate globin synthesis in the following manner.

1. Excess haeme is able to bind to the protein kinase, inactivating it. Inactivated kinase will not phosphorylate eIF2, so translation proceeds. Thus, as long as haeme is present, globin synthesis continues.
2. In the absence of haeme, a specific protein kinase will phosphorylate initiation factor 2.
3. Phosphorylated initiation factor is not active and therefore will not complex Met-tRNA, to the 40 S ribosomal subunit. Thus, no translation occurs.

The story of the translation control of globin synthesis does not end here. As stated already, there are four active α globin genes per diploid cell and only two active β globin genes. If each gene were transcribed and translated at the same rate, one would expect twice as many α globin molecules as β globin molecules. This, of course, is not the case. One finds a 1.4:1 ratio of α:β mRNA, but 1:1 ratio of the proteins. Thus, the mechanism for the equalization has to be translational. Kabat and Chappell (1977) have suggested that the equalization is done at the initiation step

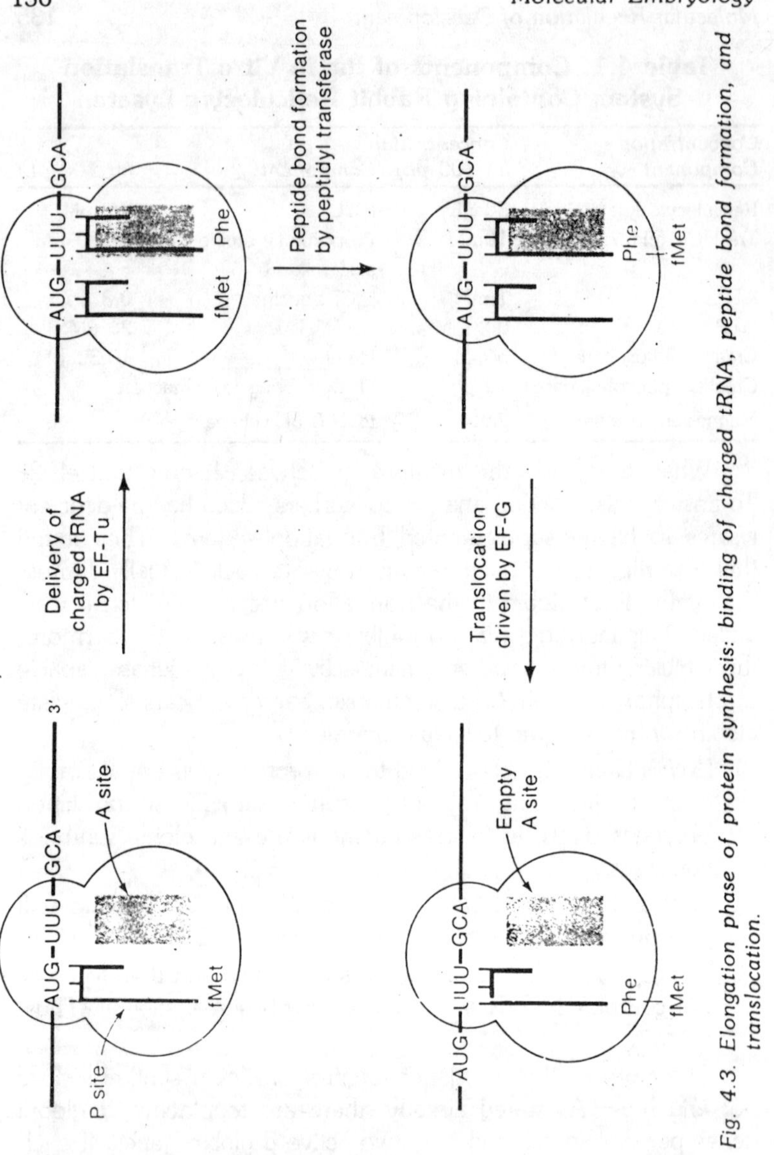

Fig. 4.3. Elongation phase of protein synthesis: binding of charged tRNA, peptide bond formation, and translocation.

of translation. They showed that the α globin mRNA competes with the β globin message for initiation factors and that the β globin message appears to be a better competitor.

The β globin message was recognized more efficiently by the initiation factors and was thus translated more frequently. The 5' ends of the two globin mRNA differ significantly, as is shown in Figure. When the two RNAs were present in equal amounts with a severally limiting supply of initiation factors, only 3 per cent of the resulting protein was α globin. However, when the unfractionated mRNA (α and β globin messages from lysed cells) were added to an excess of such initiation factors, all mRNAs were translated with equal efficiency and the resulting α to β ratio was 1.4:1. Thus, at the initiation step of translation, the proper ratios of α globin, β globin, and haeme are established. Although haemoglobin synthesis involves regulation at the transcriptional and RNA processing levels, the final molecule is constructed through fine-tuned co-ordination at the level of translation.

Translational Control of Oocyte Messages

Evidence for Maternal Regulation of Early Development

In most animal species, the diploid nucleus is not immediately expressed. Evidence that early development is controlled by factors stored in or made by the oocyte came from several experiments at the turn of the century. These experiments clearly demonstrated the dominance of maternal traits during the initial stages of embryogenesis and a switch to paternal or hybrid characteristics only later in development. Such far-reaching maternal effects have already been alluded to in our discussion of the cleavage orientation in snail embryos, in which the oocyte cytoplasm contains a factor that directs the rotations of the cleavage planes in a dextral or sinistral direction.

Table 4.2. Driesch's data on the Maternal Control of Primary Mesenchyme Cell Number in Hybrid sea Urchin Embryos

Egg	*Sperm*	*Average number of primary mesenchyme cells (range in parentheses)*
Echinus	x Echinus	55 (± 4)
Spherechinus	x Spherechinus	33 (± 4)
Spherechinus	x Echinus	35 (± 5)
Strongylocentrotus	x Strongylocentrotus	49 (± 3)
Spherechinus	x Strongylocentrotus	33 (± 3)

In 1898, Hans Driesch crossed two species of sea urchins and found that the average number of primary mesenchyme cells in the hybrid depended solely on which species provided the egg. In a related experiment, Tennant (1914) fertilized eggs from the sea urchin *Cidaris* with sperm from the sea urchin *Lytechinus*. He showed that all the chromosomes were retained in the resulting blastomeres and that the maternal pattern of development was followed completely through the invagination of the archenteron and the timing of mesenchyme formation. The paternal genes were first seen to be involved in the placement of the skeletal cells. A second type of evidence for maternal regulation of early development comes from enucleation studies.

If one could enucleate an egg after fertilization, one should be able to observe how far development could proceed before terminating. This enucleation has been accomplished by physical, chemical, and genetic means. Physical enucleation of the oocyte was accomplished by E.B. Harvey (1940). She placed unfertilized sea urchin eggs into a sucrose solution having the same density as the eggs themselves, and she found that when such eggs were centrifuged, their contents stratified and the egg was eventually split into two spheres. The lighter half contained the nucleus, whereas the heavier half was enucleated. Harvey then parthenogenetically activated the enucleated halves by placing them into hypotonic seawater. These halves ("parthenogenetic merogones") cleaved, formed abnormal blastulae (which lacked a blastocoel), and successfully hatched.

Table 4.3. Tennant's Data on the Control of Early Gastrulation Events in Hybrid Sea Urchin Embryos

Species of sea urchin	*Archenteron invagination (hours)*	*Mesenchyme formation (hours)*	*Site of origin of primary mesenchyme cells*
Cidaris o	20-33	23-26 (follows invagination)	Archenteron tip
Lytechinus o	9	8 (precedes invagination)	Archenterone base and sides
Hybrid (*Cidaris* o x *Lutechinus*) o	20	24 (follows invagination)	Archenterone base and sides

No further development was seen. Similar studies involving frog eggs have also shown that cleavage can even occur in the total absence of chromosomes (provided the egg is injected with centrioles to compensate for those normally provided by the sperm).

Here, too, an abnormal blastula is created. Physical enucleation studies suggested that certain eggs were capable of development through the midblastula stage, even in the absence of a nucleus.

Stored Messenger RNA

Evidence that the oocyte controlled early development by storing messenger RNAs was first obtained by Brachet and co-workers (1963) and by Denny and Tyler (1964). The results of these investigations demonstrated that such enucleated and activated sea urchin half eggs contained RNA and that they could synthesize proteins at a rate comparable to that of normally fertilized eggs. Because there could be no transcription from these cells, stored mRNA was implicated. Craig and Piatigorsky (1971) demonstrated that this increase in protein synthesis could not have come from the transcription of mitochondrial DNA in the oocyte. It seemed that the oocyte had stored messenger RNAs that were not translated until after fertilization. A gentler means of enucleation was made possible when it was discovered that the drug actinomycin DD inhibited RNA synthesis, and transcription could be eliminated by placing the newly fertilized egg in a solution of this drug.

Gross and Cousineau (1964) found that when sea urchin eggs were treated with enough actinomycin D to shut down 94 per cent of their RNA synthesis, the embryos still became blastulae. That this amount of development was dependent on stored messages and not on preformed proteins was shown by treating the fertilized eggs with emetin or cycloheximide. These drugs inhibit translation, and embryos fertilized in the presence of these inhibitors did not develop at all. Several investigators demonstrated a burst of protein synthesis shortly after fertilization.

The actinomycin D experiments of Gross and his co-workers demonstrated that the magnitude of the fertilized burst of protein synthesis in "chemically enucleated" embryos was exactly the same as that in controls. By 6-10 hours, however, a decline in protein synthesis was observed in the treated embryos, and they did not undergo a second burst of protein synthesis at the blastula stage. The message for this second burst of protein synthesis comes from nuclear transcription. Thus, the actinomycin D-treated embryos can develop up to the blastula stage, hatch, and then proceed no further. The stored mRNAs of the oocyte are sufficient to take development only through the hatched blastula stage. Students in amphibians have given similar results, although the stimulus for the increased rate of protein synthesis appears to be ovulation, rather than fertilization, of the egg. In activated enucleated frog eggs, the proteins synthesized shortly after fertilization are not only synthesized in the proper amounts, they are also the same type of protein as is normally made.

Here too, it appears that the oocyte cytoplasm is "preprogrammed" to carry out early developmental processes even in the absence of a nucleus. Eventually, transcription is initiated in the nuclei of the embryo: yet even his this transcription of the embryonic nuclei may be activated by the maternal factors in the oocyte. This concept is supported by the investigations of the o mutant of the Mexican axolotl. This is a maternal effect mutation wherein homozygous females produce eggs that are successfully fertilized and are totally normal until the late cleavage and early blastula stages, At midblastula, eggs shed by an o/o female are seen to have slower mitoses. These eggs will from a dorsal blastopore lip but will always arrest to gastrulation. Malacinski (1971) has shown that in wild-type blastula, new proteins are starting to be synthesized. However, the blastulae from eggs from o/o mothers do not undergo this burst of protein synthesis and have a pattern of proteins identical to that produced by enucleated zygotes.

Carroll (1974) has shown that whereas normal late blastulae show intense RNA synthesis, the mutant embryos show little or no RNA production. Briggs and Cassens (1966) demonstrated that the embryos from o/o mothers lacked a factor that activated the nuclear genome during blastula stage. Such a factor could be isolated from normal blastomere cytoplasm or from the nuclear sap of premeiotic oocytes. When this factor was injected into the eggs of o/o mothers, it repaired the gastrula arrest and allowed the embryo to develop normally. Thus, a specific factor is needed to activate the amphibian nucleus.

In the absence of such a factor, the only development that occurs is that which can be supported by the stored oocyte mRNA. In amphibians, there is enough stored oocyte material to enable the embryo to enter gastrulation. However, without new RNA synthesis, no further development can occur. Physical, chemical, and genetic enucleations of the egg made it clear that stored messages do indeed exist in the cytoplasm of the oocyte and that the products of these messages support the early development of the embryo.

Characterization of Maternal Messages

Eric Davidson and his colleagues have estimated the complexity of the oocyte mRNA in a manner similar to their analysis of DNA complexity. RNA was hybridized to denatured DNA and the half *Cot* value of the hybridization was found to be proportional to the amount of different RNA sequences present. By this analysis, they estimated that each oocyte (in numerous phyla) had enough different nucleotide sequence to account for roughly 1600 copies each of 20,000-50,000 RNA types. This is the greatest message complexity of any known cell type, and it reflects the enormous developmental potential of the oocyte. Even though this figure represents nearly 4 per cent of the total RNA of the oocyte, only a few of these messages have been characterized. The use of cell-free translation systems and cDNA probes has enabled the identification of histone mRNA and tubulin mRNA in the oocyte cytoplasm. The products of these messages are of obvious importance for the formation of chromatin and the mitotic spindle during cleavage.

Biochemical assays have permitted the identification of the messages for nucleotide reductase (an enzyme essential for the formation of deoxyribose nucleotides from stored ribonucleotides) and for the hatching enzymes, which permits the blastulae to digest their fertilization membranes. The utilization of recombinant DNA techniques discussed earlier is also being applied to maternal messages and has enabled investigators to isolate store mRNA for actin and other proteins. Interestingly enough, stored mRNAs are not uniformly distributed throughout the oocyte.

We have already seen that certain RNA sequences show cytoplasmic localization in snail and tunicate eggs, but such localizations have also been observed in sea urchin eggs. Rodgers and Gross (1978) separated blastomeres at the 16-cell stage, cultured them in actinomycin D, and found that the mesomeres had more different types of mRNA than did the micromeres. Thus, cytoplasmic localization of preformed oocyte messages can be seen in both "*mosaic*" and "*regulative*" eggs.

Mechanisms for Translational Control of Oocyte Messages

There are, at present, at least four hypotheses for the regulation of oocyte mRNA translation. Three of them involve the availability of messenger RNAs; the fourth involves the efficiency of mRNA translation. Although these hypotheses may be seen as competing with one another, it is probable that each is used by different species and that some species use more than one of these mechanisms to regulate oocyte mRNA translation.

Masked Messages

The first hypothesis contends that the oocyte messages are physically masked by proteins so that the mRNA cannot attach to ribosomes. Messenger RNA is never found devoid of proteins. However, the type of protein associated with the RNA can vary 1966. Spirin proposed that the mRNA of the oocyte was stored in Informosomes, ribonucleoprotein complexes wherein the mRNA was masked. The masked messages would be unable to bind to the ribosomes and thus would not be translated. At fertilization, the proteins masking the message would be released (possibly because of the ionic changes occurring during fertilization) and the message would be free to start translation. Support for this hypothesis followed shortly.

In 1968, Infante and Nemer found in the sea urchin oocyte ribonucleoprotein (RNP) particles that sedimented more slowly than ribosomes, and Gross and co-workers (1973) found that these particles contained various messages. Other studies showed that oocyte ribosomes and initiation factors were perfectly capable of translating exogenously supplied message, so that the control of translation did not seem to be due to the number or capabilities of the ribosomes. A series of experiments in Rudolf Raff's laboratory demonstrated that oocyte mRNA is indeed stored in a non-translatable form that is susceptible to ionic changes such as those occurring during fertilization. Nonribosomal RNP particles can be isolated from mature sea urchin oocytes and are found to contain RNA with poly(A) tails. Thus, these RNP particles seem to contain messenger RNA. These RNP particles were isolated in two different ionic solutions. Some were isolated "oocyte" buffer containing 0.35 mM K^+ and 5 mM Mg, an ionic condition approximating that of unfertilized oocytes.

Other particles were isolated into 0.35 mM Na^+ , an ion concentration that should remove the proteins bound non-covalently to the mRNA. The mRNA-containing particles prepared in the "oocyte" buffer were very poorly translated in a cell-free system. However, the RNA extracted into the sodium-containing buffer were above to support protein synthesis almost as well as isolated egg mRNA. It thus appears that the influx of sodium during fertilization may destabilize the RNP particle, allowing its mRNA to be translated. Recent certain oocyte-specific proteins have been implicated in the masking of stored messages. When globin mRNA is injected into *Xenopus* oocytes, it is readily translated. However, if this message is first mixed with proteins isolated from *Xenopus* oocyte RNP particle, very little translation is observed. Other RNA-binding proteins isolated from other tissues did not affect the translatability of the injected globin messages, so the oocyte-specific proteins of the RNP particles are implicated in the translational regulation of stored maternal messages. It is known that whereas the unfertilized sea urchin oocyte has only 0.75 per cent of its ribosomes incorporated into polysomes, the cleavage-stage blastomeres have nearly 20 per cent of their ribosomes in such structures.

It appears, then, that pre-existing messages are being recruited into polysomes. Three recent experiments show that the mRNA from the oocyte RNP becomes incorporated into embryonic polysomes. Young and Raff (1979) showed that radioactivity labelled RNA originally in the sea urchin oocyte RNPs later became associated with blastomere polysomes. A similar phenomenon is observed in the early development of the surf clam. Rosenthal and his co-workers (1980) found that when protein-free mRNA from oocytes and embryos were placed in cell-free translation systems, they both coded for the same proteins. In other words, the oocytes and embryos contained identical sets of mRNA. However, different subsets of messages are on the polysomes in oocytes and embryos. Messages for three prominent embryo-specific proteins (A, B and C) are seen as untranslated RNPs in the oocyte cytoplasm, whereas they are seen as polysomal mRNA in embryonic blastomeres.

Conversely, messages for three characteristic oocyte proteins (X, Y and Z) are found on the polysomes of oocytes, but not on those of embryos. The recruitment of mRNA can be seen from the untranslated RNP to the translationally active polysomes. Message recruitment is also seen *Drosophila* embryos. Mermod and co-workers (1980) isolated poly(A)-containing RNA from oocyte RNP, oocyte polysomes, and embryonic polysomes. This mRNA was then translated in a cell-free system containing radioactive amino acids. They found that the mRNA from *Drosophila* oocyte RNP can make certain proteins that cannot yet be synthesized by the mRNA isolated from oocyte polysomes. However, these proteins can be from the mRNA isolated from *embryonic* polysomes. Thus, mRNA formerly stored in the oocyte RNP particles has been recruited into the embryonic polysomes. In this manner, gene expression is controlled by the initiation of translation.

Uncapped Messages

Certain moths use a different mechanism for translational control within the oocyte. In order to be translated, most all eukaryotic messages need to have a "cap" on their 5' ends (Shatkin, 1976). These eukaryotic messages are characterized by 5' 7-methylguanosine. It is believed that this structure is necessary for the recognition of the mRNA with the 40 S ribosomal subunit. The stored messages of the tobacco hornworm moth oocyte have a non-methylated cap. The guanosine is present, but the methyl group has not been added to it. Such messages are not translated into proteins in a cell-free system. However, at fertilization, there is a burst of methylation in these oocytes and the cap is completed. The mRNAs with the completed caps are then able to bind to the ribosomes and initiate translation. Therefore, in two different ways, the oocyte keeps its mRNA and ribosomes separate. Whether by masking the message or not completing its 5' end modifications, the initiation of translation is inhibited. Thus, mRNAs can be stored without being translated, and fertilization triggers the events that enable their translation the oocyte.

Sequestered Messages

Some studies have disputed the evidence that oocyte RNPs are untranslatable. Rather, it is possible that the protein synthetic apparatus is compartmentalized such that mRNA (within the RNP) does not have a chance to get close to the ribosomes. The histone mRNAs of sea urchin oocytes seem to be regulated by this type of restriction. The histone messages of the oocyte are not found in cytoplasmic RNPs. Rather, they are localized in the pronucleus. It is only when the pronucleus breaks down, at the end of fertilization, that the histone mRNA gets into the cytoplasm. This may not be the case for other messages. Less than 0.1 per cent of the total oocyte messenger RNA is found in pronuclei, and those RNPs containing the messages for actin and tubulin are predominantly cytoplasmic.

Changes in Translational Efficiency

In one model of translational regulation in the oocyte the physical or chemical separation of mRNAs need not be postulated. Rather, the initial low pH of the oocyte is able to impede protein synthesis. As stated already, there is an enormous release of hydrogen ions during sea urchin fertilization, resulting in an elevation of cytoplasmic pH from 6.9 to 7.4. Winkler and Steinhardt (1981) prepared an in vitro cell-free translation system from unfertilized sea urchin oocytes.

The pH of the resulting suspension was then retained or altered by dialysis, and protein synthesis was measured by the incorporation of radioactive valine into proteins. No exogenous mRNA was added. Increasing the pH from oocyte levels (pH 6.9) to zygote levels (pH 7.4) occasioned a burst of protein synthesis mimicking that seen during fertilization. It is not known whether this increase in protein synthesis is due solely to the increased efficiency of protein synthesis or whether increasing the pH may cause other conditions (such as the unmasking of messages) that then allow protein synthesis to occur. This increased efficiency alone would not account for the specificity of translation as seen in the surf clam (where message formerly on the polysomes are replaced by others). In any event, the oocyte is able to exercise translation-level control to regulate the expression of preexisting mRNAs.

Maternal mRNA and Embryonic Cleavage

Cleavage in sea urchins requires continued protein synthesis from stored maternal mRNAs. If sea urchins are fertilized in the presence of an inhibitor of translation, no cleavage results, even though the two pronuclei fuse. No mitotic spindle forms, the chromosomes do not condense, and the nuclear membrane does not break down. Thus, it is probable that some of the proteins specified by material messages are involved in cell division during cleavage. This hypothesis has gained support from the discovery of a class of proteins called *Cyclins*. These proteins are coded for by maternal mRNAs, and they are not detectable in the unfertilized egg. However, their synthesis increases dramatically after fertilization. What is striking about cyclins is that they are destroyed upon cell division and have to be resynthesized anew from the stored messages after the completion of each cleavage. Cyclin synthesis is seen to decline as the embryo nears the end of the blastula stage.

Although originally seen in sea urchins, cyclins have also been detected in other species. Proteins A and B in the surf clam (mentioned earlier) are probably cyclins, as their synthesis correlates with the cell cycle. Although no physiological proof exists that these cyclins are responsible for regulating cell division during cleavage, the cyclical synthesis and destruction of a protein would be a likely mechanism for controlling mitosis during embryonic cleavage.

Translation Control of Casein Synthesis

Differentiated gene products are often synthesized in response to hormonal induction. In some of these cases, the hormones do not increase the transcription of certain messages but act at the level of translation. One such case involves the synthesis of casein by lactating mammals. Casein is the major phosphoprotein of milk and is therefore a differentiated product of the mammary gland. The mammary gland is prepared by the sequential actions of several hormones. Prolactin, however, is the hormone responsible for lactation, that is, actual milk production. Prolactin augments the transcription of casein messages only about twofold; its major effect appears to be the stabilization of the casein mRNA. Prolactin somehow increase the longevity of the casein message such that it exists 25 times longer than most other messages in the cell. This means that each casein mRNA can be used for many more rounds of translation. In this way, more casein molecules can be synthesized from casein message.

The Widespread use of Translational Regulation

Translational control is probably a major mechanism of development regulation. We have seen its importance in the burst of protein synthesis following fertilization in many species, in balancing the synthesis of the components of haemoglobin, and in the hormone-induced synthesis of casein. There are many more cases of translational regulation for which the detailed mechanism of operation is still unknown. Glucose, for example, is able to stimulate a tenfold increase in the synthesis of proinsulin from the pancreatic beta cell. However, when a cDNA probe was used to measure the insulin mRNA content of pancreatic beta cells, it was found that glucose-stimulation beta cells contained no more proinsulin message than do non-stimulated beta cells. Thus, the regulation of insulin synthesis by glucose appears to be under translation control.

The reactivation of dehydrated brine shrimp embryos also seems to be regulated at the level of translation. These embryos (often sold as "brine shrimp eggs") begin development and can then lie dormant for long periods of time. During this dormancy, no protein synthesis occurs. It has been found that an inhibitor of protein synthesis exists in these embryos and that upon reactivation, a 20-fold increase in the level of functional eIF2 occurs. Moreover, Sierra *et al.* (1977) showed that this inhibitor may be a protein kinase. Thus, the same mechanism that regulates haemoglobin synthesis may also be used by brine shrimp embryos. Even sperm appear to regulate the expression of certain genes by translational control. Just like eggs, sperm can store messages for later use. This can be seen in the case of lactate dehydrogenase-X in avian and mammalian sperm. Although this *protein* is only synthesized in the last stages of sperm development, the *gene* for LDH-X is transcriptionally active throughout all stages of sperm development. The synthesis of LDH-X protein probably occurs by initiating the translation of stored mRNAs. Translational control, then, is an important and widely used mechanism for regulating gene expression in development.

Posttranslational Regulation of Gene Expression

Once a protein is made, it becomes part of a larger level of organization. It may become part of the structural framework of the cell or it may become involved in one of the myriad enzymatic pathways for the synthesis or breakdown of cellular metabolites. In either case, the individual protein is now part of a complex "*ecosystem*" which integrates it into a relationship with numerous other proteins. Thus, we do not abandon the regulation of gene expression after synthesizing a protein, for several changes can still take place. First, some newly synthesized proteins are inactive without further modifications. These modifications can involve the cleaving away of certain inhibitory sections of the protein or may involve the binding of a small compound to enhance its activity. Second, some proteins may be selectively inactivated. In some cases, inactivation involves the degradation of the protein itself; in other cases, inactivation may be brought about by the binding of an inhibitory ligand. Third, some proteins must be "*addressed*" to their specific cellular destinations. The cell is not merely a sack of enzymes; and proteins will often be sequestered in certain regions such as membranes, lysosomes, or mitochondria. Fourth, some proteins need to assemble with other proteins to form a functional unit. The haemoglobin protein, the microtubule, and the ribosome are all examples of numerous proteins joining together to form a functional unit. Therefore, the expression of genetic information can still be influenced at the posttranslational level. Let us now review each of these in turn.

Activation of Proteins by Post-translational Modifications

Many newly synthesized polypeptides are inactivate unless certain amino acid residues are cleaved away. This is especially evident in the case of peptide hormones. The precursor of insulin, for instance, is a long polypeptide with three disulfide bridges. The hormone is active only after the middle and the beginning of the peptide have been cleaved away. The two remaining ends, held together by disulfide bonds, constitute the active molecule. In the case of pro-ACTH-endorphin, several small peptide hormones are made as one long precursor. This one polypeptide can be cleaved to form adrenocorticotrophic hormone (ACTH), g-lipotropin (g-LPH), and b-endorphin. Further processing of these hormones depends on the specific cell type in which they are found. In the anterior pituitary cells, ACTH is an end product that can be secreted to stimulate the production of steroids by the adrenal cortex. However, in the cells of the intermediate lobe of the pituitary, ACTH is cleaved to form a-melanocyte stimulating hormone (a-MSH).

Some protein can only work after they have been modified by the covalent attachment of smaller molecules. Phosphorylase kinase, for example, is inactive if it is not itself phosphorylated. Once it is phosphorylated, it is catalytically active. Collagen, one of the major structural proteins of the body, cannot function unless its prolines and certain lysines are hydroxylated, and failure to do so results in serious diseases. Posttranslational modification of histones may be important in regulating transcription itself. In the slime mold *Physarum polycephalum*, there are no cellular boundaries and all the nuclei are synchronously regulated within a common cytoplasm. This makes an excellent organism in which to study the biochemistry of the cell cycle. Here, histone H4 can exist in several forms, depending on its state of acetylation. Highly acetylated (2-4 acetyl groups) H4 is associated with transcriptional activity. Conversely, the phosphorylation of histone H1 appears to activate its chromatic condensing activity, thereby curtailing

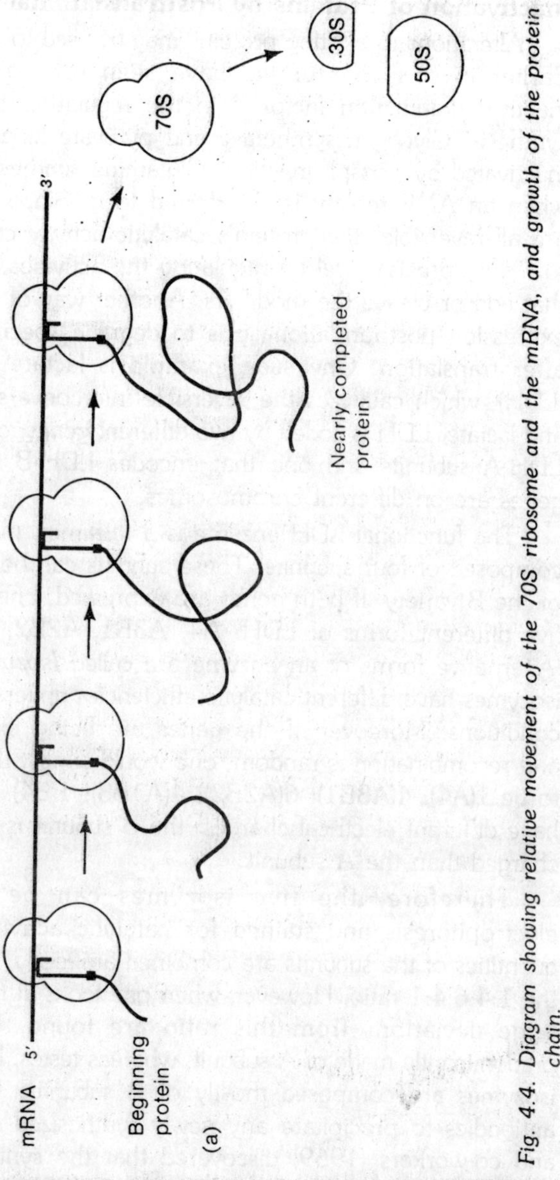

Fig. 4.4. Diagram showing relative movement of the 70S ribosome and the mRNA, and growth of the protein chain.

transcription. In *Physarum*, H4 acetylation and H1 phosphorylation are inversely expressed. Thus, the gross transcriptional activity of chromatin may be regulated by the phosphorylation or acetylation of histone proteins.

Inactivation of Proteins by Posttranslational Modifications

Alterations to existing proteins may by used to inactivate them. Earlier in this chapter we have seen how phosphorylation inactivated initiation factor 2 in the regulation of haemoglobin synthesis. Glycogen synthetase and pyruvate kinase are similarly inactivated by phosphorylation. Glutamine synthesis is inactivated when an AMP moiety is transferred to it. Since these reactions are all reversible, the protein's catalytic activity can be regulated to a very precise level by regulating the activities of the enzymes that add or cleave the modifiers. Another way of regulating gene expression posttranslationally is to degrade specific polypeptides after translation. One such example is lactate dehydrogenase (LDH), which catalyzes the reversible interconversion of pyruvate and lactate. LDH is coded by two different genes: one that encodes LDH-A subunits and one that encodes LDH-B subunits. These genes are on different chromosomes.

The functional LDH enzyme is a tetramer, meaning that it is composed of four subunits. These submits can be of either the A or the B variety. If both genes are expressed, one expects to find five different forms of LDH: A4, A3B1, A2B2, A1B3 and B4. (Alternative forms of an enzyme are called *Isozymes*.) The LDH isozymes have different catalytic efficiencies under different cellular conditions. Moreover, if the genes are being expressed equally and recombination is random, one would expect the ratio of types to be 1(A4): 4(A3B1): 6(A2B2): 4(A1B3): 1(B4). These isozymes have different electrical charges: the B subunit is more negatively charged than the A subunit.

Therefore, the five isozymes can be separated by electrophoresis and stained for catalytic activity. When equal quantities of the subunits are combined artificially, one indeed finds the 1:4:6:4:1 ratio. However, when one looks at individual organs, large deviations from this ratio are found. Muscle LDH is predominantly made of A subunit, whereas testes, heart, and kidney isozymes are composed mostly of B subunits. Using anti-LDH-antibodies to precipitate any newly synthesized A subunits, Fritz and co-workers (1969) discovered that the synthesis of LDH-A does not vary greatly from tissue to tissue. However, the rate of its degradation varies enormously. The degradation of LDH-A is 22 times greater in heart muscle than in skeletal muscle. So the control of LDH isozymes is not at the level of LDH transcription or translation, but at the level of enzyme degradation.

Subcellular Localization of Proteins by Posttranslational Modifications

After they are synthesized, proteins have to be properly positioned. Thus, the protein may ultimately be located in the soluble cytoplasm, the mitochondria, the lysosomes, the endoplasmic reticulum, or the nucleus; it may even be secreted from the cell. Those proteins destined for lysosomes, endoplasmic reticula, or secretion are passed into the lumen of the rough endoplasmic reticulum which they are still being translated. These proteins (including collagen and the peptide hormones discussed earlier) contain a signal of approximately 30 amino acids, which are recognized by a *Signal Receptor Complex* that floats in the cytoplasm. When this complex (which contains six different peptide chains and a small RNA molecule) binds to the growing peptide of a cytoplasmic ribosome, translation stops. The complex then joins to a docking protein on the endoplasmic reticulum and translation resumes.

The growing polypeptides pass through the reticular membrane and once inside the lumen, this signal sequence is removed and the remaining protein can be modified. One such modification is the covalent addition of mannose-6-phosphate to the protein. This sugar acts as an address label that packages the glycoprotein for lysosomes. The mannose-6-phosphate is recognized by receptors at specific areas of the endoplasmic reticulum, which thereby concentrate those enzymes having this address label. This region of the reticulum then buds off to form the lysosome. When binding does not occur, these proteins are secreted from the cell. Evidence for this comes from a lethal genetic syndrome, I-cell disease, where patients are unable to put the mannose-6-phosphate address label on the proteins.

The lysosomes of these individuals lack the appropriate enzymes, which have been mistakenly secreted into the blood. Some proteins have specific binding proteins, which localize them in certain regions of the cell. In mouse liver, β-glucuronidase is found both in the lysosomes and on the endoplasmic reticulum. Although the mannose-6-phosphate label can get this enzyme into the lysosomes, β-glucuronidase is bound to the liver endoplasmic reticulum by the *protein Egasyn*, which specifically recognizes this protein. In mutant mice lacking *egasyn*, there is no β-glucuronidase found on the liver endoplasmic reticulum. The localization of β-glucuronidase is different in various cell types. Tissues such as liver and kidney are rich in egasyn and have β-glucuronidase on their endoplasmic reticula. Brain and spleen cells, however, appear to be entirely deficient in egasyn, and these have all their β-glucuronidase in their lysosomes. So the intracellular localization of proteins is posttranslationally regulated and may differ from tissue to tissue.

Supramolecular Assembly

The last form of posttranslational regulation that we will consider involves the assembly of these newly synthesized proteins into a functional unit. We have already discussed several proteins that can spontaneously assemble themselves into functional units. Haemoglobin and lactate dehydrogenase, for example, both contain four polypeptide chains, which are non-covalently jointed to form the functional protein. On a slightly higher level, the contractile proteins tubulin and actin are both found to exist in polymerized or non-polymerized forms. We have seen that the polymerization of actin from globular monomers into microfilaments is essential for the extension of the acrosomal process during fertilization. Similarly, tubulin is assembled into the microtubules of the mitotic apparatus, only to be converted into monomers as mitosis finishes. These tubulin units can then be reassembled into new microtubules, which are important in establishing cell shape.

There appears to be a class of proteins (microtubule-associated proteins) that are necessary for regulating the polymerization and function of microtubules. Several proteins form assemblies with nucleic acids. Histones, for instance, are useless unless assembled with DNA to form the nucleosome particle. The eukaryotic 80 S ribosome consists of approximately 70 proteins and 4 different ribosomal RNAs. The proteins and nucleic acids have to assemble together to be functional, and it is remarkable that this assembly occurs quickly and efficiently within all cells. It should not be a total surprise, then, to discover that whole infectious viruses can be obtained by incubating the component proteins and nucleic acids together. Thus, spontaneously assembly is another important form of posttranslational control.

Collagen: An Epitome of Posttranslational Regulation

By looking at the synthesis of collagen, one of the most important structural proteins of the body, we can demonstrate the importance and variety of posttranslational control mechanisms,. First, the collagen mRNA sequence is translated into a procollagen peptide. The amino terminal of the procollagen contains a signal sequence that enable it to enter the lumen of the rough endoplasmic reticulum. This signal peptide gets cleaved from the remainder of the collagen molecule. Within the endoplasmic reticulum, the collagen molecules also encounter three enzymes that hydroxylate it. Two of these enzymes hydroxylate proline residues to 3-hydroxyproline and 4-hydroxyproline; the third coverts lysines to hydroxylysyl residues.

Only certain proline and lysine residues are in the correct positions to be recognized by these enzymes. As the collagen chains are being hydroxylated, sugar residues can be added to the hydroxylysines. The first enzyme, galactosyltransferase, adds and galactose to the hydroxylysine; the second enzyme, glucosyl-transferase, adds a glucose unit to the hydroxylsylgalactose. The region near the carboxyl end is modified by glucosamine and mannose. The next step involves the formation of intrachain disulfide bonds and the organization of three collagen polypeptides into a triple helix. Disulfide bonds are formed at the carboxyl ends between adjacent collagen molecules, establishing the preconditions for forming a triple helix.

If the prolines have been properly hydroxylated, the three chains fold over one another to produce a large triple helical region bounded on either side by globular areas. The collagen molecule is secreted in this form to the outside of the cell. Even here, processing continues. First, two proteolytic enzymes remove the globular domains at the amino and carboxyl ends. This creates a pure triple helical fragment. Next, these fragments spontaneously assemble into fibrils. These fibrils, however, lack the necessary tensile strength to function as structural proteins until the various triple helices are joined together covalently by further crosslinking. This is accomplished by enzymatically modifying lysyl or hydroxylysyl residues and linking them together. Thus, there are an impressive number of essential steps by which a protein is modified after translation. We have now looked upon the various levels regulating gene expression: transcription, RNA processing, translation, and posttranslational modification.

Each type of regulation has been found to be important in controlling gene expression during development. We are getting closer to answering the mystery of differentiation, and the new techniques of gene cloning (as Boveri predicted in 1904) should enable us to understand differentiation on a chemical level. However, the question of differentiation of individual cells is not the only problem in developmental biology. We still need to understand how the cells of the body interact to develop at the correct time and at the correct place. In the next part, we shall analyze *cellular interactions* in development that lead to the formation of tissues and organs.

5

Oogenesis Under Gene

The growing oocyte can be regulated as a special form of differentiated cell, organized to carry out a unique function. It accumulates a particular collection of gene transcripts, and by the end of its growth phase has developed a complex of unusual cytological structures. We have seen that the set of genes expressed during oogenesis bears a close, largely overlapping relation to that expressed in the early embryo nuclei. Thus from the standpoint of nuclear function, *oogenesis* is the process during which the major portion of the *zygotic* pattern of gene activity is established. This pattern is also impressed on the entering male genome, and during cleavage is propagated to the blastomere nuclei. Through there are of course genes that function only in the early embryo and not during oogenesis, and *vice versa*, in quantitative terms, both of these classes are minor, in respect to either the mass or the complexity of the transcripts for which they account.

This chapter is focused on gene expression during the growth phase of oogenesis, which begins following premeiotic DNA replication. Throughout this period newly synthesized transcripts are being delivered to the oocyte cytoplasm, and until the metaphase of the initial reduction division occurs at maturation, the oocyte remains in the first meiotic prophase. Thus, it contains a 4C rather than a 2C genome. In the typical case transcription of the genes utilized during oogenesis begins early in the growth phase, before the onset of vitellogenesis, and continues almost until maturation. At maturation there occurs the breakdown of the germinal vesicle, or oocyte nucleus. By this time all nuclear transcription has ceased, and a new set of biosynthetic activities is instituted. The molecular biology of oocyte maturation *per se* has been reviewed extensively and is not further considered here. For earlier events of oogenesis, such as germ cell migration, oogonial multiplication, and the many other biological and biochemical aspects not directly relevant to germ line transcription and gene expression during oogenesis, recent compilations such as those edited by Jones 1978 and Browder (1985) may be consulted. Follicle cell functions are reviewed in these works as well, and in articles by Mahowald and Kambysellis (1980); Wallace (1983); and Moor (1983). At the molecular level perhaps the best studied follicle cell activity is synthesis of the chorion or egg shell, particularly in lepidopteran and dipteran species. Though follicle cell function lies outside the scope of the following discussion, it is to be noted that chorion synthesis is particularly interesting in the Diptera, where it has been observed that the genes coding for the major proteins of this structure are amplified during differentiation of the ovariole.

Gene Expression in the Nurse Cell-oocyte Complex

Oogenesis is always to some extent a co-operative process, involving cells other than the oocyte itself. In most, though certainty not all forms, the growing oocyte is surrounded with follicle cells of somatic origin. Follicle cells provide essential transport, hormonal and secretory functions, but with the possible rare exceptions mentioned below they do not directly feed transcription products to the oocyte. A natural division separates forms of oogenesis in which the *germ line* gives rise both to oocytes and to nurse cells that directly provide the oocyte with cytoplasmic organelles and macromolecules, including transcripts, from those in which the oocyte nucleus is alone responsible for the synthesis of its non-yolk macromolecular constituents. Oogenesis in which the egg cytoplasm is the co-operative product of both nurse cell and oocyte genomes is termed *meroistic oogenesis*. It is known mainly in insects, though it occurs as well in other protostomial invertebrates.

Origin of the Nurse Cell-oocyte Complex

Meroistic oogenesis has been studied in most detail in certain holometabolous insect orders, the most prominent of which are Coleoptera. Hymenoptera, Hemiptera, Lepidoptera and Diptera. Many other insect orders, e.g., the Orthoptera and Odonata, carry

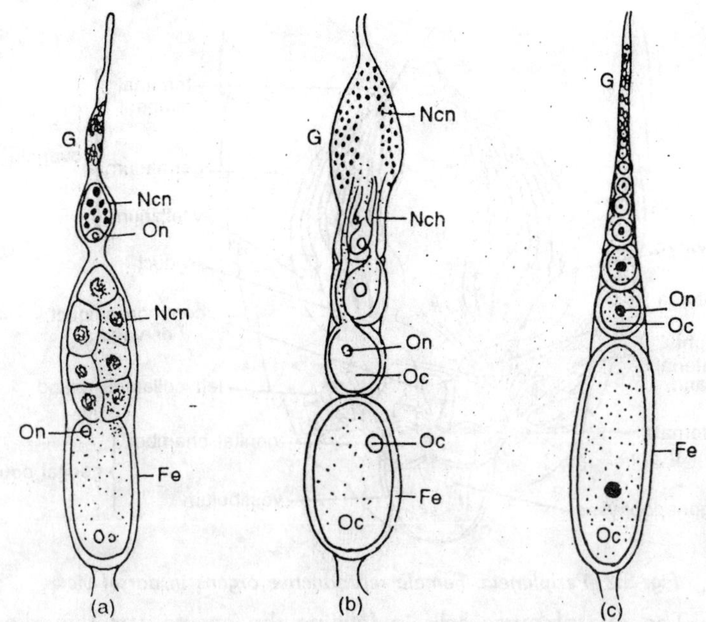

Fig. 5.1. Diagrams of the three types of insect ovary. Polytrophic and telotrophic meroistic oogenesis are portrayed in (a) and (b), respectively, and panoistic oogenesis is portrayed in (c). Fe–follicular epithelium; Ncn–nurse cell nucleus; Nch–nutritive chrod; G–germarium; On–oocyte nucleus; Oc–oocyte.

out oogenesis by the alternative route, in insects termed *panoistic* oogenesis. Nurse cells are here absent, and the oocyte growth process is autonomous, except for the assistance in yolk protein and metabolite uptake provided by the follicle cells. The chromosomes of meriostic insect oocytes typically display a condensed structure. They either do not synthesize RNA at all during oocyte growth, or do so at a very low relative rate. The RNA accumulated during the growth phase in the oocyte cytoplasm is instead transported there from the nurse cells via open cytoplasmic channels. Figure indicates the major forms of egg chamber arrangements observed. In *polytrophic* egg chambers the oocyte is directly connected to each of several nurse cells, and these to one another, by discrete intercellular channels, termed *ring canals*. In *telotrophic* egg chambers a single common duct enters each oocyte, and through this pass the combined products of a syncytial mass of nurse cell nuclei.

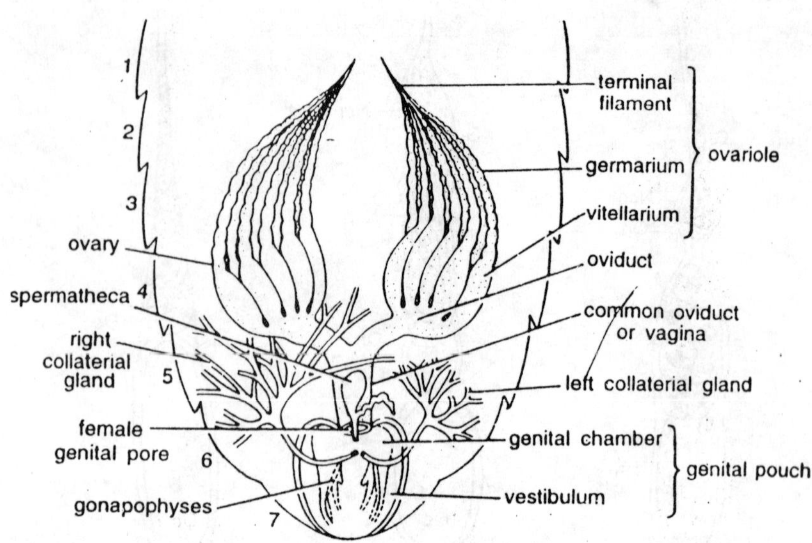

Fig. 5.2. Periplaneta. Female reproductive organs in dorsal view.

The role of nurse cells in feeding the oocyte was remarked upon by many classical writers. Early observers noted in several species that organelles as large as mitochondria pass from nurse cells to oocyte. An interesting variation exists in turbellarian flatworms, where the nurse cells, filled with oocytes after oogenesis is completed, and nurse cell contents are used to sustain growth of the embryos just as are *intracellular* components in other eggs. This unusual course of events draws attention to the essential aspect of nurse cell function, that of providing the egg with materials it will require for development. Wherever nurse cells are found, the functional nature of the oocyte nurse cell interaction is evident. In certain annelids, for example, one or two nurse cells with large polyploid nuclei are applied to each oocyte, and other oocyte-nurse cell complex is released into the lumen of the ovary relatively early in oogenesis. Oocyte growth then occurs at the expense of the nurse cells, which shrink progressively until they become small compared to the relatively enormous oocytes.

Cytoplasmic Transport in Polytrophic Egg Chambers

The structure of the nurse cell-oocyte complex of *Drosophila* as it appears at light microscope magnification. The five deeply staining nurse cell nuclei seen in the section are highly polyploid, while the oocyte nucleus of course contains only the 4C meiotic prophase genome. Three ring canals are shown, one connecting the oocyte with the follicle cells, and two connecting adjacent nurse cells. The growth phase of *Drosophila* oogenesis is divided into 14 stages (for a complete description see King, 1970), beginning with the formation of the oocyte-nurse cell complex and its positioning at the posterior end of the germarium. Oogonial replication and the establishment of new egg chambers begin during puparial life and continue in the adult. Grell and Generoso (1982) showed that the premeiotic S-phase that marks the onset of oogenesis proper occurs in the first pupal preoocytes within the interval 13-162 hr after oviposition. During this period the synaptonemal complexes are extended, and this is also the time at which meiotic recombination may occur. The entire 14 stages require 4½ days in pupae and 6 days in the adult with the difference localized mainly to the rate at which stages 1 and 2 are traversed. Polyploidization of the nurse cell nuclei begin in stage 2, during which they increase their DNA content from 4C to 8C, and this process continues through the previtellogenic stages, i.e. stages 2-7. Together these stages require about 50 hr.

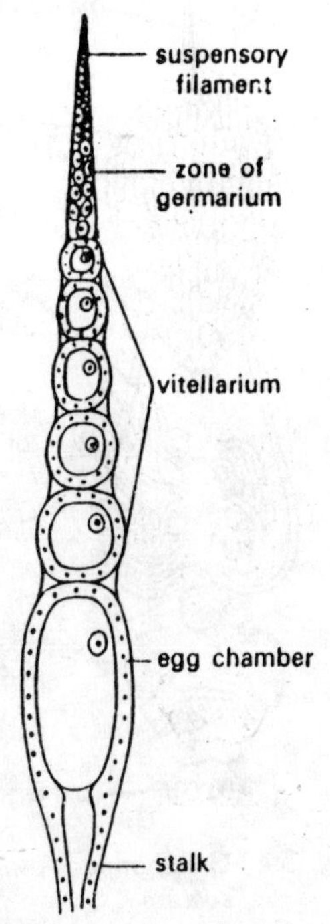

Fig. 5.3. Periplaneta. An ovariole in L.S.

After the 32 stage the nurse cell DNA does not replicate uniformly, in that satellite DNA, and some other sequences, including histone gene DNA are underreplicated. The amount of DNA present in the nurse cell genomes at stage 8, when vitellogenesis begins, is the result of 4-5 further replications of about 75% of the total DNA beyond the 32C stage. Two complete further replications have occurred by stage 10A, when the process is complete. There are then four nurse cell nuclei that contain an amount of DNA equivalent to about 1500 haploid genomes, and eleven that contain about half this amount of DNA. The mass of oocyte constituents meanwhile increases continuously, first as the result of the low of materials through the ring canals, and later by the pinocytotic ingestion of yolk as well. The mechanism of

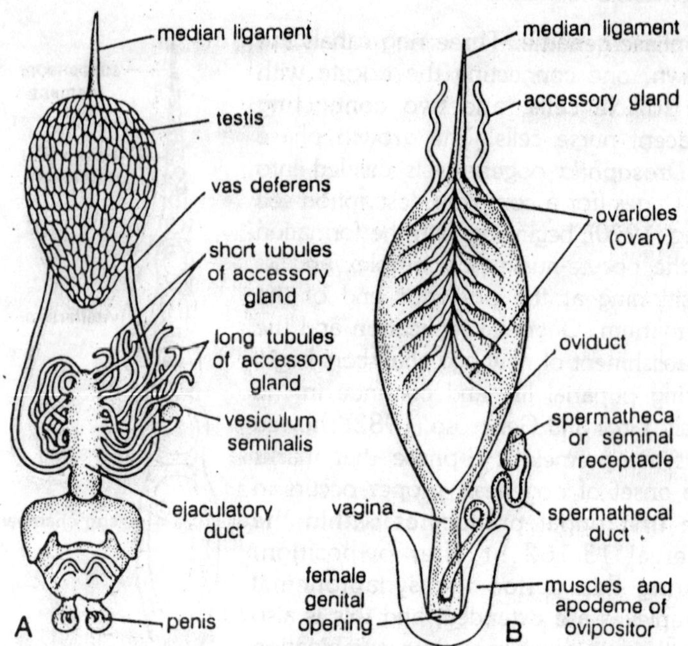

Fig. 5.4. Grasshopper. A—Male reproductive system; B—Female reproductive system.

cytoplasmic transfer through the ring canals may be electrophoretic. Though there are no equivalent observations in *Drosophila*, a potential gradient of > 1 V/cm has been demonstrated across the open ring canals of the polytrophic *Hyalophora* egg chamber, and a similar potential gradient exists in the telotrophic ovariole of the bug *Rhodnius*. Injection of fluorescein labeled compounds into both types of meroistic ovaries shows that negatively charged molecules are transported by intracellular electrophoresis from nurse cells to oocyte, and once inserted in the oocyte these molecules are diffusing backwards toward the nurse cells. The mobility within the *Hyalophora* egg chamber of fluorescein labeled lysozyme, a positively charged protein, and of its negatively charged methylcarboxylated derivative, are compared, after injection into nurse cells.

Just as originally summarised by classical observers using the light microscope, the canal area is seen to be packed with mitochondria, which have apparently been caught up in the cytoplasmic streaming that carries materials from nurse cell to oocyte. At this stage the flow from nurse cells to oocyte is accelerated as the result of a new process, the direct injection of nurse cell cytoplasm into the oocyte. As the injection proceeds, the cytoplasmic volume of the nurse cells decreases and that of the oocyte increases concomitantly. This phase continues for about 5 hr in *Drosophila* until the nurse cells have reliquinshed most of their cytoplasmic mass. They then degenerate. The velocity of the cytoplasmic stream of the ring canal has been measured during the injection process as about 2 μm sec^{-1} and once have entered the egg it dissipates in a circular motion that effectively distributes the entering cytoplasmic constituents. At its peak the total flow carried by the four ring canals entering the oocyte is calculated to be about 1.3 × 104 μm^3 min^{-1}. A similar rate of cytoplasmic flow into the oocyte was estimated by Telfer in the polytrophic ovaries of the moth Hyalophora. The end result of previtellogenic growth, followed by the direct transfer of nurse cell cytoplasm and yolk accumulation, is an enormous change in oocyte volume. In *Drosophila*, for example, the volume of the oocyte increases 90,000 fold with the few days required for oogenesis.

Origin of the Nurse Cells

In most Diptera the accessory cells that perform the function of feeding RNA and other cytoplasmic constituents to the growing oocyte through ring canals derive exclusively from the germ line. However, an exception is known in the *dipteran* family Cecidomyidae, where polyploid follicle cells that are of somatic origin fuse with a single true nurse cell to form a syncytial nurse chamber that contributes cytoplamic RNA to the oocyte. Careful analysis has shown that in the more typical example provided by *Drosophila* and in Lepidopte a and Hymenoptera, the nurse cells of a given egg chamber, together with the oocyte, constitute a clone, descended from a single oogonial stem cell. The development of the egg chamber was described for Drosophila by Koch et al. (1967), for the moth *Hyalophora cecropia* by King and Aggarwal (1965), and for the wasp *Habobracon juglandis* by Cassidy and King (1972). The structure and ontogeny of Ocyte-nurse cell complexes in various inset groups are reviewed by King (1970), Telfer (1975), and King and Buning (1985). The Ocyte-nurse cell complex in Hyalophora includes seven nurse cells and in *Drosophila* 15 nurse cells. These complexes are constructed in the terminal three and four oogonial divisions, respectively. The ring canals that connect each nurse cell to other nurse cells and/or to the oocyte are highly organized membranous structures associated with actin microfilaments. The disposition of the ring canals, which originate by incomplete cytokinesis following each oogonial mitosis, indicates the order of appearance of the nurse cells and the sequence of steps by which the egg chamber is constructed. It can be seen that except for that nurse cell which is formed first the oocyte is the cell with the largest number of ring canals (four), although oocytes with only three ring canals have been reported. Both the first nurse cell and the oocyte initially from synaptonemal chromosomal complexes. The synaptonemal complexes developing in the nurse cell nucleus later disappear, whereas in the oocyte the meiotic prophase movements proceed, and the growth phase of oogenesis ensues.

Germ Line Functions Required in Egg Chamber Differentiation

Additional insight into the process of oocyte-nurse complex formation in *Drosophila* has derived from observations on female sterile mutations that affect egg chamber formation. Morphological effects of the mutation *fes* [fs(2)B] were described by Johnson and King (1972). In homozygous *fes* females cytokinesis is often complete rather than incomplete, with the result that large numbers of abnormal cell clusters containing less than 16 interconnectect cells are formed. The presumptive oocytes are thus deprived of their complement of appropriately connected nurse cells. Normally the interconnected cystocytes all divide synchronously and this form of co-ordination is also absent in *fes* mutants. Several mutations are known that cause disorganized cytocyte divisions, resulting in masses of ovarian cells referred to as "*tumors*," including narrow (nw) and various alleles of fused, and of ovarian tumor (otu). An interesting example is the dominant mutation fs(2)D, which is heterozygous females results in inhibition of cytocyte division (heterozygous males are fertile).

Only a few egg chambers are produced, and most of these contain less than 16 oocytes, which is consistent with the concept that a condition for the switch to the oocyte as opposed to the nurse cell pathway of differentiation is the presence of four intercystocyte ring canals. Telfer (1975) drew attention to a branched cytoskeletal structure, the *fusome*, described in detail by classical cytologists and noticed more recently as well, which extends through the ring canals and interconnects the cytocyte complex. At division one pole of each spindle seems to be anchored in the *fusomes*, which may thus serve to orient each successive cleavage and ensure the appropriate geometry. The effects of the *fs*(231) mutation have been interpreted by king (1979) and King and Riley (1982), as a derangement of fusome organization. In this mutant the cytocytes often construct long linear chains, in which branches form less often than normally), and three dimensional reconstructions from serial electron micrograph sections show that fusomes are often not connected in the adjacent cells, fs(231) cystocytes frequently become polyploid and differentiate into pseudonurse cells. Incomplete cytokinesis takes place about half the time, and cells with four ring canals occur at a frequency of about 10^{-2}, while oocytes appear only at a frequency of 10^{-7}. Thus, while it may be necessary, the possession of four ring canals does not seem to be sufficient to trigger the differentiation of an oocyte. Another mutation isolated by Schupbach presumably affects this switch directly, as it causes for the formation of egg chambers that contain 16 nurse cells, but no oocyte.

The development of the oocyte-nurse cell complex can be regarded as a classical example of a clonal differentiation process. The number and orientation of the divisions by which the final 16 products are produced from the stem cell, and the selection of one of the two cystocytes that contain four ring canals as the oocyte, are all seem from these examples to be subject to specific genetic controls. The implication of the genetic evidence that function of there genes occurs only in cells that are of germ line origin, is that the differentiation of this lineage involves the activation of tissue-specific genes, just as in other cell lineages.

Transcriptional Role of Nurse Cells

The structure of the meroistic egg chamber, and the observation of cytoplasmic transport from nurse cells to oocyte, suggest that the maternal transcripts present in the mature egg at fertilization were synthesized originally in the nurse cell nuclei. An indirect argument supporting this proposition was adduced by Ribbert and Bier (1969). They compared the length of time required for oogenesis in poanoistic and meriostic oogenesis, and correlated this with the number of genomes putatively cooperating the preparation of the oocyte. In the panoistic oogenesis of the cricket *Acheta domestica*, for example, oogenesis takes more than three months, while in the meroistic dipterans, such as *Drosophila* or the blowfly *Calliphora*, oogenesis is completed within a few days. The nurse cells each achieve a DNA content of 750-1500C in drosophila, as we have seen, and of 256 in calliphora. Thus there are 1-4 $\times 10^3$ more genomes putatively involved in providing oocyte transcripts in these Diptera than in the 4C oocyte nucleus of the cricket. This argument indicates the great adaptive value of the meroistic method of oogenesis. A related speculation is that perhaps nurse cells are usually of germ line origin because some prior processes of germ line differentiation are required for expression of the specific set of genes needed during the growth phase of oogenesis, irrespective of whether the transcription of these genes occur in the germinal vesicle or the nurse cell nuclei. We now consider available molecular evidence relating directly to transcription in nurse cell nuclei and its role in the provision of maternal oocyte RNAs.

Sites of Heterogeneous RNA synthesis in the Meroistic Egg Chamber

Under normal conditions the polyploid chromosomes of most dipteran nurse cells are to *polytene*, i.e. the multiple chromatids are not aligned in register. Thus banded chromosomes such as are present in other polyploid cell types are usually not observed in nurse cells. However, the transition from polyploidy to polyteny in nurse cells seems to be fairly trivial. In some *Anopheles* species polytene chromosomes form spontaneously in nurse cells, while they are absent in close relatives. Among the unexplained effects of the *fes* mutation of *Drosophila* is the appearance of polytene chromosomes in nurse cells and giant polytene chromosomes also appear in nurse cells of some ovarian tumor (otu) mutants. In *Calliphora* polytene chromosomes can be induced in nurse cells merely by a mild cold treatment (Bier, 1960), Ribbert (1979) also found that in *Calliphora* reduction of polymorphism through inbreeding results in a high incidence of polyteny in nurse cell chromosomes. Cytogenetic studies on these *Calliphora* polytene chromosomes provide evidence that in this species all or most regions of the nurse cell genomes are equally replicated. Furthermore, the autoradiograph patterns obtained by Ribbert (1979) and accompanying observations on chromosomal puffing, show that transcriptional activity in nurse cell chromosomes is very widespread, and is also relatively intense. The amount of transcriptional activity was estimated as about an order of magnitude greater than in comparable polytene chromosomes of trichogen cells, and many *loci* were observed to remain in a puffed configuration throughout most of oogenesis. These observations imply a high complexity for nurse cell transcription, since RNA synthesis visibly occurs in a great many parts of the genome.

A classic series of autoradiographic experiments carried out by Bier (1963) demonstrated that RNA synthesized in the nurse cell nuclei is transported rapidly into the oocyte. An example, from the housefly *Musca*. Here newly synthesized RNA can be seen localized over the polytene nurse cell nuclei after a 30 min labeling period. Five hours later the labeled RNA has moved into the nurse cell cytoplasm and is apparently pouring through a ring canal into the cytoplasm of the oocyte. Note that no RNA synthesis can be observed over the oocyte at 30 min, even though the film is clearly overexposed with respect to the amount of incorporation in the nurse cell nuclei. Similar autoradiographic results have been reported for other dipteran species, and also for various coleopteran, hymenopteran, and lepidopteran species, a list which includes oocytes of both telotrophic and polytrophic construction; and in addition for the single nurse cell-oocyte complexes of a polychaete annelid.

Transport of newly synthesized poly(A) RNA from nurse cell to oocyte has been directly demonstrated by Paglia et al. (1976a,b) in manual dissection experiments carried out on the polytrophic ovaries of the silk moths *Antheraea polyphemus* and *Actias luna*. Whole follicles were labeled *in vitro*, and ribonucleoprotein particles containing newly synthesized nurse cell poly(A) RNA were observed accumulating in the oocyte cytoplasm after several hours. Ribosomes and newly synthesized rRNA are transported as well. When further synthesis in the nurse cells was blocked with actinomycin, their content of label poly(A) RNA decreased more than 10-fold within 6 hr, while that of the associated oocytes increased correspondingly. No significant incorporation of precursor into vitellogenic oocyte RNA was observed on incubation of follicles from which the nurse cells had been ablated, while isolated nurse cell complexes are transcriptionally active. Furthermore, polysomes are present only in the nurse cells, and the mRNP, tRNA and ribosomes of growing oocyte can thus be seen clearly to be stored for used later in development. By injection 3H-uridine into late pupae that were then permitted to enclose, Paglia et al. (1976a) showed that poly(A) RNA synthesized during vitellogenesis is indeed sequestered in mature chorionated eggs.

In some meroistic insects, particularly some species of midges, bugs and beetles, heterogeneous RNA of the oocyte cytoplasm is also contributed by the germinal vesicle, though most derives from the nurse cells. This can be inferred from autoradiographic evidence of germinal vesicle RNA synthesis, and has been shown more directly by other means. *In situ* hybridization with 3H-poly(U) has been used to determine the source of the oocyte poly(A) RNA in the telotrophic ovary of the milkweed bug *Oncopeltus fasciatus*. During the growth phase of oogenesis poly(A) RNA accumulates in the oocyte cytoplasm and the newly synthesized transcripts are evidently supplied by the nurse cells. Thus, these cells and the nutritive cord are heavily labeled by the probe, while the germinal vesicle contains no detectable poly(A) RNA. However, towards the end of oogenesis the nutritive cord is interrupted by the growth of the chorion, but thereafter a further net increase in oocyte poly(A) RNA nonetheless takes place. This is apparently due to late RNA synthesis in the germinal vesicle. Transcription of poly(A) RNA also occurs in the germinal vesicles of postvitellogenic oocytes of the bug *Dysdercus*. In the gall midge *Wachtliella* elimination of 16 of the 20 chromosomes present in germ line cells occurs in all other cells early in embryogenesis. During oogenesis the four somatic (S) chromosomes remain condensed in the oocyte nucleus, surrounded by a concentric fibrous lamellar structure, while the remaining (E) chromosomes are dispersed. Autoradiography shows that the S chromosomes are transcriptionally silent in the oocyte, while the E chromosomes are active. The nucleolar organizer is located on an S chromosome and nucleolar, formation and S chromosome activity are detected only in nurse cells. If the E chromosomes are experimentally eliminated from the germ line stem cells, oogenesis does not take place. Though the heterogeneous RNAs produced in the oocyte by the E chromosome may be required for oogenesis, the relative importance of nurse cell and germinal vesicle transcripts remains to be determined.

The observations reviewed so far are qualitative, and even where no germinal vesicle activity is reported it remains possible that rare maternal transcripts could be synthesized by the oocyte genomes. These might easily have escaped observation in autoradiographic experiments, focused as they are on the overwhelming synthetic activity of the thousand-fold polyploid nurse cell nuclei. The differences between those meroistic insects that utilize germinal vesicle transcripts, and the lepidopteran and dipteran examples that apparently do not could be merely quantitative. On the other hand, many of the same RNA species produced in small quantities in the germinal vesicles of meroistic oocytes may also be represented in the much greater flow of nurse cell transcripts. These uncertainties should be kept in mind in considering the locus of action of germ line mutations that display maternal effects, and the origins during oogenesis of given species of transcripts, only one of which might represent only a small fraction of the total maternal RNA.

Protein Synthesis in the Drosophila Egg Chamber

Qualitative comparisons of protein synthesis patterns in isolated nurse cells and oocytes of Drosophila have been reported by several authors. In these studies newly synthesized proteins were labelled by injection of 35S-methionine into the adult abdomen, or by incubation of whole follicles and isolated nurse cell complexes *in vitro*. Newly synthesized proteins were then displayed by 2D gel electrophoresis. These studies show that most of the several hundred proteins visualized are synthesized throughout oogenesis, though there are a few stage-specific, and that all the newly synthesized proteins found in the oocyte are synthesized as well in isolated nurse cells. On the other hand several nurse cell proteins are not detectable in the oocyte. Flies that are prevented from laying eggs retain their stage 14.00 oocytes for some days, carry out the same programme of protein synthesis as do the oocyte-nurse cell complexes of earlier stages. These observations are all consistent with a relatively simple model in which a single qualitative pattern of gene activity exists in all the active germ line cells of the polytrophic egg chamber throughout oogenesis.

Yolk is taken up by the oocyte from without, and thus only the accumulation of *non-yolk* proteins provides a useful measure of biosynthetic activity within the oocyte-nurse cell complex. The mass of total protein in the mature *Drosophila* egg is about 1.8μg of which about 22% is yolk. The rate of non-yolk protein accumulation is relatively low in stages 1-9, but after this the mass of polysomes in the egg chamber increases sharply, and the rate of non-yolk protein synthesis accelerates correspondingly. The same mass of polysomes is then present in the egg chamber from stage 10 to stage 14, and there is little further change even after fertilization. Assuming that 4 or 5% of the mass of polysomes measured in egg chambers after stage 10 is mRNA its mass would be 2.1-2.6 ng. Mermod et al. (1980) measured the mass of total poly(A) RNA per stage 14 oocyte as 2.8 ng, and an estimate of 3.8 ng of cytoplasmic poly(A) RNA was reported for early embryos by Anderson and Lengyel (1979). About 60-80% of the total poly(A) RNA is associated with polysomes throughout are oogenesis and into early embryogenesis though this can vary in mature oocytes depending on mysiological conditions. Thus these data indicate that the amounts of total as well as of polysomal mRNA found in the *aereopping embryo* are already set by stage 10 of oogenesis, though the location of the mRNA shifts thereafter from nurse cells plus oocyte to mainly oocyte. This shift probably occurs primarily as a result of the injection of nurse cell cytoplasm (including polysomes) into the oocyte.

Despite the high prosoma content in stage 12-14 oocytes no significant further net accumulation of egg chamber protein occurs after stage 12. At this point each egg chamber already contains the amount of non-yolk protein present in the mature oocyte, about 1.45μg. Current data do not suffice to distinguish between the alternatives that protein synthesis rate in the polysomes of the late oocyte has declined by a large factor due to a precipitous decrease in translational efficiency that involves change in neither polysome size or content, or on the other hand, that there is simply a sharp increase in protein turnover rate. As noted about synthesis of a large number of diverse proteins does continue in a stage 14 oocytes, though the rate is not known, and when measured very shortly after fertilization the absolute rate of proteins synthesis implies that a large function of the polysomes present are normally active. This rate is about 9.7 hg hr-1 embryo. It is near in any case that *Drosophila* differs from species such as the *sea urchin* or *Xenopus* in which the quantity of polysomes increases many fold over the period of late oogenesis, maturation and early development.

In late *Drosophila* oocytes the significance of the nonpolysomal mRNP compartment is primarily qualitative rather than quantitative. Thus certain mRNA species are preferentially localized in the non-polysomal compartment prior to fertilization and these species appear in polysomes only after development begins.

Activity of Specific Gene in the Oocyte-Nurse Cell Complex

Specific Structural Gene Products

Measurement on the accumulation of actin, tubulin, heat shock protein and histone transcripts during *Drosophila* oogenesis illustrate in particular many of the general conclusions drawn above. The tubulin measurements cited were obtained by Loyd et al. (1981), who separated stage 10B egg chambers into oocyte, nurse cell, and follicle cell frictions and determined the proteins synthesized in each compartment by 2D gel electrophoresis. Tubulin synthesis occurs in both nurse cell and oocytes at stage 10B. At this point the oocyte contains 0.6 ng of tubulin, and the nurse cells about 4-6 ng, while at the end of oogenesis the stage 14 oocyte contains about 18 ng of tubulin. After stage 11 the continued synthesis of tubulin evidently occurs in the oocyte on the templates provided earlier by the nurse cells. A more or less constant rate of tubulin synthesis obtain after stage 9, suggesting the maintenance of a constant amount of tubulin message. This could signify either repression of tubulin genes from stage 10 on, or a steady state quantity of metabolically labile tubulin mRNA. Loyd et al. (1981) also found that in stage 10B egg chambers cytoskeletal actins are synthesized by both oocyte and nurse cells, as well as by follicle cells. The accumulation of these proteins has been described by Ruddell and Jacobs-Lorna (1984). Two cytoskeletal actin genes are utilized during oogenesis, and from the rate of protein accumulation their mRNAs can be calculated to be almost as prevalent as the tubulin mRNAs. Net accumulation of actin protein ceases at stage 12, though like most oocyte mRNAs the actin message remains loaded on polysomes thereafter. Messages for three heat shock proteins also originate in nurse cells (in normal non-stressed adult), and they are released into the oocyte at stage 10-11. There are $1\text{-}2 \times 10^7$ molecules of each species of heat shock mRNA, coding for the 83 kd, the 28 kd and the 26 kd heat shock proteins, in the egg at fertilization, i.e., somewhat less than the quantities of the specific mRNAs. The histone mRNAs provide a different example, in that these messages continue to accumulate right up to stage 14. Up to stage 10A the nurse cell histone mRNA is apparently utilized primarily to provide histone for the replicating nurse cell DNA. Since polyploidization in the nurse cells is then complete and there is not further DNA synthesis in the oocyte-nurse cell complex, the translation products of the histone mRNA synthesized after this must be destined for use in the embryo. The source of the histone mRNA after stage 12 is mysterious, since the nurse cells are then walled off from the oocyte by the chorion, while the chromosomes of the late oocyte nucleus have been thought to be transcriptionally quiescent on the basis of their condensed structure and autoradiographic evidence.

The sets of genes examined are regulated differently in detail during oogenesis, and the accumulation of their protein products follows different kinetics. Nonetheless, these examples individually support the generalization that maternal mRNAs of the *Drosophila* egg are mainly synthesized in the nurse cell nuclei and injected into the oocyte, where they are found associated with polysome and engaged in protein synthesis during most or all of the ensuring periods of oogenesis.

Germ Line Functions that Affect Oocyte Structure

Several germ line mutations have been described that block oogenesis after the completion of the cystocyte divisions and formation of the egg chamber. Except that the locus of action of these mutations is the nurse cells or oocyte, for the most part the primary physiological or structural defects they cause are not known, and none are yet analyzed at the molecular level. Their significance here is to remind us that there are many additional genetically controlled aspects of oogenesis besides synthesis of maternal transcripts and proteins required after fertilization. An interesting example analyzed by Waring et al. (1983) concerns a recessive female sterile mutation, *fs*29 (other alleles are *fs*117 and *fs*445), that maps to region 12E1-12F1 on the X chromosome.

The germ line function of this gene was demonstrated by transplanting homozygous mutant pole cells into normal eggs, and mutant pole cells into normal eggs. The visible effect of the homozygous mutation is a slightly abnormal chorion structure, particularly at the anterior dorsal side, where in *fs*29 oocyte the respiratory appendages project at the wrong angle and the adjacent anterior region of the chorion is abnormally formed. A consequence is failure of fertilization, which normal occurs via the micropyle, an anterior chorion structure. Waring et al. (1983) found a sharp decrease in the amount of yolk protein taken up by *fs*29 oocytes, although synthesis of yolk occurs normally in both fat body and follicle cells. *Drosophila* follicle cells are the source of about 35% of two of the three major yolk proteins. The deficiency in fs29 animals appears to lie in the mechanism by which yolk is sequestered by the oocyte, which include an extremely specific binding to surface receptors followed by pinocytotic internalization. Probably the interruption of follicle cell-oocyte contacts that produces the fatal structural abnormality in the chorion is a peripheral side effect of decreased turgidity in the anterior region of the oocyte, due to decreased yolk content.

A second mutation, fs(1)K10, that affects chorion structure in the same region and also acts in the germ line, has been described by Wieschaus et al. (1978). In normal egg chambers morphogenesis of the dorsal appendages is carried out by nests of follicle cells apposed bilaterally to the anterior surface of the oocyte and the dorsoventral polarity of the oocyte is foreshadowed by the thicker follicle cell layer on the dorsal side. Mutations at the *fs*(1)K10 locus dorsalize the egg chamber, in that a thick follicle cell layer occurs on the future ventral side as well, and they result in eggs with enlarged doral appendages. In the few cases in which development is initiated, the embryos lack ventral structures and display a dorsalized phenotype. Function of the K10 gene is necessary only in the female germ line and normal progeny can be derived from transgenic *fs*1(K10) females bearing the wild-type gene. Another female sterile mutation that according to pole cell transplantation tests acts in the germ line, though it affects follicle cell function, is tiny.

Though it maps to the same region of the X chromosome, tiny complements *fs*29 and any of its chambers the distal follicle cell migration occurring early in oogenesis is blocked, and an abnormally thick follicular wall that prevents expansion of the oocyte is formed. This in turn produces a thickened, irregular chorion. DiMario and Hennen (1982) suggested that the primary lesion in *tiny* egg chambers is a defect in the nurse cell-oocyte surfaces over which the follicle cells are supposed to migrate.

Ribosomal RNA Synthesis

Two different mechanisms by which ribosomal RNA is supplied to the oocyte have been observed in meroistic insects. In most forms the ribosomal RNA derives wholly from the nurse cells, along with other transcript species. As this is of course the bulk form of RNA in the oocyte its transfer is easily detected. In *Drosophila* and *Calliphora* egg chamber this occurs mainly when the nurse cell cytoplasm is injected into the oocyte, while in the Hyalophora ovariole most of the rRNA is transferred to the oocyte prior to the terminal injection. Early vitellogenic Hyalophora oocytes already contain about 1 μg of total RNA, most of which is undoubtedly ribosomal, and this is also the RNA content of the seven nurse cells taken together.

During vitellogenesis the oocyte doubles its total RNA content, and this is increased to 3 μg by the terminal injection process. Ribosome transfer from nurse cells to oocyte was demonstrated in the polytrophic *Antheraea* egg chamber by Hughes and Berry (1970), and in the telotrophic egg chamber of *Oncopeltus* by davenport (1976). In these experiments the ovariole was exposed to isotopic precursor, and the subsequent appearance of labeled ribosomes in the oocyte was monitored. Entry of newly synthesized rRNA could be interrupted in Oncopeltus by ligation of the nutritive cord and in Antheraea did not occur after removal of the nurse cell cap. In many organisms the rRNA genes are amplified during oogenesis, a special mechanism required to meet the enormous demand for performed ribosomes early in embryonic development. However, significant ribosomal gene amplification, i.e., relation to the remainder of the genome, has been reported not to occur in the polyploid nurse cells of a number of meroistic insect species.

In *Calliphora* rRNA synthesis takes place in extra chromosomal nucleoli, but even here the fraction of ovariole DNA that is ribosomal is only 1.3 times that measured in diploid brain cells (Renkawitz and Kunz, 1975). In four dipteran species, *Drosophila hydei*, *Drosophila virilis*, *Sacrophaga barbat* and *Rynchosciara angelae* the ribosomal DNA is actually underreplicated by about a factor of two during nurse cell polytenization. Nor is there preferential accumulation of ribosomal DNA during oogenesis in *Drosophila melanogaster*, in the silk moths *Antheraea pernyi* and *Bombyx mori* or in Oncopeltus, where the nurse cell nuclei of the telotrophic egg chambers are polyploid only to the extent of 32-64C. These measurement imply that the total number of ribosomal genes present per egg chamber, i.e., the product of nurse cell ploidy, rRNA genes per haploid genome, and the number of nurse cells, is sufficient to provide the oocyte with the requisite quantity of ribosomes in the time available, so that ribosomal gene amplification is unnecessary. This can be seen explicitly for the *Drosophila* oocyte, where the transcription rate and other necessary parameters can be estimated.

There are about 250 rRNA genes per haploid Drosophila genome and so assuming neither under nor over-DNA replication at stage 9 when nurse cell polyploidization is complete, the number of rRNA genes per egg chamber is about 3.8×10^6. At a polymerase translocation rate of 10 nt sec^{-1} and assuming a minimum polymerase spacing of about 100 nt it would require only about 30 hr for this number of genes to produce the approximately 4.5×10^{10} ribosomal RNA molecules stored in the mature egg (there are 160 ng of rRNA per egg. Mermod et al. (1977) showed that earlier in oogenesis the rate of egg chamber rRNA synthesis is directly proportional to the degree of nurse cell polyploidization and the total number of rRNA genes. Ribosomal RNA synthesis could be a rate limiting process in *Drosophila* oogenesis. Thus in bobbed mutants in which various portions of the ribosomal gene cluster are deleted, the length of time required for oogenesis depends on the number of ribosomal genes remaining (Mohan 1971). For example, in mutant egg chambers containing only 1/3rd the normal complement of rDNA, oogenesis required 206 hr compared to 75 hr in controls. It is interesting that as result of this compensatory mechanism the mature eggs of bobbed females contain a normal quantity of rRNA.

Ribosomal DNA amplification does occur in the oocytes of some insect species, including water beetles such as *Dytiscus* and *Colymbetes*; the dipterna *Tipula*; and the neuropteran *Chrysoma*. In these examples a single "DNA body" that contains rRNA genes is found in the definitive oocyte nucleus, but not in nurse cell nuclei. Autoradiographic observations and *in situ* hybridizations with ribosomal sequence probes indicate that rDNA amplification in these bodies begins during the cytocyte divisions, and continues in the early stages of previtellogenic oogenesis. The best studied example is the water beetle *Dytiscus*. Gall and Rochaix (1974) found that the DNA body in this species contains about 3×10^6 rRNA genes, arranged in small circles, each containing one to five genes. This number is remarkably close to that present in the nurse cell chromosomes of *Drosophila*. In midoogenesis the DNA body disperses and intense rRNA transcription begins. No other transcriptional activity can be detected in the oocyte nuclei. The chromosomes are present in a condensed mass, the karyosphere, and as usual in meroistic oocytes, they appear silent in autoradiograph experiments.

The meroistic form of oogenesis is usefully considered from a logistic point of view. A great many copies of genes are put to work co-operatively in the meriostic egg chamber, all producing sequences required by the oocyte. The obvious adaptive value of this elaborate strategy is that it accelerates by a large factors the overall rate of oogenesis.

The Transcriptional Role of Oocyte Lampbrush Chromosomes

The oocytes of many animals contains spectacular meiotic prophase chromosomes that are distinguished by the presence of thousands of lateral loops. Such chromosomes were first observed by Flemming (1882) in sections taken through urodele oocytes, and they were the subject of a detailed study by Ruckert (1892), carried out on isolated germinal vesicle of shark oocytes. Ruckert recognized that they are paired chromosomes from which the numerous lateral loops project, and this aspect of their structure suggested to him the designation lampbrush chromosomes. These chromosomes are never found in the germinal vesicles of meroistic oocytes, which as we have seen are generally quiescent in regard to transcription, nor do they occur in the oocytes of many other species in which the maternal transcripts derive solely from the germinal vesicle. Though there remain some mysteries, and their patterns of transcription are in some respects unusual, we shall conclude from molecule and cytological analyses that their basic function is probably the synthesis and maintenance of the large pool of maternal transcripts resident in the oocyte cytoplasm.

Lampbrush Chromosome Structure: Transcription Units and Chromomeres

The modern era of lampbrush chromosome cytology began with the establishment of two basic structural tenets. First, lampbrush chromosomes were shown to be bivalent meiotic prophase structures in which each loop of an apposing pair contains a single DNA duplex, while the central axis contains two such duplexes. The metiotic homologues are typically united by several chiasmata. Second, the loops were perceived to be the sites of chromosomal RNA synthesis. Thus, an active locus is represented by four loops, two deriving from each chromosomal axis. The axes of lampbrush chromosomes from which the loops project were recognized classically to consist of a linear array of compacted beads referred to as *chromomeres*. GEL (1954) showed that DNA is concentrated in the chromomeres, as these are the only structures in lampbrush chromosomes that can be visualized by the Feulgen reaction.

The paired structure of the chromomeric axis was demonstrated by Callan (1955), by stretching local regions with microneedles to the point where the chromomeres would separate transversely into two distinct chromatin stands. From measurement of the kinetics of lampbrush chromosomes breakage by DNase 1. Gall (1963) deduced the presence of two DNA duplexes in the axis and a single duplexes in the loops. Important evidence also derived from the maps constructed for new lampbrush chromosomes by Callan and Lloyd (1960). These studies showed that distinctive loops can be recognized at invariant locations in the chromosomes. Particular loop morphologies are the property of species, subspecies, or individuals. Heterozygotes, or hybrids between related species, generally produce heterozygous sets of lampbrush chromosomes with display the allelic alternatives characteristic of each parent, and the frequencies at which these alternatives appear are distributed in wild populations as predicted by a Hardy-Weinberg calculation. A general conclusion is thus that the loop structure is determined by the genetic locus, i.e., the DNA sequence which it contains. Lampbrush chromosome maps have now been assembled for a number of different species, including several urodeles; the anuran *Xenopus* (Muller, 1974); and some amphisbaenian reptiles.

Transcription Units of Lampbrush Chromosomes

Early autoradiographic studies carried out with the light microscope showed that intense RNA synthesis occurs throughout the length of most loops, though there are a few giant loops that display unusual labeling patterns. The loops contain newly synthesized proteins as well as RNA. Furthermore, most loops appear to have a polarity, in that the loop matrix is thicker at one end than at the other. Electron microscopy carried out by the Miller spreading technique has not provided a convincing molecular interpretation of both the widespread transcriptional activity and the polarity observed at light microscope resolution. The transcription units of lampbrush chromosomes are packed densely with nascent RNA molecules.

Observation on both *Xenopus* and *Triturus* lampbrush chromosomes show that the polymerase molecules by which the nascent transcripts are anchored to the DNA fibril of the loop are typically only 100-200 nt apart. Though some variation is observed both within and among transcription units, the structural feature that immediately distinguishes lampbrush chromosomes is that almost all the transcription units functioning in lampbrush chromosomes are synthesizing. By contrast, when the same methods of visualization are applied to active somatic cell types, e.g., sea urchin or Drosophila embryonic cells only about 10% of the transcription units are densely packed with nascent transcripts, just as expected from consideration of the RNA prevalence distribution characteristic of such cells.

The lengths of the transcripts units observed in lampbrush chromosomes vary from about 2 kb to at least 70 kb of extended chromosomal DNA in *Xenopus*, and from about 10 to 100 kb in several urodeles. The enormous size of transcription matrices accounts easily for the polarized from of the prominent loops observed by classical methods, since the nascent transcripts associated with a given transcription unit increase in length toward the distal end. However, it is necessary to take into account the effects of the spreading procedure required for visualization. The detergent treatments involved in these preparations probably result in the removal of some proteins, and the surface tension to which the sample is exposed converts what might be imagined a solid ribonucleoprotein core of increasing diameter into the open, two-dimensional from illustrated.

A useful calculating of Macgregor (1980) shows that average dimensions of the solid ribonucleoproteins matrices form which the typical transcription complexes visualized in the electron microscope would have derived are consistent with those actually observed in the phase microscope. The relation between the number of loops and of transcription units is not straightforward, though many loops probably contain-single transcription units. However, it is not uncommon to observe two transcription units separated by a short non-transcribed region. This is particularly striking when the transcription matrices are oriented in opposite directions, indicating initiation sites on opposite strands. A given region may also contain multiple transcription units oriented similarly, and occasionally a series of matrices of identical size and orientation are observed that probably indicate the transcription of a tandemly repeated gene family. Careful light microscope observations of discontinuities in the matrices of giant loops have also provided indications that some loops bear multiple transcription units. If follows that the number of transcription units is by some unknown of loops, or of loops and associated chromomeres, was classically assumed to provide an approximation of the number of genetic loci, or at least of active loci. Given that the physical manifestation of an active locus is transcription complex, this venerable correlation is clearly incorrect.

Furthermore, while at electron microscope resolution the active transcription units of lampbrush chromosomes are easily defined, the distinction between the loop and chromomere domains inferred from light microscopy in most regions is not at all obvious. Closely apposed transcription units in doubt derive from the same loop, while at the other extreme the long inactive axial fibrils that can also be observed could represent the chromomeric DNA of the classical model. After spreading by the Miller procedure all of the inactivated regions retain a similar breaded structure, indicative on nucleosomal conformation. Nucleosomes can be perceived even within active transcription units, where the nascent fibrils and polymerase molecules are not so densely packed. An implication of the latter observation is that nucleosomal structures must be able to reform within less than a minute following the passage of a polymerase and the associated nascent transcript.

A reasonable interpretation of the structure of lampbrush chromosomes as classically observed in the light microscope is that chromomeres occur where there are long regions of inactive chromatin that *in situ* are condensed into higher order structures. These are disaggregated during the preparation of transcription spreads by the Miller procedure. Shorter inactive regions may bunch together, giving rise to a complex aggregate that would also be recognized as a chromomere is the light microscope, from which protrudes an array of multiple loops, all containing active transcription units. Hill (1979) pointed out that the alternative interpretation would require that much of the length of an average loop be occupied by inactive intermatrix spacer sequences, contrary to indications form light microscope autoradiography. A scanning electron microscope study of Angelier et al.(1984), in which the same lampbrush chromosomes were also visualized in the phase microscope, supports the view that chromosomes were also visualized in the phase microscope, supports the view that chromomeres may consist of associated regions of condensed nucleoprotein from which many loops extend. It follows that the chromomere is not an invariant structure equivalent to a single genetic locus. It is rather to be considered a compacted aggregate of transcriptionally inactive DNA, the manifestation of which depends on the spacing of the surrounding active transcription units. In different organisms among which the mode of genomic sequence organization varies, the prominence and distribution of chromomeres might also vary. Thus the differences in the number of chromomeres and the size of loops observed in comparing the lampbrush chromosomes of various species probably reflect primarily differences in the distribution and lengths of the non-transcribed DNA sequences that define the termini of the loops. This interpretation clears away the paradox that develops from the classical view that such differences would indicate fundamental variation in the number of functional genetic loci among related species.

Phylogenetic Occurrence of Lampbrush Chromosomes

For no other organisms do there exist ultrastructural or molecular observations on lampbrush chromosomes comparable to those available for amphibians. Yet these structures occur in the oocytes of many other vertebrates and in a great many invertebrates as well, and at least as perceived in the phase microscope they appear completely homologous in their cytological organization to the amphibian examples. Here are displayed lampbrush chromosomes from an orthopteran insect, *Decticus albifrons* a squid, *Sepia Officinalis Bithynia tentaculata* and a starfish, *Echinaster sepositus*. The general structural similarity of lampbrush chromosomes is illustrated in the detailed study of DeLobel (1971) on *Echinaster* lampbrush chromosomes. The map constructed for the chromosomes of this organism displays the same kinds of special structures, such as giant loops of unusual conformation, that serve as landmarks in amphibian lampbrush chromosomes and the average loop dimensions are similar to those of *Xenopus* lampbrush chromosomes.

The collected data regarding the overall distribution of lampbrush chromosomes, and where possible the duration of the lampbrush stage. It is clear that lampbrush chromosomes occur in many major groups both deuterostome and protostome. Thus, like the process of oogenesis itself, lampbrush chromosomes are probably of very ancient evolutionary origin. Lampbrush chromosomes evidently perform some fundamental function in oogenesis, since they have been retained throughout most of metazoan evolution. The list of organisms in which lampbrush chromosomes have been reported is of course limited by the choices made by investigators and the difficulty of observing them in some material. Nonetheless, it is now established that lampbrush chromosomes are not ubiquitous. A generalization suggested by their known distribution is that lampbrush chromosomes occur in large oocytes that contain relatively huge pools of heterogeneous maternal RNA, while they are absent from small oocytes.

The mouse egg, for example, contains less than 1/3000 the amount of maternal poly(A) RNA that is present in the egg of *Xenopus* and the sea urchin egg less than 1/500 this amount. The Xenopus egg, in turn, is small compared to some other anuran eggs and to urodele eggs. The same relation pertains among the echinoderms. Thus the egg of the starfish *Echinaster espositus*, which contains lampbrush chromosomes, is several hundred times the volumes of the egg of the sea urchin *Strongylocentrotus purpuratus*, which does not. The correlation suggests a simple interpretation of lampbrush chromosomes functions. This is that these structures exist where the logistic demands of supplying the growing oocyte with maternal transcripts in the allowed time require that almost all transcription units function at maximum rate. As we have seen, widespread and intense transcription is the definitive property of oocyte lampbrush chromosomes.

A logistic function for lampbrush chromosomes is also suggested by the exclusive relationship between meroistic oogenesis and the presence of lampbrush chromosomes in the oocyte nucleus. That is, lampbrush chromosomes probably perform the same function that the polytene nurse cell chromosomes do in meroistic oocytes, viz., the provision of maternal RNAs. This function appears to be exercised continuously over the relatively long period required for oocyte growth. Thus, as shown in Table the lampbrush stage lasts for weeks, months, or even longer in species for which estimates are available. In some amphibian species a mechanism has evolved that almost approaches the meroistic mode of oogenesis. Here each oocyte contains multiple nuclei. Macgregor and Kezer (1970) showed that in the tailed frog *Ascaphus*, for example, the growing oocyte has eight germinal vesicles, formed initially by oogonial nuclear divisions that are not accompanied by cytokinesis. Each nucleus is endowed with a complete set of lampbrush chromosomes, and thus there are 32 copies of every active locus functioning per oocyte.

Even more extreme examples have been described by del Pino and Humphries (1978) in studies of oogenesis in marsupial frogs. In these forms development proceeds directly from egg to juvenile, with a reduced or nonexistent larval tadpole stage, in several general of marsupial frogs the previtellogenic oocyte can be seen to contain hundreds of nuclei. Each oocyte is the product of many oogonia, formed by the disappearance of cell membranes within an oogonial cyst. During the growth phase each of the multiple nuclei is endowed with lampbrush chromosomes and all are active in RNA synthesis. Later in oogenesis all but one nucleus disappears. It may be relevant that the eggs of species carrying out direct development are relatively enormous, ranging up to 9 mm in diameter. The egg of *Xenopus*, for comparison, is about 1.2 mm in diameter. In summary, the comparative biology of occurrence, when combined with ultrastructural evidence for instance transcriptional activity in all or most of the loops, suggests that the basic function of lampbrush chromosomes is to provide the growing oocyte with the maximum possible flow of newly synthesized transcripts.

Complexity, Average Structural Characteristics, and Synthesis Rates for Lampbrush Chromosome RNAs

Size of Primary Transcripts and Association with Specific Proteins

Newly synthesized germinal vesicle RNAs have been extracted from the oocyte of several amphibian species, and the distribution of their molecular lengths determined under denaturing conditions. In one study carried out on vitellogenic oocytes of the newt *Pleurodeles poireti* transcripts as large as 10-30 kb were observed, though most RNAs recovered were smaller. RNAs up to at least 60 kb in length have been extracted from *Triturus* oocytes. Scheer and Sommerville (1982) also reported measurements on extracted germinal vesicle RNAs carried out by electron microscopy. Length distributions obtained in this study extended up to about 20 kb for Senopus oocyte nRNA. However, the number average size of the molecules recovered from the isolated germinal vesicles of all three species is around that of mature mRNA. However, the number average size of the molecules recovered from the isolated germinal vesicles of all three species is around that of mature mRNA, about 2 kb.

The RNAs examined in these studies could have suffered strand scissions during their preparations, despite all possible precautions, and whether for this reason, or because endonucleolytic cleavages occur in the course of processing while the transcripts are still nascent, the *extracted* molecules clearly fail to match the enormous dimensions of the larger transcription units observed in the electron microscope. As expected for a high complexity population of nuclear transcripts, lampbrush chromosome RNA has a DNA like base composition, except for a still unexplained bias towards unusually high uridylic acid content reported for both *Xenopus* and *Triturus* oocyte mRNAs. The nascent transcripts of the loop matrix are associated with proteins, which results in a significant compaction. Hill (1979) noted that even after preparation for electron microscopy the apparent nascent RNA length is less than half the length of the DNA form which it is transcribed, and in native RNP the degree of compaction is undoubtedly much higher. In high voltage electron microscopic images of thick sections through the loop matrices RNA fibrils can be seen connecting particles of about 20 nm diameter. The RNP assemblages have been isolated from *Triturus* oocytes by differential centrifugation and their structure analyzed in vitro.

Linear chains of the 20 nm ribonucleoprotein particles in different states of aggregation and their disaggregation into monomers after mild RNase treatment. The particles consist largely of proteins, and more than 20 distinct polypeptides have been isolated from such preparations. Antibodies prepared against these polypeptides generally react with the matrices of all loops, even though different loops often display individual morphologies, due to the various conformations into which the 20 nm particles are arranged, perhaps a function the primary RNA sequences. However, there are certain proteins that are present only on a small number of specific loop parts out of the whole germinal vesicle complement. The same proteins are included in the ribonucleoprotein complexes released into the nuclear sap from these loops when transcription has been completed. In their physical properties and constitution the ribonucleoprotein particles that contain newly synthesized chromosomal RNAs of the amphibian oocyte directly resemble the heterogeneous nuclear RNP complexes of somatic cells. This suggests that the functions mediated by these assemblages, whether transcript packaging, transport or processing, may also be similar.

Rates of RNA Synthesis in Lampbrush Chromosomes

The lampbrush chromosome stage begins in *Xenopus* in previtellogenic, early diplotene oocytes only 50-60 µm in diameter. According to the staging criteria of Dumont (1972), these are stage 1 oocytes just beginning their growth phase. Hill and Macgregor (1980) showed that the chromosomes of even these very young oocytes contain heavily loaded transcription matrices typical of those present in midvitellogenic (stage 3) lampbrush chromosomes. Transcription is initiated even earlier, in oocytes that are only 24-40 µm in diameter. Though the nascent transcripts are at first 5-10 fold more widely separated than in the later stage 1 oocytes the average length of the earliest transcription units is already the same as at the maximum lampbrush *stage*. This study yields the important inference that a high rate of chromosomal RNA synthesis is instituted very soon after the stage 1 oocyte begins to grow, rather than only at the midlampbrush stage, as earlier assumed. Martin et al. (1980) showed, furthermore, that mature lampbrush chromosome transcription units persist even in fully grown oocytes.

Lampbrush chromosomes are thus present almost from the beginning to the end of oogenesis, although in the light microscope the early and late oocyte lampbrush chromosomes are less easily resolved. The main difference in morphology between the transcription matrices of very early and very late oocytes, on the one hand, and of the dumont stage 3-5 vitellogenic oocytes classically described as "maximum lampbrush stage," On the other, reflects the degree to which the nascent transcripts are compacted in their RNP assemblages. The lateral loops of Dumont stage 6 oocytes appear shorter, and in Dumont stage 1 and 2 oocytes the nascent transcripts are about 6-fold less extended in preparations spread for electron microscopy than in stage 3 oocytes. A valuable series of direct synthesis rate measurements carried out on stage 3 and stage 6 *Xenopus* oocytes. Both stage 3 and stage 6 oocytes synthesize high molecular weight (>40S or over 7 kb). Unstable, nucleus-confined RNAs. The accumulation

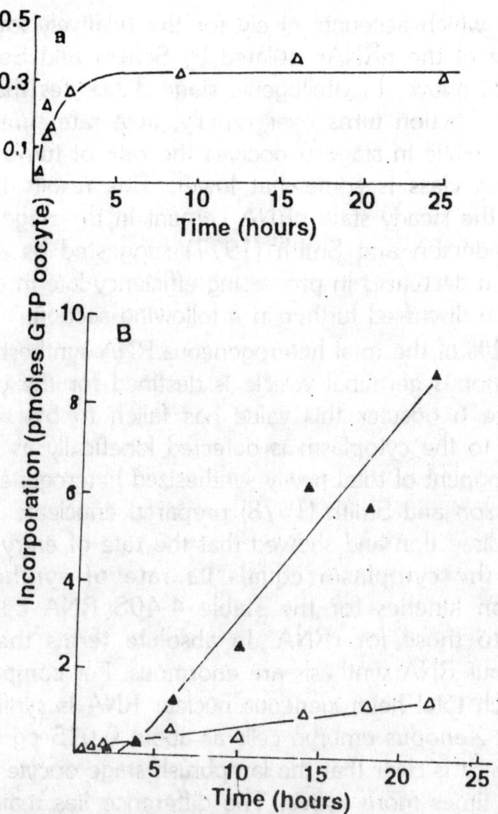

Fig. 5.5. Accumulation kinetics for newly synthesized RNAs in Dumont stage 3 Senopus oocytes. Oocytes were individually dissected from the ovary, and the follicular cell layer was removed. They were injected with 3H-GTP, the precursor pool specific activity was determined, and the radioactivity incorporated in each class of RNA was measured as a function of time. Cytoplasmic fractions were prepared by manual enucleation (a) Synthesis and turnover kinetics for >40S heterogeneous RNA. Though measured on whole oocytes this RNA fraction is confined to the nucleus. (b) Kinetics of accumulation in cytoplasmic RNA for newly synthesized 4-40S heterogenous RNA (open triangles) and ribosomal RNA (closed triangles).

kinetics of these transcripts. Transcripts of the same size range, which turn over with a half-life of about 20-45 min area of course found in somatic nuclei as well.

In addition both stage 3 and stage 6 oocytes synthesize a somewhat greater amount of unstable, nucleus-confined RNA of smaller size which accounts nicely for the relatively low number average size of the nRNAs isolated by Scheer and Sommerville (1982) noted above. In vitellogenic stage 3 oocytes the unstable 4-40S RNA fraction turns over rapidly, at a rate similar to the >40S RNA, while in stage 6 ooctyes the rate of turnover of the 4.40S nRNA class is somewhat lower. This results in a large increase in the steady state nRNA content in the stage 6 oocyte nucleus. Anderson and Smith (1977) suggested as a possible explanation a decreases in processing efficiency late in oogenesis, a proposition discussed further in a following section.

Only 11% of the total heterogeneous RNA synthesized in the stage 3 Xenopus germinal vesicle is destined for the cytoplasm, and in stage 6 oocytes this value has fallen to 5%. The RNA transported to the cytoplasm is detected kinetically as the stable 4.40S component of total newly synthesized heterogeneous RNA. Thus Anderson and Smith (1978) prepared enucleate cytoplasm by manual dissection and showed that the rate of entry of 4-40S RNA into the cytoplasm equals its rate of synthesis. The accumulation kinetics for the stable 4-40S RNA component, compared to those for rRNA. In absolute terms the rates of heterogeneous RNA synthesis are enormous. For comparison the rate at which total heterogeneous nuclear RNA is synthesized in postgastrula *Xenopus* embryo cells as about 0.015 pg min-1 per (2C) nucleus. It is clear that the lampbrush stage oocyte nucleus is a thousand times more active. The difference lies mainly in the number of polymerases transcribing each functional region in lampbrush chromosomes. As calculated earlier the density of nascent heterogeneous nuclear RNA molecules is typically less than one per 10^4 nt in somatic cells, compared to one per 10^2 nt in the active lampbrush transcription units. Transcription units may also be longer in the lampbrush chromosomes, as further discussed below, and the 4C oocyte genome contributes an additional factor of two to the comparison. However, the fractions of newly synthesized heterogeneous RNA exported to the cytoplasm from the lampbrush chromosome stage germinal vesicle, here 5-12% are not very different from those found in embryonic or other somatic cells. It follows that during the lampbrush stage the rate of flow of stable transcripts into the cytoplasm is also as much as 1000 times greater than in other cells. The significance of the observation that heterogeneous RNAs are being exported at high rates to the cytoplasm throughout the midlampbrush chromosome stage is that it directly relates lampbrush chromosome transcription to the function of supplying such RNAs to the cytoplasm. Dolecki and Smith (1979) showed, furthermore, that much of this heterogenesis RNA flowing into the cytoplasm is polyadenylated. The 4-40S RNAs are at least moderately stable, since no turnover can be detected in the form of the accumulation curve during the labeling period.

Complexity of Germinal Vesicle RNA

There have been no direct experimental estimates of the complexity of the nuclear RNA of *Xenopus* oocytes. Though the complexity and sequence content of the cytoplasmic RNAs stored in the mature egg are relatively well-known. However, the measurements of transcription rate can be utilized to calculate the approximate total length of the DNA included in the transcription matrices of the stage 3 oocyte chromosome, and from this the single copy sequence complexity may be estimated. The result is that around 30% of the *Xenopus* genome appears to be transcribed into newly synthesized nRNA. This fraction is not significantly different from that observed for many somatic cell nuclear RNAs stably accumulated cytoplasmic RNA of the *Xenopus* oocyte includes only about 4-6% of the calculated *complexity* of the nuclear RNA. This ratio falls at the lower end of the range commonly observed for somatic cells, about 5% to 20%. The conclusion that the RNA transcribed in lampbrush chromosomes is no more complex than the nuclear RNA of somatic cells is also consistent with several early hybridization studies which showed that only a minor fraction of the diverse genomic repetitive sequences is represented in the newly synthesized RNA of stage 3 *Xenopus* oocytes. Speculations regarding lampbrush chromosome function that require the whole genome to be transcribed in lampbrush chromosomes are thus not supported. Furthermore, RNAs coding for hemoglobin and vitellogenin. Both expressed specifically in terminally differentiated adults tissues, have been shown to be absent from *Xenopus* egg RNA, as are mRNAs for certain though not all of the heart shock proteins that can be expressed in somatic cells. The possibility is not excluded, however, that the rapidly decaying nuclear transcripts synthesized in lampbrush chromosomes include these particular sequences. What is clear is that a specific, though large, set of sequences is transcribed and that these in turn give rise to a specific array of stable cytoplasmic RNAs.

The basis in measurement is here almost complementary to that available for *Xenopus*. While there are no synthesis rate measurements for *Triturus* oocytes, direct attempts have been made to estimate the nRNA complexity by the single copy saturation hybridization method. Furthermore, *Xenopus* lampbrush chromosomes are small, thus rendering quantitative cytological examination at the light microscope level difficult, while extensive cytological measurements have been carried out on *Triturus* lampbrush chromosomes. A visual comparison between *Xenopus* and *Triturus* lampbrush chromosomes, photographed at the same magnification. There are about 20,000 loops in the whole chromosome set of Triturus or about 5000 loops per haploid set. From direct length measurements these are estimated to contain about 5% of the genomic DNA. The length of DNA transcribed in *Triturus* oocytes on this basis is also listed, i.e., assuming that on the average each loop consists only of transcription matrix. The value obtained, i.e., 1.5×10^9 nt, is not significantly different from the measured single copy complexity, i.e. $> 9.6 \times 10^8$ nt, taking into account that about 40% of the *Triturus* genome is repetitive.

The estimated lengths of genomic sequence represented in the nRNA of *Xenopus* and *Triturus* oocytes probably differ only by a factor of about 1.7 though the genome size of Triturus is 10 times larger than that of *Xenopus*. On the other hand as much as a six times greater *fraction* of the Xenopus genome is apparently being transcribed (30%) than of the *Triturus* genome (5%). This conclusion is directly contrary to the earlier assumption that because the lampbrush chromosomes of organisms of greater genome size have larger loops, an equivalent fraction of the genomic DNA is always being expressed. From the assumption has risen the famous mystery known as the "C-value paradox." The question posed in statements of the "C-value paradox" is why animals of similar biological organization should display manyfold differences in the length of genomic information transcribed in homologous cells, or in more extreme form, why the homologous expressed "genes" of one organism should on the average include manyfold longer DNA sequences than in a related organism. With respect to amphibian oogenesis this paradox in fact does not exist, though the enormous fraction of silent DNA accumulated during evolution in urodele genomes indeed remains a mysterious phenomenon. Thus, the six-fold greater relative activity of the *Xenopus* genome reduces the ratio of expressed genome sizes to 1.65, which might suggest a slightly greater length of sequence utilized in *Triturus* lampbrush chromosomes. This last factor, which as described in note *b* of could be a little larger probably indicates that *Triturus* transcription units on the average include about twice as much non-coding sequence that is confined to the nucleus as do *Xenopus* transcripts. However, Sommerville and Scheer (1982) found no striking difference in the fraction of oocyte nRNAs from these organisms that consists of repetitive sequences.

The significance of the difference in loop lengths among species requires re-examination as well. For one thing, contrary to the premise of the "C-value paradox." Macgregor (1980) showed that average loop lengths in fact vary far less than proportionately with genome size. It is indeed the case that in *Xenopus* the loops average only 5-10 μm, and large loops are about 10-15μm in length [though the average length has also been estimated as 12.5μm. In *Triturus* species the loops average 50 μm and the largest range up to 200 μm. In urodeles with even larger genomes than *Triturus cristatus*, e.g., *Necturus maculosus* the average loop length is slightly greater. However, since average loop length is overall a complex function of sequence organization, i.e., the distance between the relatively long regions of silent DNA that are compacted in chromomeric structures, together with the spacing, of genes active in oogenesis, the most reasonable conclusion is simply that transcription units are arranged differently large fraction of the genome would appear to consist of very long blocks of silent sequences, but the active regions bunched together in the loops include an amount of single copy genomic sequence that according to factor of two the same as in expressed in the genome to *Xenopus*.

Transcription of Specific Sequences in Lampbrush Chromosomes

Transcription of specific sequences has been visualized by *in situ* hybridization of labeled DNA probes with the nascent matrix RNA of lampbrush chromosomes loops. Such hybridization is abolished by prior treatment with RNAse, and the signal obtained depends directly on the dense packing of the nascent transcripts. The first application of this procedure to lampbrush chromosomes was as an attempt to identify the locus of the active 5S rRNA genes in *triturus* lampbrush chromosomes . The probe consisted of labeled 5S DNA that had been purified from genomic DNA by isopycnic centrifugation. As it may have contained some other sequences, the results of this particular investigation remain equivocal. However, the *in situ* hybridization method, when utilized with cloned probes, has provided useful information in regard to both the locus of particular genes, and of greater importance the general properties of transcription in lampbrush chromosomes.

Transcription of Repetitive Sequences

A great variety of repetitive sequence is transcribed in amphibian oocyte nuclei, and is also included in the interspersed poly(A) RNAs that by mass form the dominant fraction of the heterogeneous maternal transcripts stored in amphibian egg cytoplasm. Heterogeneous nuclear RNAs that have an interspersed sequence organization have also been extracted directly from the germinal vesicles of several amphibian species. When renatured and spread for electron microscopy these RNAs form partially duplexed multi-molecular structures similar to those observed in studies of renatured oocyte cytoplasmic poly(A) RNA. In addition many low molecular weight RNAs are transcribed from repeated genes. Aside from 5S rRNA these include U1 and U2 snRNAs which are synthesized in the *Xenopus* oocyte, and the 181 nt cytoplasmic species known on OAX.

There are other, yet unidentified tandemly repeated sequences transcribed in lampbrush chromosomes as well, among them the clusters of short, densely packed transcription units visualized in the electron microscope by Scheer (1981). Molecular measurements carried out on genomic DNA show that the members of most short repetitive sequence families in amphibian genomes are widely interspersed, rather than tandemly repeated. Since many different interspersed repeat families are actively transcribed during oogenesis, *in situ* hybridizations to loop matrix RNA carried out with probes representing such sequences as a class would be expected to reveal reactions at multiple loci. Just this result was obtained by Macgregor and Andrews (1977), whose probe consisted of a low C_0t DNA fraction labeled *in vitro*, and also by Sommerville and Scheer (1982) who utilized an RNA probe prepared from renatured RNAse digested nRNA, labeled *in vitro* with ^{125}I. An unexpected observation in the study of Macgregor and Andrews (1977) was that only certain regions of some loops reacted with the repeat sequence probe. These loops are interpreted to contain more than one transcription unit, and some of these lack sufficient repetitive sequence to react detectably.

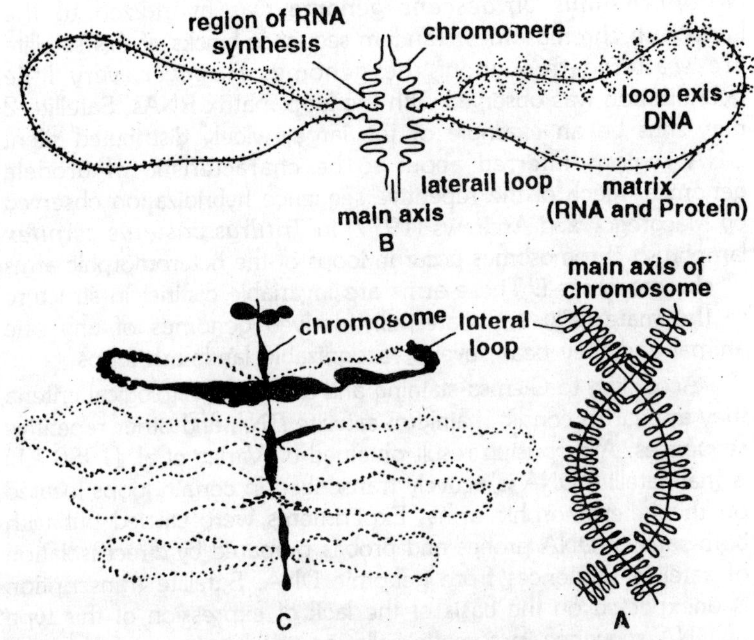

Fig. 5.6. Diagrammatic representation of lampbrush chromosome.

Exactly the same pattern of labeling is present in the homologous loops of chromosomes from different oocytes, from both the same and different animals. This particular observation refutes an early theory of lampbrush chromosome function, according to which the genomic DNA is constantly being spun out into the loops from the chromomeres, so that at any specific time a different set of sequences is being transcribed in given loops. As expected a widely distributed pattern of hybridization is also observed with cloned probes representing *single interspersed repeat sequence families*. From a study of Jamrich et al. (1983) carried out on *Xenopus* lampbrush chromosomes. The probe represents a sequence present in about 10^3 copies per haploid genome. These appear to be so extensively interspersed that they are present in a significant fraction of the lampbrush loops visible. A similar result was reported by Kay et al. (1984), who found that about 100 pairs of lampbrush chromosome loops react with a different cloned repetitive sequence probe.

In another experiment a probe representing satellite 2 of the *Notophthalmus viridescens* genome was hybridized to the lampbrush chromosomes. Tandem sequence blocks of this satellite are scattered throughout the genome. However, very little hybridization was observed with the loop matrix RNAs. Satellite 2 may thus be an example of the large, widely distributed silent DNA regions inferred about to be characteristic of urodele genomes. Much of the repetitive sequence hybridization observed by Macgregor and Andrews (1977) in *Triturus cristatus carnifex* lampbrush chromosomes occur in loops of the heteromorphic arms of chromosomes I. These arms are invariable distinct in structure in the maternally and paternally derived genomes of any one animal and they bear several recognizable landmark loops.

According to Giemsa staining and additional cytological criteria they appear to consist mainly of satellite DNA and other repetitive sequences. A surprising result obtained by Varley et al. (1980,a,b) is that satellite DNA is actively transcribed in certain loops located on the heteromorphic arms. Experiments were carried out with both satellite DNA probes and probes prepared by direct isolation of satellite sequences from genomic DNA. Satellite transcription is unexpected on the basis of the lack of expression of this type of DNA sequence in somatic cells, its usual presence in inactive heterochromatic regions of the chromosomes; and the extremely low complexity and tandem repetition of these sequences. Thus, it would appear that in lampbrush chromosomes sequences may be transcribed that are not expressed in their cell types, though of course this result in itself provides only a suggestion to this effect, as measurements on transcription of the same satellites in somatic *Triturus* cells have not been carried out. Further exploration of satellite transcription in lampbrush chromosomes has provided a new insight into the functional characteristics of these unique structures.

Transcription of satellite Sequences at the Histone Gene Loci of Newt Lampbrush chromosomes

A series of investigations on histone gene transcription in oocytes of *Notophthalmus viridescens* has provided direct evidence that initiation from histone gene promoters can result in the transcription of downstream satellite DNA sequences. There are about 600-800 copies of each of the histone genes per haploid genome, organized primarily in 9 kb clusters that contain all five of the genes plus intragenic "spacer" DNA in the order H1, H3, H2b, H2a, H4. The 9 kb gene clusters are separated by long stretches of satellite 1 DNA, some of which extend for more than 50 kb. This satellite consists of tandem repeats of a 222 nt sequence. Gall et al. (1981) showed by *in situ* hybridization to the DNA of both somatic mitotic chromosomes and oocyte lampbrush chromosomes that the histone genes are located predominantly in two major clusters, on chromosomes 2 and 6. Additional minor sites may exist as well.

The number of histone genes at each major site is approximately equal, i.e., each includes 300-400 of the 9 kb clusters. *In situ* hybridization of the histone probes to lampbrush chromosomes matrix RNA reveals clusters of labeled loops at the corresponding regions of these chromosomes. An example of histone probe hybridization at the chromosome 2 locus. from the size and number of the labeled loops it may be surmised that in this oocyte a significant fraction of the histone genes present on chromosomes 2 are being transcribed. This may not be the case at the other histone gene locus or in different animals, since the DNA/ DNA hybridizations revelaed some chromomeric labeling. A significant correlation is observed between the chromosomal locations of the histone gene clusters and the presence of landmark bodies, known as "spheres", which are attached to the chromosomal axis. The spheres are composed of an acid protein, conceivably accumulated for the special purpose of transporting or processing histone transcripts. The sphere loci are the sites of histone proe hybridization in *Triturus cristatus* and *Triturus alpestris* as well as in *Notoph-thalmus*, though in each case they are found on different chromosomes. Sphere loci also exist in *Xenopus* lampbrush chromosomes, but in this organisms the location of the histone genes is not yet known.

Lampbrush chromosomes loops at the *Notophthalmus* sphere loci react with cloned satellite 1 probes as well as with histone probes. The most important observations have been obtained with asymmetric probes transcribed from M13 templates. These experiments demonstrate that within certain loops there are multiple transcription units, each many kilo-bases in length, since only sharply defined regions of these loops are labeled by the probe. Thus transcription units of opposite polarity can be seen to about directly within the same loop. Both strands of the satellite sequence are represented in the loop RNAs in different transcription units. Transcription initiates at a histone promoter within the gene cluster, and continues in the downstream direction across the remaining histone genes within the cluster and on into the flanking satellite DNA. The large size and the morphology of the transcription matrices implies that transcription fails to stop until another gene cluster is reached, or until a transcription unit initiated in the opposite direction from an adjoining cluster is encountered. Thus, in *Notophthalmus* lampbrush chromosomes, termination fails to occur at the end of the histone gene sequences.

Unfortunately, nothing is known in regard to histone gene transcription in somatic cells of this newt, and thus it is not yet demonstrated that inefficient termination is a pecularity of the transcription of these genes in lampbrush chromosomes. Mature histone messages accumulate in amphibian oocyte cytoplasm during oogenesis and must be derived from such read through transcripts. Processing of these transcripts thus must occur, probably by a strand scission at the 3' terminus of the initial histone mRNA in each transcript, since the satellite 1 sequences are not found outside the nucleus. The satellite transcripts together with the read through histone gene transcripts are presumably degraded within the germinal vesicle.

It remains to be determined whether read through transcription is a special feature of the histone genes or a general explanation for the enormous size of some urodele transcription units. Since there occur in lampbrush chromosome loops well defined transcription units separated by sort regions of silent DNA, there probably exist some means of termination within at least some loops other than encounter with an adjacent transcription unit. Nonetheless, it is possible that read through is a common feature of transcription in lampbrush chromosomes, perhaps an indirect consequence of the generally high rate of initiation, i.e. relative to the rate at which termination complexes could form. Readthrough transcription initiated at a normal gene promoter could account for the satellite sequences detected in the heteromorphic arms of the *Triturus c. carnifex* lampbrush chromosomes by Varley et al. (1980 a,b). This region of the genome must contain some functional genes, since homozygosity for the heteromorphic arms is lethal in this newt species. Ribosomal RNA sequences are also transcribed in chromosome I, though this is not the site of the true nucleolus organizers, where the vast majority of ribosomal genes are located, and which appear silent in lampbrush chromosomes. The transcription of the ectopic rRNA sequences in lampbrush chromosomes is mediated by polymerase II rather than polymerase I, and therefore it is likely that it too is initiated at a functionally unrelated upstream promoter. It is reasonable to suppose in the absence of additional knowledge that transcription of any sequence that is found by *in situ* hybridization to be represented in lampbrush chromosome RNA could have been initiated at the normally regulated promoter of another nearby gene. Note that there is no evidence in the best studied example, the histone genes, that transcriptional initiation occurs at abnormal locations in lampbrush chromosomes; that is, the promoters at which initiation occurs are those of the various histone genes.

Processing of mRNA Precursors in the Xenopus Oocyte Nucleus

Several of the observations thus far reviewed reflect on the capacity of the amphibian oocyte nucleus to process and selectively transport mRNA precursors. For example, there is the presence in the mature egg of a large quantity of non-translatable poly(A) RNA that differs from most mature mRNA in its content of interspersed repetitive sequences. Further evidence reviewed below shows that in *Xenopus* oocytes RNA of this nature is actively exported to the cytoplasm during the lampbrush stage. It is clear, on the other hand, that a stringent selection of those transcripts destined for the cytoplasm does take place, since as the complexity of germinal vesicle RNA is 15-25 fold greater than that of cytoplasmic RNA. These observations completely exclude the possibility that the oocyte germinal vesicle randomly "leaks" heterogeneous nuclear RNA into the cytoplasm. The measurements in addition that the flow of newly synthesized heterogeneous RNA into the cytoplasm includes only 5% (stage 6 oocytes) to 125 (stage 3 oocytes) of the nucleotides initially polymerized into heterogeneous nuclear RNA.

Processing has been demonstrated in the *Xenopus* oocyte nucleus in several studies on the molecular fate of specific precursors, either synthesized in the germinal vesicle from injected plasmids, or injected directly. One such experiment that may be relevant to the possible disposition of the read through histone transcripts was carried by Krieg and Melton (1984). A histone precursor RNA was synthesized *in vitro* from a plasmic containing a chicken H2b histone gene, and introduced into the oocyte nucleus. Generation of the native histone message from this precursor requires a correct 3; endonucleolytic processing reaction, and this was observed. Furthermore, the appearance of a correctly terminated histone mRNA was not affected by the presence of several hundred nucleotides of additional vector sequence distal to the proper termination site. Birnstiel and associates also have demonstrated the production of correctly terminated histone mRNAs after injection of cloned sea urchin histone genes into *Xenopus* oocyte nuclei. Both a highly conserved 3' terminal sequence element in the histone gene, and a small ribonucleoprotein appear necessary for the formation of normally terminated messages. These experiments indicate the existence of the mechanisms that would be required for production of mature histone messages from the 5' terminal gene sequences of the readthrough histone transcripts synthesized in the lampbrush chromosomes.

Direct evidence that *Xenopus* oocytes also possess the capacity to carry out RNA splicing reactions has been obtained by injecting cloned genes that include introns. In several such experiments synthesis of the protein coded by the exogenous sequences is reported to occur. Examples include the ovalbumin gene, in which correct processing involves the precise removal of no less than seven introns, and the SV40T antigen, requiring precise excision of one intron (Rungger and Turler, 1978). Accurate splicing of a yeast $tRNA^{tyr}$ precursor in the *Xenopus* oocyte nucleus has also been demonstrated. Processing of the yeast $tRNA^{tyr}$ precursor involves removal of 5' leader sequence, and also of extra nucleotides at the 3' end and a sequential series of base modifications, as well as the splicing reactions. All of the enzymatic machinery required for these activities is confined to the nucleus.

The observations of Anderson and Smith (1977, 1978) that in late oocytes there seems to be an increased retention of nuclear RNAs and a lower intranuclear turnover rate suggest that the efficiency of the processing reactions might be relatively low. By "efficiency" is meant the fraction of precursor molecules that are converted into mature messages. A kinetic study of the processing of SV40 transcripts in the stage 6 *Xenopus* oocyte nucleus suggests that the efficiency with which processing is carried out is indeed poor. Miller et al. (1982) injected SV40 DNA into the oocyte nuclei, and found that 90% of the primary transcript is degraded within the nucleus. The half-life of these SV40 primary transcripts is only 20-40 min, just as for the endogeneous nRNA. However, a small fraction of the transcription products appears to be converted into correctly processed 19S poly(A) mRNA that accumulates in the cytoplasm. An even smaller fraction of the precursor is converted into 16S mRNA.

It seems unlikely that these results are artifactual consequences of overloading the processing capacity of the oocyte with SV40 transcripts, since the amount of SV40 transcript synthesized in the study of Miller et al. (1982) was less than the amount of endogenous nRNA transcript, and the total amount exported to the cytoplasm was only about 10-20% of the endogenous newly synthesized poly(A) RNA flow into the cytoplasm. A further investigation of the processing of late SV40 transcripts by Wickens and Gurdon (1983) demonstrates that the requirements for transport to the cytoplasm are endonucleolytic cleavage at the site of the 3' terminus of the mature message, and polyadenylation. Both spliced and unspliced derivatives with these terminal features are found in the cytoplasm, while non-polyadenylated transcripts that include sequences distal to the mature 3' terminus are retained in the nucleus. These results suggest that the amount of processing may be a function of the relative rates at which transcription, degradation, preparation of the 3' terminus, and the splicing reactions occur. Where the rates of the later are relatively insufficient, unprocessed transcripts will either be degraded, or if they have acquired mature 3' termini, exported to the cytoplasm where further processing reactions cannot take place.

An investigation of Green et al. (1983) in which human β-globin precursor RNA was injected into oocyte nuclei shows that molecules lacking the 5' cap structure are rapidly degraded. The β-globin precursor molecules were synthesized enzymatically *in vitro*, and only if capped prior to injection were they found to be stable. Accurate splicing out of both β-globin precursor introns occurs in the oocyte germinal vesicle, and for this a terminal poly(A) sequence is not required. However, these splicing reactions are also inefficient, in that the majority of the globin precursor molecules are never processed. A sequence specific form of selective nRNA processing has been demonstrated in the oocyte nucleus as well. Thus Bozzoni et al. (1984) showed that two of the nine introns included in primary transcripts produced from an injected ribosomal protein gene are not spliced out, while all nine introns from a different ribosomal protein gene transcript are removed. Most of the incompletely processed precursor molecules are confined to the germinal vesicle, but again some are recovered from the cytoplasmic compartment.

A role for snRNPs in at least some splicing reactions carried out in *Xenopus* oocytes has been demonstrated by Fradin et al. (1984). These experiments showed that injection of various antibodies against U1 RNP sharply inhibit splicing of the SV40 late transcript, resulting in an accumulation of unspeliced precursor. Many of these antibodies had no effect on the splicing of the early transcript, however, suggesting that a variety of other snRNP cofactors could be required for the processing of different precursor molecules. It may be relevant that the *Xenopus* oocyte germinal vesicle *lacks* several snRNA species that are synthesized in embryos after the blastula stage. Among these are two particular U1 snRNA species found in somatic cells, though other U1 snRNA are synthesized during oogenesis. The mature *Xenopus* egg contains about 8×10^8 molecules of the oocytes types of U1 snRNA, sufficient to provision the several thousand nuclei that have been formed by the time transcription resumes at the blastula stage. It remains to be determined whether deficiencies in specific snRNA pool are in any respect responsible for the efficiency of nRNA processing in the oocyte.

In summary, it is clear that the oocyte nucleus can carry out all known RNA processing reactions; that it does so inefficiently, however, and that if they have been polyadenylated, transcripts that are not mature messages may be exported to the cytoplasm. These conclusions provide a reasonable interpretation for the presence of the interspersed cytoplasmic poly(A) RNAs. However, this proposition must remain inferential, since the results have all been obtained with exogenously introduced sequences, many of which are foreign to the *Xenopus* oocyte. The post-transcriptional disposition of an endogenous primary transcript of the lampbrush chromosomes, read-through or otherwise, has yet to be described in similar detail.

Messenger RNAs and Interspersed Heterogeneous RNA of the oocyte Cytoplasm: The logistic Role of Lampbrush Chromosomes

Accumulation and Turnover of Cytoplasmic Poly(A) RNA

We have so far regarded the newly synthesized oocyte transcripts mainly from an intranuclear perspective. However, the key issue, in considering the ultimate function of lampbrush chromosome, is the relation between the chromosomal, transcription products and the cytoplasmic RNAs accumulated in the course of oogenesis. This relation is by no means an obvious one. In Xenopus, the only organism for which extensive measurements exist, the quantity of total poly(A) RNA present at the end of oogenesis is about 80-90 ng and the same amount is to be found in midvitellogenesis stage 3 oocytes. Relative measurements carried out at a series of different stages of oogenesis by Rosbash and Ford (1974) and Golden et al. (1980) show that in fact the final quantity of oocyte poly(A) RNA has already accumulated before the end of Dumont stage 2. Golden et al. (1980) also reported that the patterns of accumulation of several cloned maternal sequences follow that of the total poly(A) RNA, all achieving their final levels, as judged from RNA gel blots, by the beginning of vitellogenesis. The only exceptions observed were mitochondrial poly(A) RNAs, which continue to accumulate throughout oogenesis.

The mature *Xenopus* oocyte contains about 10^7 mitochondria, and about 4 ng of mitochondrial DNA or ~12 mitochondrial genomes per mitochondrion. However, mitochondrial transcripts account for only about 15% of the newly synthesized cytoplasmic poly(A) RNA of the stage 6 oocyte. The early appearance in oogenesis of the final content of cytoplasmic poly(A) RNA seemed particularly paradoxical before the demonstration by Hill and Macgregor (1980) that even in growing stage 1 oocytes only 60 μm in diameter the chromosomes contain densely packed transcription units similar to those found in Dumont stage 3 oocytes.

A certain amount of confusion has been caused by the fact that continuous transcriptional activity takes place in the lampbrush chromosome in stage 2-6 oocytes, while the net mass of the cytoplasmic poly(A) RNA remains constant. Several discussions

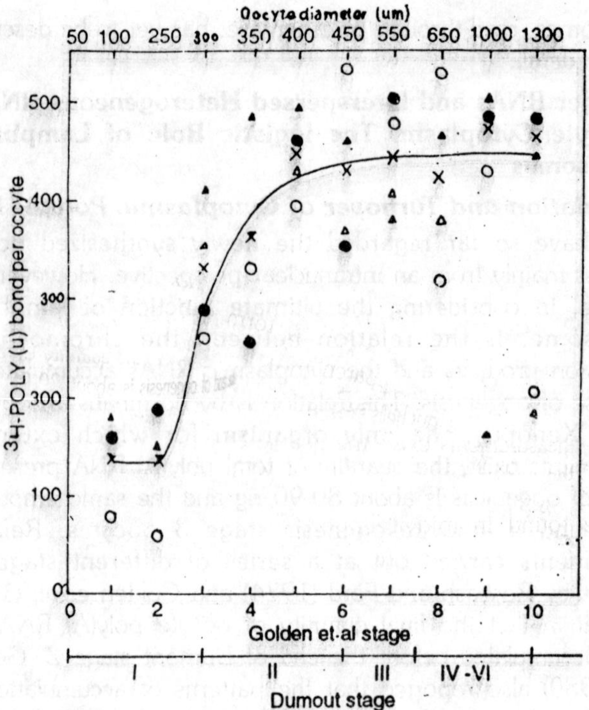

Fig. 5.7. Poly(A) content of Xenopus oocytes, estimated by ^{3}H-poly(U) hybridization.

appear in the literature in which it is concluded that because there is no further *accumulation* of cytoplasmic poly(A) RNA, newly synthesized poly(A) RNA must not be exported to the cytoplasm during most of the lampbrush chromosome stage. This is obviously a faulty deduction, however, in that it is directly contradicted by the measurements of Anderson and Smith (1977, 1978) and Dolecki and Smith (1979). These demonstrate a continuous flow of newly synthesized heterogeneous RNA into the cytoplasm in Dumont stages 3 and 6 oocytes, most of which is polyadenylated. The poly(A) RNA accumulation function means simply that the poly(A) RNAs emerging from the germinal vesicle accumulate until a steady state content is reached, and thereafter this is maintained by the balance between new synthesis and turnover. The lampbrush chromosomes provide the synthetic activity that throughout oogenesis drives the kinetic process. The following calculations show that the maximum rate of transcription in virtually all the transcription units represented in maternal RNA is required in order for this function to be carried out.

There are on the average about 4×10^6 copies of each maternal mRNA sequence in the *Xenopus* egg, representing 1-2 $\times 10^4$ different transcription units. The polymerase II translocation rate in Xenopus oocyte nuclei is 10-20 nt sec^{-1}. Given that the transcribing polymerases are packed only 100 nt apart (i.e., as in lampbrush chromosome), to synthesize 4×10^6 molecules of an average transcript on four single copy genes per nucleus would require about 80 days ($4 \times 10^6 \times 7$ sec/transctipt) - (4 genes/ chromosome set $\times 8.64 \times 10^4$ sec/day)]. The average duration of the growth phase of stage 1 and stage 2 when net poly(A) RNA is accumulating has been estimated to be as long as 8-9 months. However, 2-3 months is a reasonable minimal estimate for young frogs. Thus, vitellogenic oocytes appear four months after metamorphosis is complete while at three months after metamorphosis no oocytes have yet advanced into stage 2. A complementary calculation based on the complexity of the maternal RNA was presented by Anderson et al. (1982). In stage 3 and stage 6 Xenopus oocytes about 1.7-2.2 pg min^{-1} of stable newly synthesized 4-40S RNA, most of which is polyadenylated, enters the cytoplasm. Of this perhaps 10% consists of repetitive sequence transcript and ≤15% of mitochondrial transcripts. Thus, a minimum value of 1 pg min-1 can be taken as the flow into the cytoplasm of newly synthesized transcripts representing single copy genomic sequence. The complexity of this RNA is about 4×10^7 ntp and we assume for the limit calculation that all of the maternal species are continuously being synthesize. Thus the number of copies of each sequence exiting from the oocyte nucleus is about 45 per minute (1 pg min^{-1}, 1.8×10^9 nt $min^{-1} \div 4 \times 10^7$ ntp), or 11 transcripts per minute attributable to each of the four active genes per nucleus. However, this rate requires one initiation per 5.5 sec. which is almost the same as calculated independently from the measured polymerase translocation rate, if one assumes only 100 nt between transcribing polymerases (i.e., 100/15 sec = 7 sec/initiation). Therefore, virtually all of the transcription units active in synthesizing cytoplasmic poly(A) RNA must be maximally packed with transcribing polymerases. Since the mean size of *Xenopus* oocyte poly(A) RNA is about 2.2 kb, the number of such transcription units must be close to the total number included in the lampbrush chromosome loops, i.e., almost 2×10^4 (4×10^7 - 2.2. $\times 10^3$). It follows that to account for the measured flow of stable heterogeneous RNA into the oocyte cytoplasm the large majority of the transcription units active in the oocyte nucleus must be densely packed with polymerases. We may conclude that the transcriptional structure of lampbrush chromosome is in fact required to explain both the rate at which the poly(A) RNA of the oocyte initially accumulates (stages 1-2) and the rate at which it is synthesized and exported at steady stage (stages 2-6).

Given a steady state pool as large as 80 ng of poly(A) per oocyte, and a flow rate of 1 pg min^{-1} (1.44 ng day^{-1}), the half life of the newly synthesized cytoplasmic poly(A) RNA would be about 35 days (ln 2 x 80 ng ÷ 1.44 ng day^{-1}). Of course a fraction of the poly(A) RNA may be turning over at a higher rate. Since no detectable turnover was observed in the kinetic experiments of Andrson and Smith (1978) and Dolecki and Smith (1979). The half-life of such a fraction would still have to be greater than several days. In any case, it is clear that many if not all of the newly synthesized cytoplasmic transcripts ultimately turn over at a rate which these calculations predict would be too low to permit detection in labeling experiments, and yet would still be sufficient to provide a kinetic steady state.

Transport of Interspersed RNA to the Cytoplasm at the Midlampbrush Stage

The interspersed cytoplasmic RNAs were those that are stored in nature oocytes, where they constitute about 70% of the poly(A) RNA mass. This class of RNA can be experimentally defined by its ability to renature, forming multi-molecular networks held together by repetitive sequence duplexes. Anderson et al. (1982) showed by renaturing poly(A) RNA extracted from stage 3 oocytes that interspersed RNAs have already been deposited in the cytoplasm by the mid-lampbrush stage. Electron micrographs of the partially duplexed multimolecular structures. These structures included about 50% of the total poly(A) RNA mass examined. The oocytes were labeled by microinjection of precursor, and after 2 days the newly synthesized poly(A) RNA was extracted from the cytoplasm. About 90% of the labeled molecules fractionated as interspersed RNA. Thus in quantitative terms, the major product of lampbrush chromosome transcription that is exported to the cytoplasm at least at this late stage of oogenesis is interspersed poly(A) RNA. These experiments confirm the view that newly synthesized lampbrush chromosome transcripts continue to enter the cytoplasm even in late oogenesis. The fraction and the ultimate disposition of this large class of heterogeneous maternal RNAs are still to be determined.

The Quantity of Functional mRNA in Xenopus Oocytes

During oogenesis of the rate of protein synthesis per oocyte increases over 100-fold from about 0.18 ng hr^{-1} to about 22 ng hr-1, as illustrated. The quantity of ribosomal RNA per oocyte increases to about the same extent. Thus the fraction of ribosomes engaged in protein synthesis remains approximately constant. L.D.Smith et al. (1984) report this as 1.6% at stage 1 and 2.4% and 2.3 % at stage 3 and 6, respectively, in good agreement with the earlier estimation of about 2% made by Woodland (1974). Since the mass of total poly(A) RNA remains constant after stage 2, an increasing fraction of the poly(A) RNA pool must be involved in the translational apparatus as oogenesis proceeds. The amount of mRNA engaged on the oocyte polysomes at stage 3 and 6 is compared to the total amount of translatable message estimated

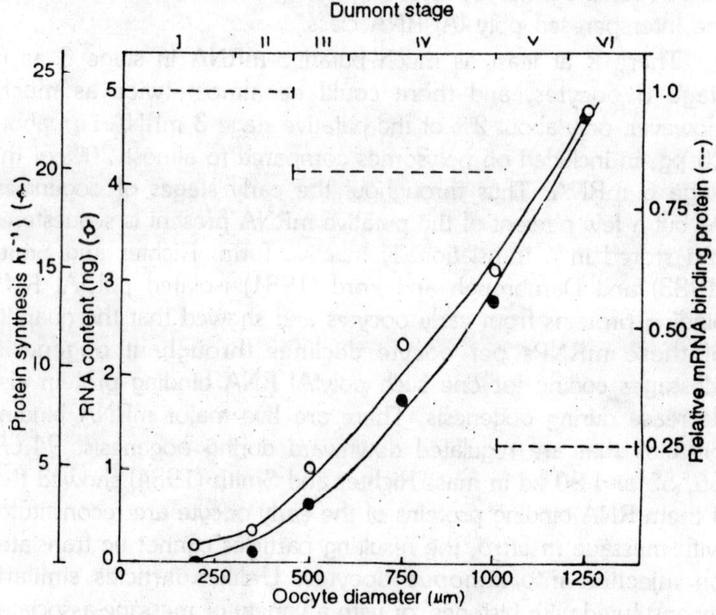

Fig. 5.8. Protein synthesis and mRNA binding protein in Xenopus oocytes.

to be present in the oocyte, calculated from the total poly(A) RNA content, after deducting mitochondrial poly(A) RNA and interspersed poly(A) RNA. The latter is known to be largely non-translatable. Thus, the mass of true maternal message in the stage 6 oocyte is estimated to be about 21 ng. To account for the quantity of polysomes present in the embryo up to the blastula stage, when mRNA begins to be synthesized a new would require about the same quantity of bona fide message as by these calculations is apparently present in the stage 6 oocyte, i.e., assuming that the mRNA is stable after fertilization. Thus during maturation the polysome content increases about 2-fold and a further gradual increase in the fraction of polysomes occurs as additional maternal mRNAs are recruited in early embryogenesis. By blastula stage about 15% of the ribosomes are engaged in polysomes. The mass of polysomes at blastula stage is therefore about six times that in the stage 6 oocyte (15%/2.3%), while the mass of putative mRNA in the oocyte is about 5-fold in excess of that being utilized (21.4 ng/4 ng). The conclusion is that the unused bona fide mRNA in the stage 6 oocyte just constitutes a sufficient supply of maternal mRNA for the embryo, assuming no significant contribution from the intersperesed poly (A) RNA class.

There is at least as much putative mRNA in stage 3 as in stage 6 oocytes, and there could be almost twice as much. However, only about 2% of the putative stage 3 mRNA i.e., about 20 pg, in included on polysomes compared to almost 20% of the stage 6 mRNA. Thus throughout the early stages of oogenesis. All but a few percent of the putative mRNA present is sequestered and stored in a translationally inactive form. Richter and Smith (1983) and Darnbrough and Ford (1981) isolated poly(A) RNA binding proteins from early oocytes and showed that the quantity of these mRNPs per oocyte declines throughout oogenesis. Messages coding for one such poly(A) RNA binding protein also decrease during oogenesis. There are five major mRNA binding proteins that are regulated downward during oogenesis. 94.68, 56, 55 and 50 kd in mass Richter and Smith (1984) showed that if them RNA binding proteins of the early oocyte are reconstituted with message *in vitro*, the resulting particles cannot be translated on injection into *xenopus* oocytes. Unlike particles similarly reconstituted with histones, or with a variety of message-associated proteins extracted from other cells. As the amount of the translational repressor declines through oogenesis, and the quantity of message engaged on polysomes increases mRNA molecules stored early in oogenesis are probably released for translation at later stages. Thus, the sequestration proteins could serve the function of controlling the rate of biosynthesis during the growth phase of oogenesis.

The mechanism responsible for the failure of the remaining 80% of the putative stage oocyte mRNA to be translated during oogenesis remains unclear. The bulk of this fraction of the poly (A) RNA clearly is capable of protein synthesis when extracted and tested *in vitro*, or injected into *Xenopus* oocytes. Perhaps within the later oocyte it remains translationally repressed by the same mRNA binding proteins as present earlier. However, this is an unnecessary hypothesis. A sufficient that have demonstrated limitation in some component(s) of the stage 6 oocyte translational apparatus, other than the supply of accessible mRNA. The key observation is that in stage 6 oocytes injected mRNA can compete for translational components with endogenous mRNA, or with other injected messages. One limiting factor required for translation of membrane-bound mRNA is associated with rough endoplasmic reticulum, since coinjection of a rat liver cytoplasmic messages membrane fraction enhances the ability of the oocyte to translate injected of this class. The translational component(s) that are limiting for cytosol messages remain unknown. Limitation of translation capacity is a specific feature of late oocytes, however. Thus injection of globin mRNA into stage 3 and stage 4 oocytes causes an increase in total protein synthesis, and no competitive depression of endogenous protein synthetic activity is observed, while the same exogenous mRNA injected into stage 6 oocytes competes with the endogenous messages. The highest rate of protein synthesis that can be induced by i.e., 21 ng hr–1. This suggests that by stage 4 the maximum translational capacity of the oocyte has been set (i.e., until it is expanded at maturation), but that this capacity is not fully untilized until late in oogenesis, perhaps because of the prevalence of repressive mRNA binding proteins in younger oocytes.

mRNAs for Specific Proteins

Many diverse proteins have been identified in late *Xenopus* and *Rana* oocytes, stored in relatively enormous quantities for use after fertilization. Examples include histone, of which a sufficient supply exists to assemble the chromatin of over 20,000 cells; nucleoplasmin, which is apparently involved in the mobilization of histones for chromatin assembly and which constitutes 10% of the total nuclear protein RNA polymerases, equal in amount to that contained in about 10^5 larval amphibian cells, DNA ligases tubulins also resident in a huge pool amounting to about 1% of all soluble oocyte protein; some even more prevalent actin species that constitute >8% of the total soluble protein in stage 6 oocytes the ribosomal proteins included in the 10^{12} ribosomes of the mature oocyte; proteins binding specifically to 'snRNAs and a variety of other nucleoproteins. Several of the later are considered in the following section, in connection with 5S RNA synthesis during oogenesis, and others, eg., those associated with cytoplasmic mRNAs have already been mentioned. DNA polymerase provides a particularly striking example, since there is no nuclear DNA synthesis during the growth phase of oogenesis. The number of molecules of $\alpha_1 + \alpha_2$ DNA polymerase increase continuously from about 10^8 in the stage 1 oocyte, to about 10^9 in the stage 3 oocyte and 2×10^{10} in the unfertilized egg. For comparison, in the hatched tadpole, a level of about 7.5×10^4 molecules of $\alpha_1 + \alpha_2$ DNA polymerase is maintained per cell by new synthesis. The egg thus stores sufficient a-polymerases to provision over 2×10^5 embryonic cells.

Though sequences represented in the maternal RNA continue to be transcribed throughout the lampbrush stage there are only a few examples that provide evidence on the origins of specific maternal mRNAs. Anderson et al. (1982) showed that transcripts labeled to high specific activity by *in vitro* incubation of isolated stage 6 oocyte nuclei reacted with six different cloned cDNAs derived from the maternal poly(A) RNA. However, there is yet no direct demonstration, in these or any other experiments, of the entrance of particular newly synthesized mRNAs into the cytoplasm in late oocytes. The problem is technically difficult because specific activities adequate for the detection of individual mRNA species cannot be obtained in whole oocytes, even after injection of high specific activity precursor nucleotides, and thus only the overall flow rates have so far been measured. Quantitative studies of protein synthesis during oogenesis have shown by two dimensional gel electrophoresis that most protein species continue to be synthesized at the same relative rates throughout, which would be consistent with later utilization of a pool of mRNAs stored from stage 1-2. Though a considerable number of stage-specific changes have been reported in these studies, they are apparently the result of alterations in the fraction of mRNA of given species that are loaded on the oocyte polysomes. Thus, kind and Barklis (1985) reported that the *in vitro* translation products of poly(A) RNA extracted from stages 2, 3, and 6 oocytes are essentially identical.

We turn now two sets of identified proteins for which cloned probes have been generated the histones, and the ribosomal proteins (rP), the *Xenopus* oocyte contains about 4×10^8 molecules of each of the rP mRNAs investigated, though the mRNA quantities for individual rP species differ up to four-fold with respect to one another. The greater content of histone mRNAs is more than accounted for by the difference in the respective gene copy numbers. Thus there are about 40 genes for each histone species per haploid genome in Xenopus 2-5 copies of each rP gene. The number of RNA polymerase mRNA molecules, estimated indirectly from the subunit synthesis rate measurements of Hollinger and Smith (1976), is several fold lower. A similar number of molecules of the mRNA for the 70 kd heat shock protein is also reported, i.e., about 3×10^6. For comparison, the prevalence of the average poly(A) RNA species in the egg or stage oocyte is about 4×10^6 molecules.

Both the rP mRNAs and the histone mRNAs the maximum level of accumulation is achieved by the beginning of vitellogenesis at Dumont stage-2, just as for the bulk of the poly(A) RNA Pieranderi-Amaldi et al. (1982) and Baum and Wormington (1985) noted that the rP mRNAs have declined in quantity by stage 6, and Dixon and Ford (1982) also reported that synthesis of rP is much decreased in late oogenesis. On the other hand, the quantity of histone mRNA remains about the same throughout the later phases of oogenesis. This difference in behaviour is directly interpretable in terms of the temporal requirements for the respective proteins. The assembly of ribosomes occur most rapidly in mild-oogenesis, and is largely completed by the end of oogenesis, mRNAs coding for ribosomal proteins achieve their peak concentrations just prior to institution of the maximum rate of rRNA synthesis, which occurs at stage 3. Thus, by stage 2, 50% of the rp mRNA is loaded on polysomes, compared to only a few percent of total oocyte mRNA and the high level of translational utilization of the rP messages persists throughout oogenesis. Much of the rP mRNA s degraded after fertilization, and some species of ribosomal proteins are not synthesized again until the neurula stage, though the mRNA begins to be produced several hours earlier at gastrulation. In contrast, both the large pool of histone protein synthesized during oogenesis and the mass of maternal histone mRNA are destined for use in early development

The histones begin to be synthesized at a high rate immediately upon maturation, and maternal histone mRNAs thereafter support all new histone synthesis until the blastula stage. Sufficient measurements on accumulation of both histone mRNA and histone proteins during oogenesis in *Xenopus* exist so that a quantitative image of the kinetics of this process emerges. We have seen that accumulation of the maternal histone mRNA pool occurs very early in oogenesis. Assuming that the polymerases synthesizing histone transcripts are packed only 100 nt apart; that the translocation rate is 15 nt sec^{-1} (Anderson and Smith, 1978); and that there are 40 genes for each core histone each of the four haploid genomes of the 4C oocyte nucleus; a period of 4-5 months would be required to synthesize the 3.4×10^8 histone mRNA molecules that according to table 4.6 constitute for each core species the maternal histone mRNA pool. As discussed earlier, this is similar to the estimated time required for accumulation of the steady stage content of an average poly(A) RNA coded by a single copy gene, about 3 months. This calculation requires the histone transcription units to be in the densely packed lampbrush chromosome configuration during early oogenesis.

In addition, it requires that the histone mRNA be stable during its initial accumulation phase, and thereafter be sufficiently long-lived so that it decays no faster than it can be replaced by new synthesis. Woodland and Wilt (1980b) showed that sea urchin histone mRNA injected into oocytes turns over with a half-life of only a few hours. There are at least two known factors that could contribute to the stability of *endogenous* histone mRNA in the oocyte. Unlike these injected messages the endogenous histone mRNA might be complexed with protective proteins, such as the mRNPs analyzed by Darnbrough and Ford (1981) and Richter and Smith (1983). In addition, 50-75% of the histone mRNA in the *Xenopus* oocyte is polyadenylated. It has been demonstrated directly by Huez et al. (1978) that prior polyadenylation increase the stability of HeLa cell histone mRNA injected into *Xenopus* oocytes.

The four nucleosomal core histones are synthesized at a rate of about 50 pg hr^{-1} during oogenesis. This is only a very small fraction of the rate that could be supported by the mass of stored histone message, and indeed during maturation the rate increases 50-fold to about 2500 pg hr-1. The final histone content is about 135 ng. This includes about 3-4 p moles of each of the core histones, and 0.5 p moles of H1 Van Dongern et al. (1983b) found that accumulation of Hla, the major H1 species, is completed during stage 3. Though some H1 synthesis continues after this it must be balanced by turnover. At the rate of 50 pg hr-1 for the stored core histone, it would take about 100 days to accumulate the amounts of stored core histone protein in the egg. Again, this is not inconsistent with the time usually required for *Xenopus* oocytes to progress from stage 2 onwards as discussed earlier.

The stored histones are mainly accumulated within the germinal vesicle of the oocyte, where they are stored in stoichiometric complexes of several kinds with the nucleophilic proteins. N_1, N_2 and nucleoplasmin N_1 and N_2 are synthesized exclusively in germ line cells, including lampbrush stage oocytes, but not after fertilization in the embryo, or in somatic cells. The embryonic role of these maternal proteins may be to bind, neutralize, and transport into the blastomere nuclei the newly synthesized histones that are translated from maternal mRNA during maturation and in early cleavage. *Nucleoplasmin* and N_1 and N_2 are not the only important proteins destined for use after fertilization that are stored in the germinal vesicle. For example most of the α DNA polymerase the RNA polymerase, the actins and the snRNPs are similarly localized in the enormous nucleus of the stage 6 oocyte. Germinal vesicle breakdown at maturation releases all of these components into the cytoplasm, where they remain until after fertilization, when they are redistributed to the newly formed nuclear compartments of the embryo.

The Functional Role of Lampbrush Chromosomes

A coherent interpretation of the data so far reviewed is as follows. Lampbrush chromosomes are the cytological manifestation of a transcriptional apparatus in which the distinguishing—and entirely unique—feature is that almost all the active transcription units are operating at the maximum possible initiation rate. The basic role lampbrush chromosomes is to provide the transcription products loaded into the oocyte cytoplasm. This function can be considered in two phases. Early in the lampbrush stage (in *Xenopus*, Durmont stages 1 and 2) net accumulation of maternal poly (A) RNA occurs until the levels that will be maintained thereafter are achieved. This has been illustrated in many different ways, e.g., by measurements of the totally poly(A) RNA content, analyses of the spectrum of proteins synthesized at different stages, and observations on several specific transcripts. The second phase begins with vitellogenesis. Throughout this process the oocyte is synthesizing the enormous supplies of proteins that will be required during development, of which the best understood example is provided by the histones.

The role of the lampbrush chromosomes during this phase is evidently to maintain in kinetic steady state the required level of the maternal transcripts. This is demonstrated directly by measurements of the flow of newly synthesized the poly(A) RNA into the cytoplasm in stage 3 and stage 6 oocytes. We have seen that when the observed flow rates are considered in conjunction with the complexity of the RNA, the maximum possible transcription rates would be required merely to preserve the levels of transcripts present, even where the half-lives of these in excess of a month. Thus it may be concluded that *lampbrush chromosomes provide the transcriptional solution to the logistic problem of building and maintaining very large oocytes*. The same problem is solved in a completely different manner in meroistic oogenesis. As the following sections of this chapter illustrate there are still other solutions utilized in the same oocytes in answer to the special logistic demands posed by the requirements for 5s RNA and 18 and 28S rRNA.

There could exist adaptive advantages for a system of oogenesis in which the maternal constituents are continuously renewed, even though this entails (in larger eggs) the maintenance of an extremely high rate of synthesis right to the end of oogenesis. Many amphibians, for example the tropical temporary water breeder *Engystomops pustulosus*, store mature oocytes within the ovary until an appropriate environment for egg laying becomes available and this probably also is characteristic of *Xenopus*. Continuous molecular turnover and renewal could provide a means of escaping a rigid requirement to complete oogenesis and shed eggs on a certain schedule, there by improving adaptive flexibility in response to uncertain environmental conditions. In any case, it should not seem peculiar to assume that the transcripts of the oocyte are not absolutely stable. In all living cells that have been examined, except for special examples of dormancy such as cryobiotic systems, seeds, and spores, mRNAs are found to decay, usually stochastically, and almost always with kinetics far faster than implied by the minimum half lives required in the foregoing considerations of amphibian oocyte transcripts.

The subject of lampbrush chromosome function has a curious intellectual history. Lampbrush chromosomes have provoked a remarkable series of colourful, but wrong interpretations, among which are the once widely accepted idea that the loops are the products of the chromomeric genes the "master-slave" theory; the proposition that new DNA sequences are progressively unwound into the loops from the chromomeres; the "C-value paradox"; and most recently, the assertion that the intense RNA synthesis in the loops is all non-productive, in the sense that none of the newly synthesized RNA is exported to the cytoplasm. The value of these proposals has lain mainly in the efforts they have stimulated to disprove them and none seem at this point likely avenues to further understanding. But it is scarcely the case that we have reached the end of this curious progression. Though we may not understand in logistical terms what lampbrush chromosomes do, we clearly do not understand why they are doing it. Thus the majority of the heterogeneous transcripts exported from the lampbrush stage oocyte nucleus are non-translatable, interspersed RNAs, of unknown function.

Perhaps these are unless products of readthrough transcription, exported to the cytoplasm as a side effect of some yet unknown characteristic of the post-transcription processing apparatus of the oocyte. In this case there remains to be discovered an unexpected feature of the mechanism by which transcripts are handled in the oocyte germinal vesicle, a mechanism so valuable that it is retained even at the expense of loading the cytoplasm with 70% non-functional poly(A) RNAs destined for modification and translational use after fertilization, or perhaps they constitute some totally different class of maternal RNA required in developing systems about which we yet understand nothing. A further solution to the problem of lampbrush chromosome function, and perhaps a basic new understanding of oogenesis, could lie somewhere among the answers to these newly arisen mysteries.

The Accumulation of Ribosomal RNAs

The mechanisms by which 5S rRNA and the 18S and 28S rRNAs are accumulated are for *Xenopus* relatively well-described. About 10^{12} ribosomes must be assembled per oocyte, and this logistical requirement is met by three completely different strategies. The synthesis of ribosomal proteins has already been discussed, and in this section we compare the means by which the heavy ribosomal RNAs and 5S rRNA are synthesizes and accumulated. Though they can only summarized briefly here, the molecular details of the processes by which these two classes of ribosomal genes are controlled provide interesting new insights into the mechanisms by which the genes transcribed by polymerases I and III are developmentally regulated.

5S rRNA Synthesis in Xenopus Oocytes

Low molecular weight RNAs, mainly tRNAs and 5S rRNA, are the major transcription productions of Dumont stage 1 and 2 oocyte. Thomas (1974) showed that after 24 hours of labeling about 24% of the radioactive RNA in the cytoplasm of these previtellogenic oocytes is tRAN and 39% is 5S RNA, and data of rosbash and Ford (1974) indicate that 5S RNA and tRNA account for about 70% of the mass of total RNA then present. In previtellogenic oocyte the molar ratio of newly synthesized tRNA to heavy ribosomal RNA is as high as 25, while for 5S RNA continue to be synthesized, but as the rate of synthesis of 18 and 28S rRNA accelerates after stage 2, these ratios shift dramatically, previtellogenic oocytes are thus very unusual in the large fraction of their total accumulated RNA accounted for by stable low molecular weight species.

About 50% of the 5S RNA synthesized in previtellogenic oocytes is store in a 7S particle, where it is complexed with a single 38.4 kd protein. The remaining 50% is found in a multimeric 42S particle, which contains in addition to the same 38.5 kd protein and 5S RNA and two other protein molecules of 50 kd protein and 5S RNA and two other protein molecules of 50 kd mass. The later proteins are believed to be bound to tRNA, which is present are believed to be bound to the tRNA, which is present in the aminoacyl form and in the 42S particles of previtellogenic oocytes turn over with half-lives of 4-36 hr, depending on the tRNA species. Both forms of storage particle are prominent only in previtellogenic oocytes. The proteins of 42S particle are reported to be involved in transport of 5S RNA to the nucleolar site of ribosome assembly. This occur as newly synthesized 18s rRNA become available. In the ribosome the 5S RNA is complexed with a ribosomal protein that differs in molecular weight and amino acid composition from the storage particle proteins. The proteins of the 7S and 42S particles are synthesized at the highest rate in stage 1 oocytes. The same general mechanism, viz., early synthesis of 5S RNA and storage in 7S and 42S RNP particles, followed during vitellogenesis by incorporation of the stored 5s RNA into ribosomes may occur in many large vertebrate oocytes that also bear lampbrush chromosomes. Thus homologous 42S and 7S particles have been reported in previtellogenic oocytes of anurans, urodeles, and teleost fish.

There are two classes of 5S RNA genes in *Xenopus*, those active in oocytes and those active in both somatic cells and oocytes, termed the somatic 5S RNA genes. Oocyte-type 5S RNA differs from somatic 5S RNA at 6 out of 120 nt. Only the oocyte-type of 5S RNA is stored in the 7S and 42S RNP particles. The products of the somatic 5S genes are apparently unstable in the oocyte, since though they are synthesized these transcripts fail to accumulate. After fertilization and up until the midblastula stage, both classes of 5S RNA genes are silent as are all other nuclear genes. When transcription resumes some oocytes 5S RNA synthesis occurs, as does somatic 5S RNA genes are expressed consitutively. The very high rate of oocyte 5S RNA synthesis. After the gastrula stage, however, synthesis of oocyte-type 5S RNA dwindles to an insignificant level. Thus the oocyte-type genes are regulated, in that they are mainly transcribed only during oogenesis, while the somatic 5S RNA synthesis in early oogenesis is accounted for by the large number of oocyte-type genes. Thus there are about 20,000 oocyte 5S RNA genes and about 400 somatic 5S RNA genes per haploid genome. During oogenesis each oocyte-type 5S RNA gene is responsible for the synthesis of about 10^7 5S RNA molecules.

Chromatin containing the 5S RNA genes can be transcribed with fidelity *in vitro*, by added polymerase III. Exploitation of this advantage has led to a relatively advanced understanding of the chromatin regulatory factors involved in developmental control of these genes. A DNA sequence within the gene, extending from nucleotide +45 to +97 and known as the "internal control region," is required for the formation of a transcription complex and for initiation of transcription by polymerase III.

The active transcription complex contains three protein species, *Tellia*, B and C. Aside from the polymerase itself. One of the most important properties of the transcription complex is that once formed it remains stable and functions for many rounds of transcription. Though little is so far known of the transcription factors B and C, it is established that a single molecule of TF IIIA bound directly to the internal control region of each active 5S RNA gene. This has been shown explicitly by mapping the DNA protein contact point, and by determination of the binding stoichiometry. In the absence of TFIIIA a transcription complex cannot be assembled and the genes instead may form stably repressed complex with histone. This is apparently the fate of the oocyte-type 5S RNA genes in the nuclei of somatic cells such as adult erythrocytes or kidney tissue culture cells, since these genes can be reactivated and assembled into active transcription complexes following treatments that remove histone H1.

Furthermore, oocyte-type 5S DNA chromatin, i.e. the DNA in nucleosomal form, can be reconstituted *in vitro* together with histone H1 to reproduce the repressed structure in tissue culture cells the somatic 5S RNA genes are instead included in stable transcription complexes, and they require only the addition of polymerase III for transcription. The oocyte 5S RNA genes are found in such complexes only during oogenesis. A simple binary choice between TFIIIA binding or histone H1 binding may constitute a sufficient explanation for the regulation of the 5S RNA genes. *In vitro* experiments with TFIIIA demonstrate a 4:1 preference for somatic over oocyte-specific genes, apparently the result of the three nucleotide difference in the oocyte and somatic gene sequence with an important reason of the internal control region. However, when a mixture of oocyte and somatic type 5S gene derivatives is injected into fertilized *Xenopus* eggs a preference for the somatic 5S gene sequences of up to 200-fold greater activity observed. *In vivo* the discrimination between the endogenous somatic and oocyte-type genes for a limiting amount of regulatory factors may indeed by the causes of the 1000-fold greater activity observed (on the average) for each of the somatic-type 5S RNA genes in somatic cells, relative to each of oocyte-type genes. This type of explanation depends on the decrease in TFIIIA concentration that occurs during embryogenesis. In the fertilized egg there are about 3×10^9 molecules of TFIIIA; in the gastrula stage embryo about 9×10^4 molecules per cell, and in tadpole stage embryos and adult kidney about $1\text{-}2 \times 10^4$ per cell.

The relative *kinetics* with which there forms either an active transcription complex with TFIIIA or an inactive complex with histone may be determinant, since the somatic cell content of TFIIIA is still sufficient to bind about 1/4th of all the cellular 5S RNA genes, i.e., including the oocyte-type genes. Thus Gargiulo et al. (1984) showed that on coinjection of excess oocyte and somatic 5S RNA genes into the *Xenopus* oocyte nucleus, the greater affinity of the somatic genes for a limiting factor, presumbly TFIIIA, results in the irreversible formation of inactive histone complexes with the oocyte type genes. This result mimics the events that occur in newly formed somatic cells, and demonstrates that in the presence of histones the binding preference of TFIIIA for the somatic 5S RNA gene control sequences is apparently sufficient to account for the activation of somatic type 5S RNA genes and the stable repression of oocyte-type 5S RNA genes.

Furthermore, Brown and Schlissel (1985) demonstrated that injection of TFIIIA itself into syncytial *Xenopus* embryos that have been prevented from undergoing cytokinesis results in a sharp increase in the fraction of endogenous 5S rRNA transcript deriving from the oocyte-type genes, i.e., after transcriptional reactivation at the blastula stage. This experiment provides a direct indication that a determining parameter controlling the selective expression of the somatic type 5S RNA gene is the limited concentration of TFIIIA in late embryonic and somatic cells. The observation noted above that in normal late blastula stage embryos when there are still about ten molecules of TFIIIA per 5S RNA gene there is a transient reactivation of the oocyte-type 5S RNA genes is also consistent with this interpretation.

The changing prevalence of the positive regulatory factor TFIIIA also helps to explain directly the course of 5S RNA synthesis during oogenesis. A key observation made by Pelham and Brown (1980) and Honda and Roeder (1980) is that TFIIIA is the 38.5 kd protein that is complexed with the 5S RNA in the 7S particles of previtellogenic oocytes. In other words, this protein binds to both the anticoding strand of the 5S RNA genes are all actively transcribed. However, as the available TFIIIA is progressively sequestered together with the newly synthesized 5S RNA in the 7S and 42S storage particles, a decreases of 5S RNA transcription occurs. When these particles disappear later in oogenesis the total amount of TFIIIA in the oocyte has decreased by a factor of about two. The TFIIIA genes has been cloned, and estimates of the level of TFIIIA mRNA provided by Ginsberg et al. (1984).

There are about 5×10^6 molecules of TFIIIA message in the early oocyte, classifying this transcript as a member of the low abundance sequence class. This falls to about 1×10^6 molecules by stage 4-6 of oogenesis, and to only about 1 per cell in the swimming tadpole. In summary, the 5S RNA of the oocyte ribosomes is provided by the active transcription of about 80,000 genes in the 4C oocyte nucleus, occurring mainly during the previtellogenic period of oogenesis. The transcription of these genes is regulated by the formation of a specific transcription complex including three protein factors plus polymerase III. As the concentration of 5S RNA builds, one of these factors. TFIIIA, becomes bound instead to the 5S RNA product. Competitive interactions between a limiting amount of TFIIIA and histones are probably involved in the regulatory events that result in the extremely accurate discrimination between the oocyte and somatic types of 5S RNA genes later in development.

Ribosomal RNA Synthesis during Oogenesis

Though the same enormous number of the large rRNAs are required to be synthesized during oogenesis as of 5S rRNA, there typically no more than a few hundred genes for these rRNAs per haploid genome. In organisms whose oocytes contain lampbrush chromosomes the ribosomal genes are amplified early in oogenesis, and are present as extrachromosomal circular DNA molecules that are packaged into nuclear structures. Transcription of ribosomal DNA during oogenesis occurs in the amplified rDNA. Ribosomal DNA amplification has been observed in a number of urodele and anuran-amphibians, in teleosts, and in orthopteran insects all of which utilized lampbrush chromosomes in their oocytes. Ribosomal gene amplification is also reported in coleopteran and in a few dipteran insects, which carry out meroistic oogenesis, and to low level in an echiuroid worm and in lamellibranch molluscs. As reviewed in the following section ribosomal gene amplification probably does not occur in the mouse, though it does in primates, and it is absent in those sea urchin oocytes that have been examined, as well as in some other invertebrates, e.g., the nematode *Pangarellus*. Replication of the rDNA occurs mainly at the pachytene stage in amphibian oocytes though the process being in the premeiotic oogonia.

In orthopteran oocytes amplification also takes place during the pachytene stage. The original copies used for rDNA replication are of chromosomal origin and usually but probably not always the amplified rDNA of each original circle derives from a single genomic rRNA gene. This interference is supported by the observations that spacer sequence heterogeneity, which is observed frequently in adjacent genomic rDNA repeat units is only occasionally seen within the tandem repeats of a given *amplified* rDNA molecule. Most of the amplified rDNA is synthesized by some form of rolling circle replication of pre-existing extra-chromosomal rDNA circles.

In *Xenopus* oocytes the extrachromosomal rDNA formed by the amplification process is packaged in a large number of extrachromosomal, nucleoli, which can be observed in the germinal vesicle sap throughout oogenesis. On germinal vesicle breakdown the extrachromosomal rDNA is released to the cytoplasm, where it remains during early embryogenesis. By the gastrula stage it has disappeared. Thiebaud (1979) showed that the number of extrachromosomal nucleoli per oocyte varies from 500 to 2500 in oocytes of the same female. The maximum number is attained at stage 4, and after this some nucleoli apparently fuse resulting in a lower value in later oocytes. The number of rRNA genes per nucleolus also varies, ranging from 500-10000. On the other hand, the *total amount of amplified rDNA per nucleus is always the same*, about 30-35 pg or approximately 2.5×10^6 rRNA genes. Since there are about 450 rRNA genes per haploid genome this represents for the oocyte about a 1400-fold increase in gene number. However, even after amplification a long period is required for the accumulation of the large rRNAs, and rRNA synthesis is likely among the rate-limiting processes in the growth of the amphibian oocyte, just as in the growth of the *Drosophila* oocyte.

In the earliest previtellogenic oocytes there is little rRNA synthesis, though extrachromosomal replication of the rDNA has been completed. Scheer et al. (1976b) showed that in the newt

Triton alpestris previtellogenic oocytes (equivalent to Dumont stage 1 in *Xenopus*) synthesized rRNA at only about 0.01-0.5% of the rate measured in vitellogenic oocytes. This relatively low synthetic rate is correlated with sparse packing of transcripts in the extrachromosomal nucleolar genes, and with the presence of many totally inactive gene regions. In contrast, 90-95% of the nucleolar ribosomal genes are being transcribed in midvitellogenic oocytes, and the transcripts visible on these are tightly packed, with as many as 130 transcripts per gene region. From the known length of the amphibian 40S rRNA precursor, the spacing between polymerases in the extrachromosomal nucleolar genes is calculated to be only about 100 ntp.

Several estimates the rate of rRNA synthesis for stage 3 *Xenopus* oocytes are shown. The rate apparently varies greatly depending on the stage of gonadotropic hormonal stimulation of the female. As ribosomes are a major constitute of the cytoplasms, hormonal determination of their rate of accretion, and also of the rate of yolk uptake appear to provide direct physiological control over the kinetics of oocyte growth. The difference in synthesis rate between stimulated and non-stimulated females is due to the frequency of initiation in the active extrachromosomal nuclear genes, and in the fraction of these genes that are transcribed at all, since hormone treatment does not affect the fundamental polymerase translocation rate. Under optimal stimulation the rate of rRNA production in midvitellogenic oocyte is close to that expected were all of the 2.5 × 106 extrachromosomal nucleolar genes functioning at the maximum rate. Thus over 3×10^5 rRNA molecules are being synthesized per second, compared to 10-100 sec^{-1} in somatic cells. The instantaneous rate of rRNA synthesis exceeds the rate of total heterogeneous RNA synthesis in the lampbrush chromosomes, and is many times the rate of stable heterogeneous RNA flow in the cytoplasm. The newly formed rRNA molecules exit into the cytoplasm through "pore complex," which occupy 25% of the area of the nuclear membrane. From the number of these structures Scheer (1973) calculated that 1-2 molecules of rRNA pass through each pore per minute.

Net rRNA accumulation continues at least under some conditions in stage 6 *Xenopus* oocytes. Under conditions of hormonal stimulation synthesis of 40S ribosomal precursor occurs in these oocytes at an average rate about 50% of that in stage 3 oocytes. However, this rate varies greatly from female to female, controlled by unknown physiological factors. The 40S ribosomal precursor synthesized in stage 6 oocytes seems not to be processed with the normal efficiency. On germinal vesicle breakdown the precursor are released into the cytoplasm where they persist into early cleavage. In species other than *Xenopus* more sharply decreased levels of rRNA synthesis are observed towards the end of oogenesis. For example, in postvitellogenic mature oocytes of *Triton alpestris* the nucleolar genes contain only about 13% of that observed in midoogenesis. In mature oocytes of hibernating *Rana pipiens* at least 70% of the extrachromosomal ribosomal gene units appear completely devoid of nascent transcripts when examined in the electron microscope. In summary, the rate of rRNA synthesis on the extrachromosomal nuclear genes increases from a low initial level to maximum at midvitellogenesis and then fall again at maturity, to an extent that varies according to species. These regulatory changes reflect large differences in the frequency of chain initiation and in the fraction of the rRNA genes utilized. It has been noticed that in either very young or mature oocytes when not all the extrachromosomal genes are functioning, adjacent transcription units may display large differences in activity. The rate of initiation of each gene in the tandem nucleolar arrays thus appears to be regulated independently during oogenesis.

Transcription of the ribosomal genes is controlled by two upstream regions, viz. a proximal promoter, and a set of regulatory elements located distally in the intergenic spacer. The promoter located immediately at the 5' end of the gene has been mapped by observing the effects on transcription of various 5' deletions. These experiments were carried out either by injecting synthetic deletion mutants into the *Xenopus* oocyte nucleus, or by transcribing them in a cell-free polymerase I system that consists essentially of an oocyte nuclear homogenate. Transcripts of exogenous origin were distinguished by utilization of mutant *Xenopus laevis* gene constructs in the *Xenopus borealis* transcription systems. The proximal promoter is found to extend from nt -142 (i.e., 142 nt upstream of the start of the transcript) to nt -6. The 13 nt sequence from -7 to -6 is the major requirement for accurate transcription in injected oocyted nuclei, and this sequence element is conserved among the rRNA genes of several *Xenopus* species. This region is evidently directly utilized for polymerase I initiation. At -27 is a T_6 sequence that occurs in an analogous position to the "TATA Box" of gene transcribed by polymerase II. The remainder of the promoter region is apparently required at least under conditions of limited initiation in order to obtain maximal transcription rates. There is some evidence that portions of the promoter region extending out to -142 serve as the binding site for a transcription factor.

An interesting indirect light on the mechanism by which these genes are activated derives from an analysis of the effects of topological constraint on rDNA transcription. A supercoiled configuration was found to be required for continued expression of injected plasmids, but not for transcription of the endogenous rRNA genes in the oocyte nucleus. Thus on injection of restriction endonucleases, both classes of gene are digested, but while this abolishes transcription from the injected plasmids it does not effect endogenous rRNA synthesis. The torsional strain to which the supercoiled plasmids are subjected may thus mimic the effects of the assembled transcription complex on the endogenous rDNA. A novel aspect of regulation in the tandem rRNA gene sets is the role of sequences, located between about 200 and 2500 nt upstream in the non-transcribed spacer. The importance of these spacer sequences for transcription of the rRNA genes has been shown in several ways.

Moss (1983) found that in oocyte nuclear injection experiments deletion of a portion of the spacer region reduces the amount of transcription 20-fold compared to a coinjected wild-type gene, and a similar conclusion was drawn by Busby and Reeder (1983) from experiments in which deletion mutants lacking far upstream sequences were injected into *Xenopus* eggs, so as to observe the effect on developmental regulation in the embryo. After the midblastula transition the injected genes are activated, and deletion of these spacer sequences reduces their transcription in the embryo 5-10 fold. Additional show that when different rDNA plasmids are coinjected into the oocyte nucleus in sufficient quantity to induce competition, those plasmids the most transcript. The key spacer elements responsible for these effects are the repetitive sequences identified in Figure 4.5(d) as the 60/81 base pair repeat. Both forms of this repeat include a 42 ntp sequence also found in the proximal promoter at position -73 to -114. The 42 npt elements have the properties of enhancer sequences. Thus, they endow a plasmid carrying them with the ability to complete successfully with a second plasmid that lacks them, and their effect occurs exclusively on genes to which they are in *cis* relation (Labhart and Reeder, 1984). Furthermore, they operate equally well in either orientation, or several kb away from the gene. This interpretation also explains an old mystery of *Xenopus* species hybridization experiments. When X. *laevis* and X. *borealis* are crossed, the X. *laevis* rRNA genes are always utilized predominantly in the hybrid embryos, irrespective of the direction of the cross (Honjo and Reeder, 1973; Casdidy and Blackler, 1974). X. *laevis* rRNA spacers turn out to contain more than 20 copies of the 42 ntp enhancer elements while the X. *borealis* genes contain only four.

Thus "nucleolar dominance" can be recreated experimentally by injecting genes from these two *Xenopus* species into oocyte nuclei of either, under conditions where competition will occur. The intergenic spacer regions also contain several nearly perfect duplications of the complete proximal promoter sequence. These upstream promoters are capable of initiating transcription themselves under certain circumstances, both experimentally and naturally occurring, and they are thus to be regarded as potentially functional. However, they are followed by a "failsafe" terminator at position -243 that prevents readthrough from them into the gene proper.

The role played by the duplicated upstream promoters is not yet apparent, though it is known that the requirements for their function are subtly different than for the proximal promoter. They could serve in some way as an ciliary elements necessary for the function of the contiguous enhancer sequences. These detailed researchers into the regulation of the genes that synthesize the rRNAs may provide a preview of the immense molecular complexity of the transcriptional apparatus that functions during oogenesis. Though major in amount the 5S and 40S rRNAs are but two of the 1-2 × 10^4 diverse RNA species that are being transcribed in the lampbrush chromosome stage nucleus. A lession that we have encountered several times in this chapter, as well as earlier, is that the detailed mode of regulation is particular to each gene, or functionally related set of genes. It is likely that we can anticipate the discovery of many new regulatory structures of diverse properties when equivalent studies are carried out on individual single copy transcription units of the lampbrush chromosomes.

NON-MEROISTIC OOGENESIS WITHOUT LAMPBRUSH CHROMOSOMES

In the oogenesis of many species neither nurse cells nor lampbrush chromosomes are utilized. Nurse cells and lampbrush chromosomes are best interpreted as alternative devices for the supply of a large variety of transcripts at elevated rates, as we have seen. The requirement for high synthesis rates is not an invariant feature of oogenesis, however. Such requirements do not exist in animals in which the final quantities of maternal RNAs are relatively low, and the allowed time for completion of oogenesis is sufficient. In such animals the structure of the oocyte transcriptional apparatus as well as the kinetics of RNA synthesis are very different from those we considered earlier. Yet in qualitative terms the end result, viz. the transcript pool of the mature oocyte has the same distinctive characteristics as in the species discussed earlier. Thus, where investigated, oocytes that develop without either nurse cells of lampbrush chromosomes also contain sufficient ribosomes and stored maternal mRNAs to support protein synthesis after fertilization, and at least in sea urchins, interspersed maternal poly(A) RNA as well. The characteristics of transcription in oocytes that do not utilize lampbrush chromosomes provide interesting comparisons, which illuminate those aspects of oocyte nuclear function that are intrinsic to the process of creating an egg, irrespective of any particular logistic strategy.

RNA Synthesis in Urechis and Sea Urchin Oocytes

The echiuroids represent an unsegmented lower protostome grade of organization, while the echinoderms constitute a major branch of deuterostome evolution. Yet the strategies utilized in these most unrelated creatures for the preparation of the maternal transcript pools are strikingly homologues. Although the evidence is still incomplete it appears that the oocytes of neither form possess lampbrush chromosomes. In the light microscope the diplotene meiotic prophase chromosomes of growing *Urechis* oocytes appear to assume a diffuse configuration that is difficult to resolve. However, it does not resemble a true lampbrush chromosome structure. Electron microscope observations on chromatin spread from Urechis oocytes have to been reported. Lampbrush chromosomes were thought to be present in sea urchin oocytes on the basis of light microscopy. However, despite determined attempts to visualize them, typical lampbrush chromosome transcription matrices could not be demonstrated in electron microscope preparations of *Strongylocentrotus purpuratus* oocyte chromatin. Except for densely packed ribosomal gene matrices, only sparsely distributed nascent transcripts were observed. Since the prevalence of maximally packed transcription matrices provides a definitive ultrastructural characteristic for true oocyte lampbrush chromosomes, it can be concluded that such structures are absent from the growing oocytes of at least this sea urchin species.

Both the *Urechis* egg and the sea urchin egg are relatively small, and they contain amounts of rRNA that are proportional to their respective volume. The amount of poly(A) RNA or of heterogeneous RNA stored in the *Urechis* eggs is not reported, though from measurements both of hybridization kinetics and of the content of poly(A) tracts it would appear to be less than 1% of the total RNA. However, Rosenthal and wilt (1986) found that most maternal mRNAs in fully-grown *Urechis* oocytes lack poly(A) tracts, though after fertilization they are rapidly adenylated. In the egg of the sea urchin *Strongylocentrotus purpuratus* there is about 30 pg of mRNA, about half of which is polyadenylated, plus approximately twice this amount of interspersed poly(A) RNA that is not translatable. The *average* prevalence of each RNA species of the heterogeneous sequences class is only $1\text{-}2 \times 10^3$ molecules per egg. The data suffice to illustrate the quantitative contrast between the *Strongylocentrotus* and *Urechis* eggs on the one hand, and on the other, eggs such as that of *Xenopus*. The volume of the *Xenopus* egg is about 3000 and 1000 times greater than the volumes of the *Strongylocentrotus* and *Urechis* eggs, respectively; the RNA content is about 1500 and 500 times greater, respectively; and the prevalence per egg of an average complex class maternal poly(A) RNA species in *Xenopus* is again about 1000 times the corresponding value for the sea urchin egg. Yet the complexities of the maternal RNA in *sea urchin*, *Urechis*, and the *Xenopus* eggs, as well as the length of time required to complete the growth phase of oogenesis, are all very similar. A difference of about three orders of magnitude therefore exists in the quantitative demands exerted on the transcriptional apparatus in *Xenopus* compared to sea urchin or *Urechis* oocytes.

In echiuroid worms and several other groups of marine organisms, oogenesis is of the solitary type, in which follicle and nurse cells are ab ent, and the oocytes develop autonomously in the coelomic cavity. In the case of labeling, accessibility of oocytes of diverse stage, and the possibility of removing and then reimplanting growing oocytes in the coelom, these forms offer some unique advantages for the study of gene activity during oogenesis, though as yet they have been relatively little exploited. For example, immature oocytes of the polychaete annelid *Platynereis dumerilii* can be transferred between the coeloms of animals of different genotype and mature eggs recovered that are able to undergo embryological development. Fischer (1977) utilized this method to demonstrate maternal inheritance of larval eye colour, which is evidently the consequence of the autonomous expression during oogenesis of genes affecting the pigmentation pathways. Measurement carried out by Lee and Whiteley (1984) on growing oocytes of another polychaete annelid. *Schizobranchia insignis*, show that the rate of RNA synthesis during oogenesis decreases sharply as the oocyte grows, and that the egg RNA appears to include high complexity components similar to those observed in other forms. In addition, a variety of molecular measurements that are directly relevant to the transcriptional processes carried out by relatively small solitary oocytes have been reported for *Urechis*.

Patterns of Transcription during Oogenesis in Urechis

The course of RNA synthesis in *Urechis* oocytes differs in several ways from that in amphibian and other oocytes that bear lampbrush chromosomes. As in the amphibians the ribosomal RNA genes of Urechis oocytes are amplified, but only to the moderate extent of about six times the number of rRNA genes included in the 4C chromosomal complement. The excess rDNA is located in a single very large nucleolus present in the germinal vesicle throughout oogenesis. In *Urechis* oocytes rRNA is the major transcript species synthesized at all stages of oogenesis. The rate of rRNA synthesis actually increases 4-5 fold in late vitellogenic oocytes, and rRNA continues to be produced, though at a lower rate, even in mature unfertilized eggs. There is no evidence for preferential accumulation of 4S and 5S RNA in previtellogenic oocytes, as in amphibian and teleost oogenesis. These low molecular weight species are instead synthesized continuously in Urechis oocytes at all stages. In Urechis about 8% of the growing ribosome pool is utilized for protein synthesis throughout the growth phase of oogenesis. Thus as the oocyte develops it engages an increasing mass of mRNA. Heterodisperse RNA that is detectable after 2 hr of labeling also continues to be synthesized at all stages of oogenesis, and poly(A) tracts, presumably associated with the heterogeneous RNA molecules, accumulate. Some specific examples have been reported by Rosenthal and Wilt (1986), in a study of the representation in total RNA of a series of transcripts identified by cDNA clones. Certain of these transcripts accumulate continuously throughout the period of oocyte growth, though the quantity of others, presumably those required only for translation during oogenesis, has decreased sharply by the final stage. At least for many of those species that remain polyadenylated throughout, this provides another contrast with *Xenopus*, where as we have seen the final net quantity of poly(A) RNA is attained very early in oogenesis. In quantitative terms, the rate of accumulation of poly(A) RNA in the cytoplasm of *Urechis* oocytes is probably less than 1% of the steady state flow into the cytoplasm of newly synthesized poly(A) RNA in midlampbrush chromosome stage *Xenopus* oocytes.

Transcription in Sea Urchin Oocytes

Most sea urchins have an annual reproductive cycle and from observations on oocyte size and number during this cycle it is possible to estimate the approximate lengths of the various phases of oogenesis. Oogonial multiplication occurs a new each year. Intertidal populations of *Strongylocentrotus purpuratus* spawn intermittently from the late November to May or June, and in these animals oogonial multiplication takes place primarily between April and November, when the latest oogonial DNA synthesis is observed in each cycle. The number of previtellogenic oocytes increases during the winter months to a maximum value that is attained in March, and remains constant until the following September, when the first group of oocytes enters the vitellogenic growth phase. Between March and September the newly formed oocytes grow slowly in volume, from about 10-15 μm diameter to about 30 μm. Their rate of growth then increases about 2-fold, and mature oocytes 80 μm in diameter are found about 2-3 months later. Oocytes in all stages of the growth process can be observed in the ovary at any one time during the winter months. It may be concluded that the previtellogenic growth phase requires at least six months (March to September), and the more rapid vitellogenic growth phase about 2-3 months (September to November-December). Oocytes that begin vitellogenesis later in the spawning period must spend and even longer time in previtellogenic stages, though the length of the previtellogenic growth phase may depend largely on environmental conditions such as nutrient supply. The previtellogenic oocytes are located in the walls of the ovarian tubules, where they exist in close contact with other cells, though no specific follicular structures are evident. As the oocytes increase in size they lose their associations with the epithelial wall, and the mature and late vitellogenic oocytes accumulate in the lumen of the ovarian tubules. Both meiotic reduction divisions are completed in this location, i.e., prior to release from the ovary.

RNA synthesis has not been studied in any detail in previtellogenic oocytes, and it is known only that they synthesize transcripts of all classes, including low molecular weight RNAs, 18 and 28S rRNAs, and heterogenous RNAs. Hough-Evans et al. (1979) showed that the nuclear RNA of previtellogenic oocytes has complexity of at least 1.6×10^8 nt i.e., close to 30% of the single copy genome. Note that this is about the same as the complexity of embryo and adult sea urchin nuclear RNAs. Furthermore, the particular single copy sequence set represented in the previtellogenic oocyte nuclear RNA is within the limits of experimental detection wholly represented in gastrula stage embryo nuclear RNAs. The significance of this experiment is shows that *most if not all of the transcription units operative in midembryogenesis have already been activated at the previtellogenic stage of oogenesis.*

In the sea urchin stable heterogenous maternal transcripts accumulate *sequentially* during oogenesis. The cytoplasmic fraction of previtellogenic *S. purpuratus* oocytes thus contains only about 45% of the ultimate maternal RNA sequence set. The prevalence of typical maternal sequences is about $1\text{-}2 \times 10^3$ molecules per egg, and even for sequences present at ten times this concentration less than an hour would be required for their transcription from single copy genes and accumulation, were these genes being initiated at the maximum possible rate. This calculation assumes the polymerase translocation rate measured for sea urchin embryo nuclei, about 9 nt sec^{-1}. However as discussed earlier, few of the transcription units in the oocyte seem to be heavily loaded with nascent transcripts. The actual time course of cytoplasmic heterogeneous RNA accumulation during sea urchin oogenesis has yet to be measured.

Diverse RNA species are also synthesized in vitellogenic oocytes, including both heterogeneous and ribosomal RNAs. Ruderman and Schmidt (1981) showed that at this stage the majority of precursor incorporated per oocyte in a 2 hr labeling periods is located in mitochondrial rather than nuclear transcripts, just in unfertilized eggs and early embryos. Messenger RNAs for the histones were identified among the newly synthesized transcription products of the germinal vesicle, and a large fraction of the newly synthesized histone mRNA is rapidly assembled into polysomes and translated. There are about 10^6 copies of each core nucleosomal histone mRNA of the early or α-subtypes per egg. However, Angerer et al. (1984) showed by *in situ* hybridization with early histone gene probes that the α-histone mRNAs do not accumulate in vitellogenic oocytes, but appear only after maturation, when they are synthesized and stored within the egg pronucleus. Were all 400 histone genes coding for each early histone species in the haploid pronucleus as active as are these genes in cleavage stage embryos the accumulation of $\sim 10^6$ mRNA molecules of each histone species would require less than two days. The histone mRNAs synthesized in vitellogenic oocytes are thus likely to be of the cleavage stage (CS) type. CS histone are stored in the unfertilized egg in sufficient quantity to accommodate the DNA of a number of embryonic division cycles. In contrast to the α-histone messages, the total poly(A) RNA, and several other specific mRNAs, are distributed throughout the cytoplasm of the mature egg, and total poly(A) RNA is accumulated both in terminal vesicle and the cytoplasm during the vitellogenic phase of oogenesis.

It is likely, by analogy with the case in *Xenopus*, that the α-histone mRNAs are sequestered in the pronuclear compartment by some specific protein to which they are bound. Messenger RNA for actins and tubulins are also prevalent in growing oocytes, and these proteins are actively synthesized during oogenesis. However, at maturation these proteins cease to be translated at a high rate, only a small amount of maternal mRNA coding for these proteins is stored in the unfertilized egg. A different picture emerges in considering rRNA accumulation during sea urchin oogenesis. Measurement of absolute rRNA synthesis rate were carried out in *Tripneustes gratila* oocytes by Griffith et al. (1981). The average rate of synthesis per oocytes is about 1.1×10^5 molecules of rRNA precursor per hour. The mature egg contains about 4×10^8 ribosomes, and thus 4 to 5 months would be required for accumulation. Since all stages of oocyte were included in these experiments, it is possible that previtellogenic oocytes synthesize rRNA less actively, and that an even longer time is necessary, e.g., the whole of the oocyte growth period. In echinoids rDNA amplification does not occur. The 4C *T. gratila* oocyte genome contains about 200 rRNA genes and in order to account for the measured rate of synthesis initiation must occur on each rRNA gene about every 6.5 sec. These rRNA synthesis measurements lead to the conclusion that essentially all of the rRNA genes are operating at close to the maximum possible rate throughout oogenesis. Thus if the polymerases were packed only 100 nt apart, a translocation rate of about 15 nt sec^{-1} would be required.

Aronson and Chem (1977) determined rates of 13 nt sec-1 for both polymerase I and II in feeding *S. purpuratus* larvae, and 6-9 nt sec^{-1} for embryos of this species, which are cultured at a temperature several degrees lower than are *T. gratilla* embryos. It follows that for sea urchins as for other diverse creatures considered in this chapter, e.g., *Drosophila* and *Xenopus*, the rate of accumulation of rRNA may be among the parameters that determine the overall pace of oogenesis. Seventy per cent of the poly(A) RNA stored in sea urchin eggs consists of non-translatable, intersperesed transcripts, just as in amphibian eggs.

The structural characteristics of the interspersed RNAs that can be recovered from mature sea urchin eggs, and their disposition in the embryo, are reviewed in the discussion of maternal RNAs. From their complexity it is clear that the interspersed RNAs are the stored products of a great many discrete transcription units operative during oogenesis. Posakony et al. (1983) estimated that the prevalences of a set of individual interspersed poly(A) RNAs fall in the range 10^3 to 10^4 molecules per egg, and accurate molecular titration data for two cloned examples indicate a few hundred to about a thousand copies per egg. Thus, the abundance of these sequences is about 2-3 orders of magnitude lower than for the interspersed poly(A) RNAs of *Xenopus* eggs, just as are the respective abundances in these two systems, of histone mRNAs, rRNAs total poly(A) RNA, or average complex class transcripts. Yet both the structure and the amount of interspersed transcripts relative to total egg poly(A) RNA is the same in the eggs of these two species. This surprising correspondence clearly does not favour the specific explanation that interspersed poly(A) RNA are incidentally released to the cytoplasm because processing in the germinal vesicle cannot keep pace with the exceptionally high rate of transcript production, e.g., as in lampbrush chromosomes. That is, the rate of heterogeneous RNA transcription in sea urchin oocytes is probably quite low. Perhaps the flow of interspersed poly(A) RNAs into the cytoplasm is a tangential consequence of a basic inefficiency of processing in the oocyte nucleus that is yet unexplained, though very general in occurrence. Or perhaps their storage in the cytoplasm in the course of two such different forms of oogenesis implies that like other maternal transcripts species, these also perform a functional role after fertilization.

Transcription and the Accumulation of Maternal RNAs during Oogenesis in the Mouse

Overall Rate of Heterogeneous RNA synthesis and the absence of Lampbrush chromosomes

Oogonial multiplication, premeiotic DNA synthesis, and the initial stages of the meiotic prophase occur in eutherian mammals during fetal life. In the neonatal mouse the primary oocytes are arrested at early diplotene in a non-growing state, termed the dicyate phase. Beginning a few days after birth and throughout the reproductive life of the female small groups of oocytes periodically withdraw from the dictyate pool, and enter the growth phase of oogenesis. Dictyate mouse oocytes may thus persist in a quiescent though viable state for anywhere from a few weeks to 18 months or more, and in the human where a similar course of events, obtains, for over 40 years. Once activated the process of oocyte growth in the mouse requires only two weeks, during which the diameter of the oocyte increases from about 12 μm to 85 μm, a greater than 300 fold change in volume. The maternal RNAs found in the mature egg are synthesized and accumulated during this period, and many cytological and biochemical changes that lie outside the scope of this discussion occur as well. Within several days after growth is completed the follicular structure surrounding the oocyte undergoes an enormous expansion, and the meiotic prophase terminates at ovulation with the completion of the first reduction division.

Midgrowth phase oocyte denuded of follicle cells can be cultured in the presence of a feeder layer of somatic follicular cells or in excised follicles, under conditions in which growth continues at about 70% of the normal rate. The synthesis and turnover rates of RNAs synthesized in cultured oocytes that had been labeled for various periods of time were measured by Bachvarova (1981). As in *Xenopus* lampbrush chromosome oocytes, by far the major fraction of the newly synthesized heterogeneous RNA is unstable, displaying a typical nuclear RNA half-life estimated at about 20 min. Less than 2% of the heterogeneous RNA synthesized is destined to become stable transcription products (0.0075 pg min^{-1} ÷ 0.44 pg min^{-1}). This value is similar to those estimated from the similar comparisons for amphibian oocytes. The low ratio of stable RNA transcript production to total heterogeneous RNA transcription is thus clearly not dependent on the overall rate of chromosomal RNA synthesis, but is rather an intrinsic aspect of transcription and processing in the oocyte nucleus. The major result is for our purposes the low absolute rate of heterogeneous RNA synthesis. The comparable rate was estimated 18.3 pg min^{-1} in the lampbrush chromosome stage *Xenopus* oocyte, which is over forty times the rate measured for the mouse oocyte. Yet these organisms have about the same haploid genome size, approximately 3 pg.

A light and electron microscope investigation of the cytological state of mouse and Rhesus monkey oocyte chromosomes during the growth phase was carried out by Bachvarova et. al. (1982). At light microscope magnification the oocyte chromosomes appear fuzzy and diffuse, with central proteinaceous cores surrounded by loose bundles of fibrils. In basic organization these structures bear no homology with the lampbrush chromosomes of amphibian oocytes. The key observation is that when examined in the electron microscope the transcription units of growing mouse oocyte nuclei are generally populated only sparsely with nascent RNA molecules. A discussed earlier, this provides direct evidence that true lampbrush chromosomes are absent where the rate of heterogeneous RNA transcription is relatively low. Lampbrush chromosomes are to be considered the cytological manifestation of a condition in which the maximal transcriptional initiation rate is established in essentially all active regions of the oocyte genome.

The complexity of the stored maternal RNA has not been measured for any mammalian egg. If it is assumed to be approximately that of sea urchin and *Xenopus* maternal poly(A) RNAs, and thus that something over 10^4 diverse transcription units are represented the rate of synthesis is of 0.005 pg min^{-1} per nucleus for stable poly(A) RNA (or of 0.0075 pg min^{-1} per nucleus for total stable heterogenous RNA) implies that on the average no more than one or a few nascent transcripts would be found on the active genes of the growing oocyte. Thus 0.005 pg min^{-1} would represent a production of about 5000 molecules min^{-1} per nucleus, assuming a mean length of 2000 nt for the poly(A) RNA distributed among at least 40,000 transcription units in the 4C oocyte nucleus. It follows that initiation would occur on each productive transcription unit, on the average, only once every eight minutes, and thus for a 37°C polymerase translocation rate on the order of 20-50 nt sec^{-1}, the nascent transcripts would be spaced 1-2.5 × 10^4 nt apart. This result predicts exactly the low transcript frequency though of course were processing very inefficient, or if there are transcription units that produce RNAs wholly confined to the nucleus, the transcript density in some regions could be markedly higher.

Accumulation of Stable Maternal RNAs

The general pattern of RNA synthesis appears not to change during the growth phase of oogenesis. A number of investigations have shown that at all stages mouse oocytes synthesize 18 and 28S rRNAs, 5S rRNA, tRNAs and poly(A) RNAs, and that throughout, the ratio of these various components in the newly synthesized RNA remains constant. An early growth stage characterized primarily by synthesis of tRNA and 5S RNA thus does not exist in the mouse which in this respect resembles the sea urchin and Urechis examples, and differs from those vertebrates that utilize lampbrush chromosomes. By day 14-15 the oocyte has accumulated its final maternal complement of rRNA, about 300 pg, as well as of stable poly(A) RNA, though its volume is only 50-60% of that a mature 21 day oocyte. Though it continues to grow and to accumulate protein for the following 7 days, 14-15 day oocytes are already functionally mature, in the sense that they have attained competence to resume the meiotic division process. RNA synthesis, but not accumulation, also continues in late oocytes, and become undetectable only at germinal vesicle breakdown and meiotic maturation.

For the first 9 days of the growth phase of rRNA synthesis is about 2/3 of that or about 0.01 pg min^{-1}, while the rate of 0.015 pg min^{-1} obtains between days 9 and 14-15. Ribosomal gene amplification has not been reported in the mouse, though a four-fold rDNA amplification has been noted both in diplotene baboon oocytes and in human oocytes. However, ribosomal gene amplification is not required to account for the measured rate of rRNA accumulation in mouse oocytes. Thus Kaplan et al. (1982) calculated that the rate of 0.015 pg min^{-1} represents the synthesis of about one rRNA molecule per 15 sec for each of the 1200 genomic rRNA genes present in the 4C oocyte nucleus. This rate is below that measured in somatic mouse tissue culture cells.

The mouse egg is unusual in the large quantity of maternal poly(A) RNA that it contains, relative to total RNA. Bachvarova and De Leon (1980) labeled oocytes at all stages of the growth period by injection of radioactive precursor into the bursal sac, and on recovery of matured eggs from 4 to 19 days later found that about 8.3% of the mass of labeled RNA retained is poly(A) RNA. The length of the interval between injection of precursor and collection of ova did not affect this result. Calculations based on the mass of poly(A) per egg indicate the quantity of stored poly(A) RNA is about 25 pg, or 5.5% of the total RNA. This represents about 2.4×10^7 poly(A) RNA molecules of average length. Since the bulk of the RNA accumulating during growth is rRNA, which is completely stable, the close agreement between the calculated mass of poly(A) RNA and the fraction of total egg RNA labeled during oogenesis that is poly(A) RNA shows that the poly(A) RNA synthesized in the oocytes is largely stable as well. This has also been demonstrated directly in pulse-chase experiments carried out on cultured follicles by Brower et. al. (1981). As it accumulates the poly(A) RNA is deposited in the cytoplasm of the growing oocytes, according to *in situ* hybridizations carried out with ^{3}H-poly(U).

About half the poly(A) RNA accumulated in the course of oogenesis disappears during the maturation period, though it is not clear whether in general it is degraded or in only deadenylated. The oocyte actin messages, however, have been shown to be deadenylated rather than degraded during maturation though total stable. RNA content does decline during this period by about 20%. In any case, the fraction of *newly synthesized* oocyte RNA that is poly(A) RNA is about twice the fraction of mature, unfertilized egg RNA that is poly(A) RNA. Thus about 18-20% of precursor incorporated in the relatively stable transcripts of *growing oocytes* is found in poly(A) RNA. The newly synthesized poly(A) RNA of the oocyte distributes into two functionally distinct compartments. About 23% is assembled onto polysomes and utilized directly for oocyte protein synthesis, and the half-life of this fraction is about 6 days, according to measurements carried out on cultured oocytes in contact with follicle cells or in whole follicles. The remainder is stored, and is displays no detectable turnover. The poll of poly(A) RNA present in the oocyte at termination of growth is thus the sum of the steady state quantity of polysomal poly(A) RNA, and the quantity of non-translated poly(A) RNA that has been synthesized and accumulated during oogenesis. Given that the rate of synthesis of total poly(A) RNA is 0.005 pg min-1, the steady state amount of polysomal poly(A) RNA would be about 14 pg, and the final quantity of stored poly(A) RNA about 78 pg, for a total of 92 pg, a fifth of all the RNA in the oocyte. This value may be compared with the equivalent fraction in the mature *Xenopus* oocyte, which is about 2.2% (90 ng poly(A) RNA - 4000 ng Total RNA).

Protein Synthesis during Oogenesis in the Mouse

The absolute rate of protein synthesis increases proportionately with oocyte diameter from shortly after the beginning of the growth phase to its termination. Protein synthesis rate measurements were obtained on cultured oocytes, and are based on determination of the methionine pool specific activity. The maximum rate of synthesis is about 42 pg hr^{-1} in full grown oocytes of about 80 μm diameter, which is close to forty times the rate observed in 10-15 μm oocytes. Of the total newly synthesized protein only 1-2% represents proteins encoded by the 10^5 mitochondrial genomes present per egg. De Leon et al. (1983) showed that throughout the period of oocytes growth about 35% of the ribosomes are engaged in polysomal structures. The remainder may be aggregated in large assemblages utilized for storage of ribosomes in an inactive form, as they can be sedimented from homogenates at very low centrifugal forces.

Assuming a ribosomal translocation rate of a 3 codons sec^{-1} per ribosome (balues for 37°C) vertebrate systems range from about 2-8 codons sec^{-1} per ribosome; the rate of protein synthesis measured in fully grown oocytes is close to that expected, providing that 35% of the ribosomes are included in the translational apparatus (i.e., 2.6 × 10^7 robosomes). However, this rate of translation is 4-8 fold *lower* than expected for the steady state quantity of *polysomal* poly(A) RNA i.e., 14 pg. The major cause would seem to be a low rate of translational initiation, since the ratio of the mass of mRNA to the mass of polysomal rRNA is about four times greater than the 4% value obtained for fully-loaded polysomes (i.e. 14 pg mRNA ÷ 0.35 × 300 pg rRNA = 0.13). The rate of ribosomal translocation may also be depressed about 2-fold. A caveat pointed out by Kaplan et al., (1982) is that the protein synthesis rate measurements of schultz et al. (1976b) were carried out under culture conditions inadequate to support net growth, and this could conceivably have adversely affected the translational activity of the oocytes.

In any case, assuming the rates measured, it can be calculated that if all the protein synthesized during oogenesis were stably accumulated, this would account for only about 60% of the 25% ng of protein in the fully grown oocyte. This is in fact an overestimate, since only about 60% of the newly synthesized protein is stable. The oocyte this probably absorbs the balance of its protein from the blood, by endocytosis, just as yolk is taken up in lower vertebrates. Qualitative analysis by two-dimensional gel electrophoresis has indicated that most of the ~400 species resolved are synthesized throughout oogenesis, and that the newly synthesized species are the same as the species detected by sensitive staining procedures. However, there are a few specific proteins whose rate of synthesis increases sharply in fully grown oocytes.

At meiotic maturation, the overall pattern of protein synthesis changes noticeably, and these changes are programmed at a post-transcriptional level since there is little new mRNA synthesis at this time. The stored maternal mRNA accumulated during oogenesis does not appear to be utilized until maturation, when a qualitative pattern of protein synthesis is established that persists until well after fertilization. Synthesis of a number of known protein species has been studies in mouse oocytes. Though the estimates for the number of functioning mRNA molecules are of limited accuracy they suffice to indicate the minute accumulations of mRNA in the mouse oocyte compared, for example to the specific message contents in Xenopus oocytes. These estimates do not include stored maternal mRNAs coding for the same protein species. However, in all the cases shown, the rate of synthesis falls rather than increases during maturation, when the maternal mRNA begins to be utilized. The histones and ribosomal a proteins, which were also considered in respect to protein synthesis in *Xenopus* oocytes, again provide interesting contrasts. In the mouse, the 60 pg of core histones accumulated during oogenesis would suffice for the chromatin of only about 10 cells. As will be recalled, mature *xenopus* oocytes contain histone protein sufficient for over 10^4 cells. The leisurely pace of cell division in the mouse imposes no urgency on new histone synthesis, and whereas in *Xenopus* there is a 50-fold increase in the rate of histone synthesis during maturation, in the mouse the rate actually declines about 40% during maturation. Thus it is unlikely that there is a very large store of maternal histone mRNA, such as is found, e.g., in both sea urchin and *Xenopus* eggs.

An interesting aspect of ribosomal proteins is that many of these are encoded by mRNAs that are distinctly of the rare message class. For example, LaMarca and Sassarman (1979) showed that synthesis of several individual ribosomal proteins represents less than 10^4 of total protein synthesis, and on the basis of the calculation there would be no more than about a thousand molecules of the messages for such individual species functioning in the oocyte. The quantity of each functional ribosomal protein mRNA species in the mouse oocyte in thus equivalent of ≤ 10 molecules per typical somatic cell (i.e., normalizing 1.3×10^7 polysomal poly(A) RNA molecules in the full-grown mouse oocyte to the amount of poly(A) RNA in a somatic cell. $\sim 10^5$ molecules). The molar rates of synthesis of the individual ribosomal proteins differ by as much as a factor of four. However, this imbalance is compensated by a differential accumulation of newly synthesized molecules of the various species in the germinal vesicle, so that within this compartment an equimolar ratio is established.

The total amount of newly synthesized protein that is distributed to the germinal vesicle is about 0.9 pg hr^{-1} in the full-grown mouse oocyte, and of this histone accounts for about 10% and the ribosomal proteins about 25%. After maturation ribosomal proteins continue to be produced in the absence of any new rRNA synthesis. It is not clear whether the maternal messages utilized for this synthesis are the same as were already functioning in the polysomes prior to maturation, or alternatively, are recruited from the previously inactive stored poly(A) RNA pool. Three secreted proteins, the major constituents of the zona pellucida, account for at least 10% of the newly synthesized proteins in growing oocytes. The zona pellucida is the thick glycoprotein coat that surrounds the oocyte, and its constituents are synthesized exclusively during the growth phase of oogenesis. Denuded oocytes actively incorporate amino acids or labeled fucose into these proteins.

An O-linked oligosaccharide component of these glycoprotein species, designated ZP3 (~83 kd) apparently serves as the specific sperm receptor molecule. A second protein ZP2 (~120 kd), which constitutes over 50% of the total zona pellucida protein, is involved in the hardening of the egg coat on fertilization. This process is mediated by proteolytic enzymes released from cortical granules, just as in the sea urchin. The message for these proteins must be relatively very prevalent, and the genes for ZP1-3 are likely to be transcribed much more intensely than the genes coding for typical maternal poly(A) RNAs required after fertilization. A complementary example has been described by Cascio and Wassarman (1982). This is a set of proteins designated the fertilization proteins. FPI^{-14} are synthesized at very low levels, each accounting for about 0.055% of total protein synthesis in growing oocytes, and FP5 and FP6 are not detectable at all. However, after fertilization. FP1-6 become major species, together representing 3-5% of the total protein synthesis in one and two cell embryos. The dramatic change in their rate of synthesis which occurs as a result of maturation and fertilization. The fertilization proteins are among those coded primarily by the stored, *inactive* poly(A) RNA of the oocyte. This is demonstrated in the cell-free translation experiment. The partition between polysomal and stored poly(A) RNA in the oocyte thus encompasses some sharp *qualitative* delineations, illustrated respectively by the zona pellucida proteins and the fertilization proteins.

6

GENOME AND DIFFERENTIATION

The process of cellular differentiation in developing multicellular organisms culminates in the formation of specialized somatic cell types. Once cell types acquire a specific cellular phenotype for performing a specialized function, no further changes usually occur. We call this the stability of the differentiated state. It is now widely accepted that the process and stability of cell specialization are under genetic control. But the frequency and extent to which this genetic control involves irreversible genetic changes among eukaryotic animals remain unanswered. Irreversible genetic changes involving DNA losses or whole chromosomes, so far as we know, occur as a regular event in only a few animal species. Examples of such losses are present in some protozoans, nematodes, crustaceans, insects, and mites of the lower animal phyla. Among vertebrates, DNA losses occur in the holocephalan fish, the mammalian Marsupialia, as well as in the genetic rearrangement of the immunoglobulin genes in chicken (Weill and Reynaud, 1987) and mammalian B-lymphocytes and of certain receptor genes in mammalian T-lymphocytes.

Most of the evidence available today indicates that a large portion of the genome is retained in specialized somatic cells. For example, the activation of dormant genes in specialized cells is a relatively common phenomenon, consistently demonstrated in several experimental systems (DiBerardino *et. al.*, 1984), including cell cultures, transdifferentiation, heterokaryons, cell hybrids (H. Harris, 1985), cancer and nuclear transplantation into oocytes and eggs. In these cases, the activation of dormant genes is accompanied by changes in the cell phenotype of specialized cells. Therefore, the evidence indicates that many silent genes are maintained in the genome in the absence of expression and that under favourable conditions they are stimulated to function.

Nevertheless, a fundamental problem concerning the developmental process in multicellular organisms still remains unsolved, i.e., whether most differentiated somatic cell types retain the genomic totipotency of the zygote nucleus. The genomic potential of differentiated cells can best be evaluated by the transplantation of nuclei oocytes and eggs, because this procedure has the potential to test the entire genome. The idea that the gene content of a cell might be tested by injecting a living nucleus into an enucleated egg was suggested by Hans Spemann (1938). The rationale was that the type of development that results is a reflection of the gene content of the introduced nucleus. This idea materialized when Briggs and King (1952, 1960) succeeded in obtaining metamorphosed frogs following transplantation of blastula and early gastrula nuclei into enucleated eggs of *Rana pipiens*. Later, McKinnell (1962), Gurdon (1962), and subsequently others obtained fertile adult frogs from embryonic nuclei, demonstrating, for the first time, the genomic totipotency of embryonic nuclei. These studies were extended to the insect *Drosophila* and the teleost fish; their young embryonic nuclei were also shown to be totipotent.

Recent studies in mice indicate that totipotency is restricted to early cleavage stage nuclei. Tsunoda *et. al*. (1987) obtained several fertile mice from nuclei of the 8-cell stage, but no development ensued from nuclei of the inner cell mass of blastocysts; however, Illmensee and Hoppe (1981) reported fertile mice from these latter-stage nuclei. Among the larger mammals, three lambs and two calves have resulted from 8-cell and 9- to 15 cell-stage nuclei, respectively. So far, the success of nuclear transfers in mammals is limited to stages that are much earlier than that in amphibians however, mammalian nuclear transfer studies are still in their infancy, and we await future developments in this area. This chapter deals mainly with nuclear transfers from terminally differentiated somatic cells, i.e., cells in which the genome is imprinted to specify a specific and final cellular phenotype.

So far, successful nuclear transfer studies from terminally differentiated somatic cells have been performed in frogs. The results of these studies summarized below demonstrate that the genomes of several differentiated somatic cell types can (1) undergo widespread activation, (2) direct the formation of a multiplicity of cell types, and (3) display genetic multipotentiality, but (4) evidence for genomic totipotency of differentiated somatic cells in still lacking.

Nuclear Transfers from Differentiated Somatic Tissues

The critical experiment posed by Spemann (1938) concerned the genetic repertoire of differentiated somatic cells. The term-differentiated cell has undergone multiple uses. For example, the unfertilized egg is considered by some to be a highly differentiated cell (gamete), yet at fertilization it becomes totipotent. The critical test of nuclear equivalence can best be assayed on genomes that have been imprinted to specify a specific and final cellular phenotype, i.e., genomes of cells that normally have no other potential in the organism. Here we consider such cell types as definitive tests of the genomic potential of differentiated cells. Nuclear transfer experiments that assess the genetic potential of differentiated cells must include evidence for two conditions: (1) the donor cells are differentiated, and (2) the nuclear transplants are derived from the donor nucleus, and not from the egg nucleus. We shall consider two series of experiments that differ in the certainty that the donor cells were derived from differentiated cells.

The first series comprises the initial attempts to work with differentiated cells at a time when the techniques usually precluded distinction between stem cells and differentiated cells in the tissues employed. In the second series the specialized properties of the donor cells were documented. In both series only those nuclear transplants that developed to postneurula stages, larval (tadpole) stages, and adults will be considered. Postneurula embryos have developed primitive organ systems, which in the latter stages display muscle and heart function.

During the early larval stages, the major cell types, tissue, and organ systems have differentiated and are functional. Thus, transplanted nuclei capable of programming eggs to develop to these stages are considered genetically multipotent. Nuclei that direct the formation of fertile frogs are interpreted to be totipotent. Verification that the nuclear transplants developed from the transplanted test nucleus and not form the egg nucleus that remains due to faulty enucleation have been performed in the following ways. In *Xenopus*, the egg nucleus is inactivated by ultraviolet (UV) irradiation but, since this procedure is successful in only 61-92% of cases, a genetic marker is required. The most used marker in *Xenopus* has been the 1-nu (nucleolar) mutation that can be seen in cells from the gastrula stage and beyond. However, as pointed out by DuPasquier and Wabl (1977), it is necessary to verify that one nucleolus is accompanied by diploid set of chromosomes and does not result from haploidy or aneuploidy. In *Rana pipiens*, triploidy has served as a convenient cytogenetic marker because the triploid set of chromosomes can be identified throughout the life cycle of the frog.

Another verification procedure has also been used in *Rana pipiens* because microsurgical removal of the egg nucleus can be accomplished in 98-100% of cases. The exovate formed at the time of enucleation adheres to the vitelline membrane outside of the egg. This membrane is later removed, sectioned, and stained with the Feulgen procedure that specifically stains DNA and exhibits the presence of the egg nucleus in the exovate. This information together with the determination of chromosome number of the nuclear transplant and the time of the first cleavage insures that development ensued from the test nucleus.

Differentiated Somatic Tissues

The advanced development of nuclear transplants obtained from differentiated somatic tissues of *Xenopus* larvae and adults is summarized in Table 6.1. Five adult frogs were obtained from five original larval nuclei, four from nuclei of the larval intestine, and one from a nucleus of larval epidermis. Three of these adult frogs were fertile- two from intestinal nuclei, and one from a non-ciliated cell of the epidermis. The donor cells from the intestine were selected on the basis of large size. The epidermal cell from the experiments of Kobel *et al.* (1973) was identified as non-motile (non-ciliated). Nuclei from epidermal cells that were motile (ciliated), and therefore specialized, did not give advanced development.

The authors concluded that the single case showing totipotency was derived from a non-specialized cell. In addition to the studies tabulated in Table 6.1, endodermal nuclei of the primitive gut taken from postneurula embryos and initial larval stages before tissue differentiation directed *Xenopus* eggs to develop into 20 fertile frogs (~3%), most of which bore the 1-nu genetic marker of the donor nucleus. At least four metamorphosed frogs developed

Table 6.1 A. Nuclear Transfers from Differentiated Somatic Tissues[a]

Donor			Total nuclei	Total transfers[b] at stage									
				Postneurula		Larva		Metamorphosis		Adult			
Stage	Tissue	Identification of donor cells	tested	N	%	N	%	N	%	N	%	Verification	Reference
Larva	Intestine[x]	Size	726	36	5.0	31	4.3	-	-	4[2F]	0.6	1-nu	Gurdon and Uehlinger (1966)
	Intestine[x]	Size	522	8	2.0	8	2.0	1	0.2	-	-	98% haploids[c]	Marshal and Dixon (1977)
	Epidermis[x]	Nonciliated	440	2	0.4	2	0.4	-	-	1	0.2	1-nu	Kobel et. al (1973)
	Minced Tadpoles[x]	Cell culture	3686	23	0.6	9	0.2	3	0.1	-	-	1-nu[d]	Gurdon and Laskey (1970)
Adult	Liver[x]	Cell line	365	2	0.6	2	0.6	-	-	-	-	-	Kobel et. al. (1973)
	Skin, lung kidney[x]	Cell culture	2322	26	1.1	7	0.3	-	-	-	-	1-nu[d]	Laskey and Gurdon (1970)
	Intestine[x]	Size	1112	10	0.9	6	0.5	-	-	-	-	95% haploids[c]	McAvoy et. al. (1975)

a Superscript letters: X, Xenopus; F, fertile.
b Includes the result of serial transfers.
c Denotes efficieny of UV irradiation in controls.
d Chromosome number determined in nuclear transplants. 1-nu, one nucleolus.

Table 6.1 B. Nuclear Transfers from Differentiated Somatic Cells[a]

Donar				Total transfers[b] at stage							
				Tadpole							
		Differentiated state	Total nuclei	Postneurula		Prefeeding		Feeding			
Stage	Tissue/cell	of donor cells	tested	N	%	N	%	N	%	Verification	Reference
Larva	Melanophores cell	Pigmented	257	2	0.8	-	-	-	-	1-nu	Kobel et. al (1973)
Juvenile frog	Erythrocytes[R]	Shape, hemoglobin	51	4	8.0	4	8.0	3	5.9	Triploid[d]	DiBerardino et. al. (1986)
Adult	Erythrocytes[R]	Shape, hemoglobin	130	11	8.5	6	4.6	-	-	Exovate[d]	DiBerardino and Hoffner (1983)
	Erythroblasts[X]	Shape, hemoglobin	442	8	2.0	8	2.0	-	-	1-nu[d]	Brun (1978)
	Skin, cell culture[X]	Keratin antibody	129	6	4.6	4	3.1	-	-	1-nu[d]	Gurdon et. al. (1975)
	Spleen[X]	Immunogen antibody	100	6	6.0	6	6.0	-	-	1-nu[d]	Wabl et. al. (1975)

a Superscript letter: X, Xenopus laevis; R. Rana pipiens

b Includes the results of serial transfers.

c Stages of hindlimb bud.

d Chromosome number determined in the nuclear transplants.

from eggs injected with larval nuclei—one from an intestinal cell selected on the basis of large size and three from cell cultures. The latter three were obtained from cell cultures, initiated by mincing whole tadpoles that were in the stage of beginning circulation in the gills, but it is known if all the donor cells were differentiated. The above tissues (intestine, epidermis, and cell cultures of tadpoles) also contained nuclei capable of directing eggs to develop into 69 postneurula embryos, and 50 of these proceeded into larval stages.

Nuclear transfers of adult somatic nuclei into *Xenopus* eggs resulted in 38 postneurula embryos, and 15 of these developed into larvae. The somatic nuclei were derived from a cell line established from liver from cell cultures of skin, lung, and kidney, and from *in vivo* intestinal cells selected on the basis of size. Several conclusions can be made from the above studies. First, cells from both larval and adult tissues contain some nuclei that can direct the formation of the diverse cell types found in postneurula embryos and tadpoles. Therefore, these nuclei display genetic multipotentiality. Second, genetic totipotency has been shown for a few nuclei from advanced embryonic and young larval stages, since these nuclei directed the formation of fertile frogs. Third, the interpretation of these results remains equivocal, because with the exception of the non-ciliated epidermal cells, no valid criteria were used to distinguish between stem cells and differentiated cells in the tissues employed.

Differentiated Somatic Cells

In this section we consider the advanced development obtained from nuclei of donor cells that were unequivocally differentiated. Among the cell types listed in Table 6.2, melanophores, erythrocytes, and erythroblasts were directly identified at the time of nuclear transfer. For tests of melanophore nuclei, short-term monolayer cultures from prefeeding *Xenopus* larvae were used, and the melanophore donor cells were identified on the basis of pigment inclusions. Their nuclei promoted enucleated eggs to develop into postneurula embryos 5 (0.8%) that contained the 1-nu genetic marker of the donor cells.

Nucleated erythroid cells of amphibians are particularly ideal for nuclear transfer experiments: (1) they are unequivocally differentiated; (2) they can be conveniently isolated from the peripheral blood by low speed centrifugation; (3) the cells exist

Table 6.2 Nuclear Transfers from Germ Cells[a]

Donor				Total transfers at stage								
			Total nuclei	Postneurula		Feeding larva		Metamorphosis		Adult		
Stage	Gonad	Cell	tested	N	%	N	%	N	%	N	%	Reference
Larva	Genital ridge	Primordial germ cells[R]	410	35	8.5	31	7.6[b]	-	-	-	-	Smith (1965)
Larva to metamor phosis	Genital ridge, un-differentiated and differentiated gonads	Primordial germ cells, gonocytes[P]	2986	46	1.5	12	0.4	9	0.3	6	0.2[2F]	Lesimple et. al. (1987)
Adult	Testis	Enriched for sper-matogonia[R]	116	4	3.5	1	0.9	-	-	-		DiBerardino and Hoffner (1971)

a Superscript letters: R, Rana pipiens; P. Pleurodeles waltl; 2F, two adult were fertile.

b Ten larvae raised, final stage not indicated, one in the climax of metamorphosis.

singly in the plasma, and therefore, no chemicals are required to achieve cell dissociation as is the case when using donor cells from solid tissue; and, most importantly, (4) individual erythroid cells can be directly identified with the stereomicroscope at the time of nuclear transfer because of their oval shape and presence of hemoglobin. Erythrocyte nuclei from juvenile *Rana* frogs promoted the highest yield of tadpoles and expressed the greatest genomic potential among the differentiated somatic cell types tested: 8% of the original nuclear population promoted the formation of prefeeding tadpoles and 5.9% directed the development of feeding tadpoles that advanced to hindlimb bud stages, and in the best cases survived up to a month. All the tadpoles bore the triploid marker of the donor nucleus.

Nuclear transfer studies performed on adult erythrocyte nuclei from *Rana* resulted in 4.6% prefeeding tadpoles. The origin of the latter tadpoles was verified by recovery of the egg nucleus in the exovate and determination of chromosome number. Another study of adult erythroid nuclei was performed in *Xenopus* (Brun, 1978). In the latter study, erythroblast nuclei (2.0%) promoted development as far as the first tadpole stage at which hatching is initiated, and all these nuclear transplants bore the 1-nu genetic marker. Two other studies examined the genomic potential of adult skin and spleen cells. The donor cells were derived from a nearly homogeneous population. In the case of the skin cells, correlative studies showed that over 99% of the skin cells from short-term culture were producing immunoreactive keratin.

Nuclear transfers from similarly cultured skin cells resulted in 3.1% prefeeding tadpoles that carried the 1-nu genetic marker. In the case of the spleen cells, their nuclei directed the development of prefeeding tadpoles (6.0%) that bore the 1-nu genetic marker of the donor cells of which 96.1-98.7% were judged to be producing immunoglobins (Ig). In summary, nuclei from five documented differentiated somatic cell types promoted the formation of 37 postneurula embryos. With the exception of cultured melanophores, nuclei from the other 4 cell types promote the formation of 28 tadpoles. Thus, there is evidence for genomic multipotentiality of differentiated somatic cells from the larva, juvenile frog and adult. However, *to date, no nucleus of a documented differentiated somatic cell nor of any adult somatic cell (stem or specialized) has yet been shown to be totipotent.*

Nuclear Transfers from Germ Cells

The genomic totipotency of germ cell nuclei certainly does not require formal proof via the nuclear transplantation assay, because the germ cells of an individual retain the repertoire of zygotic genes, except for an occasional mutation. Nevertheless nuclear transfer tests of germ cells have been informative because they showed that their genome after transplantation into eggs expresses different degrees of developmental potential depending on the stage of differentiation of the donor cells. Smith (1965) tested nuclei of primordial germ cells taken from *Rana pipiens* tadpoles at the initial feeding stage.

The primordial germ cells are large and yolky and can be easily distinguished from the small somatic cells of the genital ridge. The investigator's efficiency in microsurgical removal of the host nucleus was 99.5%, as judged by the production of androgenetic haploids. Among the 31 normal tadpoles that developed from 7.6% of the injected eggs, 10 were reared for periods of 1-3 months, were then autopsied and found to be normal. Although the precise larval stages were not reported, a photograph of one metamorphosing tadpole was illustrated. Lesimple *et. al.*, (1987) tested germ cell nuclei of the salamander *Pleurodeles walti* taken from progressive stages of larval development up to metamorphosis.

The larger germ cells were distinguished from the smaller somatic cells. Among the total injected eggs, 12 (0.4%) attained the feeding larval stage, 9 (0.3%) metamorphosed and 6 (0.2%) developed into adults. Two of the adults were mated and produced normal progeny. The adults arose from experiments in which the donor nuclei were not genetically marked. The authors argue that it is unlikely that the adults arose from the egg nuclei, because in *Pleurodeles* gynogenesis results in 50% ZZ males and 50% WW females, but they obtained 3 ZZ males and 3 ZW females and no WW females. Nevertheless, proof of this interpretation is lacking. One study examined the developmental potential of adult germ cell nuclei, since their nuclei are derived from differentiated germ cells. The cell suspension of donor cells from adult testes was enriched for spermatogonia, and the donor cells were selected on the basis of size.

Enucleated *Rana pipiens* eggs injected with these nuclei yielded 4 (3.5%) postneurulae and 1 (0.9%) feeding tadpole. Recovery of the egg nucleus in the exovate indicated that development was directed by the transferred nucleus. Failure to reveal the genomic totipotency of differentiated germ cells in nuclear transplantation experiments indicates that their genome and/or chromatin undergo modifications required for specifying the phenotype of a differentiated cell.

Under the experimental conditions, the modifications are stable and prevent the full expression of their genomic potential, but certainly they are not irreversible. Lesimple *et. al.* (1987) reported a decrease in the percentage of hatching larvae when the germinal nuclei were taken after the mid-larval stages. They suggested that differentiation of germinal stem cells into spermatogonia or oogonia might account for the decrease in successfully nuclear transplantations, while the retention of a few stem cells (gonocytes) might account for the successful nuclear transplants.

Imprinting of the Genome

Recent nuclear transfer studies in the mouse have shown that normal development requires the presence of both the female and male pronuclei. Embryos with two female or two male pronuclei arrest early in embryogenesis. Those embryos with solely maternal or paternal chromosomes exhibited different phenotypes. Embryos with a diploid set of maternal chromosomes developed into well formed but small 25-somite embryos; however, the extraembryonic tissues were underdeveloped. By contrast, embryos with a diploid set of paternal chromosomes differentiated poorer embryonic structures but better extraembryonic tissues than the embryos with maternal chromosomes. These results indicate that the maternal and paternal genomic contributions to embryogenesis are functionally different and that both are essential form normal development.

The investigators have concluded that a differential imprinting of the genome occurs in the male and female gametes during gametogenesis and suggest that different subsets of genes are inactivated in the parental genomes but in a complementary pattern. This imprinting of the genome is reversible, however, because homozygous uniparental mouse embryos in chimeric combination with normal embryos can develop into fertile adults that produce normal gametes derived from the uniparental genome. These latter investigators interpret the cases of totipotency to stem from reimprinting the genome during gametogenesis. In amphibians, gynogenetic diploids produced by suppression of the second polar body can develop into fertile adults. These cases include *Rana pipiens* and *Xenopus laevis*. Thus, if differential imprinting of amphibian parental genomes occurs during gametogenesis, such imprinting must be easily reversed in amphibian embryos.

Amphibian embryos also differ from mouse embryos in their onset of zygotic RNA transcription. In *Xenopus*, the first detectable RNA transcription occurs at the sixth cleavage stage, whereas in murine embryos RNA synthesis is initiated during the first cleavage stage. The later stage of onset of RNA transcription in frog embryos would permit the genome of a transplanted frog nucleus additional cell cycles to adjust to a new programme of gene expression compared with the one cell cycle in the mouse. This differences in the onset of RNA synthesis might account for successful nuclear transplantations from advanced stages of frog cells, whereas successful nuclear transfers in mammals are so far limited to cleavage stage nuclei.

Whether the temporal programme of gene expression, parental imprinting of genomes or other factors account for the difference in nuclear reprogramming between mammals and frogs remains to be elucidated. Although the successful cases of amphibian nuclear transplants are those that are frequently emphasized, many frog nuclei from determined regions of embryos before overt tissue differentiation displayed developmental restrictions after transfer to enucleated eggs. Somatic nuclei tested from different embryonic stages, germ layers and primitive organs from various anuran and urodelen amphibians showed a progressive decrease in the percentage of normal nuclear transplants when the donor nuclei were taken from progressively more advanced stages of embryogenesis.

Other main points derived from these studies were (1) a few of nuclei still displayed genetic totipotency, and this was especially evident in the endodermal nuclei of hatched larvae of *Xenopus*; (2) but most nuclei did not promote normal development of the test eggs; (3) the developmental restrictions analyzed in endodermal and neural nuclear transplants were stable, since they could not be reversed through serial transplantation; (4) furthermore, the restrictions expressed by endodermal nuclei were intrinsic since they were not corrected by parabiosing nuclear transplant embryos with normal ones, nor could they be corrected when a haploid set of egg chromosomes was combined with the diploid set from the endodermal nucleus. Among the *Rana* nuclear transplants derived from endodermal and neural nuclei, a small number displayed a pattern of developmental abnormalities consistent with the origin of their nuclei.

During postneurula stages some nuclear transplants derived from late gastrula endodermal nuclei exhibited deficiencies and degenerative changes in ectodermal and mesodermal organs, but not in endodermal tissues. This phenotype was not reversed by serial cloning, alleviated by parabiosis with normal embryos, or associated with obvious chromosomal abnormalities. Similar studies of nuclei from the presumptive medullary plate regions of late gastrula and the definitive medullary plate of early neurula and mid-neurula embryos showed a small group with a phenotype different from the endodermal nuclear transplants.

The neural nuclear transplant group (13%) displayed during larval stages cellular deficiencies mainly in mesodermal and endodermal derivatives but showed good differentiation in the organs and tissues of ectodermal origin and did not present detectable chromosomal abnormalities. These distinct phenotypes suggested that by late gastrulation some endodermal and neurula nuclei have acquired stable properties for specific pathways of cell differentiation. Although these phenotypes were described more than 20 years ago, the mechanism probably is analogous to the imprinting of the genome recently described in murine nuclear transplants. In the mouse differential imprinting of parental genomes, presumably during gametogenesis, leads to two distinct phenotypes, depending on whether the conceptus developed from two male or two female pronuclei. In some of the amphibian nuclear transplants, we probably observed the effect of genomic imprinting initiated during embryogenesis.

The limitation of these studies is that the data are not extensive. Perhaps, an analysis of the pattern of DNA methylation in the endoderm and neural nuclear transplants would clarify this interesting problem. Presumably, the normal frog nuclear transplants were derived from stem cells present among the other differentiating donor cells and would differ from the abnormal nuclear transplants in their pattern of DNA methylation.

Reversal of Genomic Imprinting

The genome of differentiated cells has been imprinted by molecular mechanisms to specify a particular cellular phenotype.

Evidence available indicated that this imprinting results from the interaction of chromosomal proteins with the genome and modification of its DNA by methylation. Here we consider the extent to which the genomic imprinting in differentiated somatic frog cells can be erased under differing cytoplasmic environments.

Nuclear Transfer into First Meiotic Metaphase Oocytes

Nuclei of *Rana* erythrocytes have expressed the greatest genomic potential among the genomes of differentiated somatic cells so far tested. The main difference in the experimental procedure was that the erythrocyte nuclei were initially injected into oocytes, whereas the other nuclear types were injected directly into eggs. The standard host for nuclear transplantation has been the egg, the host originally proposed by Spemann (1938). However, we hypothesized that the expression of the genetic potential of differentiated somatic cells might be enhanced if the donor nuclei were first exposed to the molecular components in the oocyte that normally prepare the oocyte chromosomes to participate in fertilization. Initially, we tested whether somatic nuclei of embryonic cells would, in fact, function properly in the egg after residing in oocyte cytoplasm.

Oocytes at the stage of first meiotic metaphase were used for hosts. Such oocytes display in the animal pole the first black dot, indicative of the first meiotic metaphase. A single broken embryonic cell, together with it nucleus, was injected into the animal hemisphere near the equator region. At this time, the oocytes are not activatable, but becomes so about 24 hrs. later when they have matured in *vitro* into eggs (18°C). The matured oocytes (eggs) were activated by penetration of a glass microneedle. Within 10 min., the second black dot, indicative of the second meiotic metaphase of the egg, appeared and was removed microsurgically with a glass microneedle. We found that embryonic somatic nuclei exposed to cytoplasm directing meiotic events could still support development through embryogenesis.

Cytological studies performed on a parallel series of nuclear transplants showed that transplanted embryonic nuclei formed metaphase chromosomes on newly induced spindles in concert with the behaviour of the oocyte nucleus during first meiotic metaphase. When the matured oocytes were activated, the injected nuclei transformed into pro-nuclei. Thus, somatic nuclei can reversibly respondent to the molecular cues directing meiotic and mitotic events in a sequential manner and still direct the host to develop through embryogenesis. These findings indicated that nuclei of differentiated cells could now be tested in oocytes to determine whether the imprinting of their genome could be erased by exposure to the molecular components of maturing oocytes.

Genomic Potential of Erythrocytes

The erythrocyte satisfies the most stringent criterion of a differentiated cell, it is non-cycling and terminally differentiated cell, one that is almost transcriptionally quiescent. Its oval shape and the presence of hemoglobin (Hb) permit its direct identification under the stereomicroscope at nuclear transfer. Thus, any development ensuing from the transplantation of an erythrocyte nucleus would unequivocally result from a differentiated genome. The developmental potential of the adult erythrocyte genome was tested in two ways. In one series, the nuclei were injected into enucleated eggs; in another series, the nuclei were injected into oocytes at the stage of first meiotic metaphase. Some of the blastulae derived from the hosts of each series were used to provide donor nuclei for a second transplant generation.

The blastula nuclei were injected singly into enucleated eggs and nuclear clones were produced. None of the transplants from the egg series developed beyond the early gastrula stage. However, among the 12 clones derived from the original oocyte series, eight clones (67%) contained members that developed into postneurula embryos, and six clones (50%) had members that differentiated into prefeeding tadpoles. When erythrocyte nuclei from juvenile frogs were tested under the same conditions, they directed the development of feeding tadpoles that formed hind limb buds in 75% of the four nuclear clones (DiBerardino et al., 1986). These tadpoles survived for up to 1 month and are the longest surviving and most advanced nuclear transplants derived from documented differentiated somatic cells.

The interval from zygote to metamorphosis is approximately 3 months. Thus, the month old tadpoles developed and survived for about one third of this interval. Several conclusions resulted from the nuclear transfer studies of the non-cycling and terminally differentiated erythrocytes of *Rana*. First, the erythrocyte genome retained the genes for directing the differentiation of the cell types, tissues, and organ systems of tadpoles. In fact, the specification of the various cell types in these tadpoles requires the activation of dormant genes that never functioned in the cell lineage of an erythrocyte. Second, exposure of the erythrocyte genome to the cytoplasm of maturing oocytes induced the most widespread activation of the erythrocyte genome ever obtained in an experimental system. In the regard, the imprinting of the genome that formerly specified the phenotype of an erythrocyte was erased to the extent of specifying the cell phenotypes of tadpoles. Third, serial transplantation of erythrocyte nuclei through eight transplant generations showed that once the erythrocyte genome is activated by its progression through the cytoplasm of the oocyte and activated egg, it maintains its potential not only for replication, but also for directing the complex functions of organogenesis in excess of 100 cell cycles.

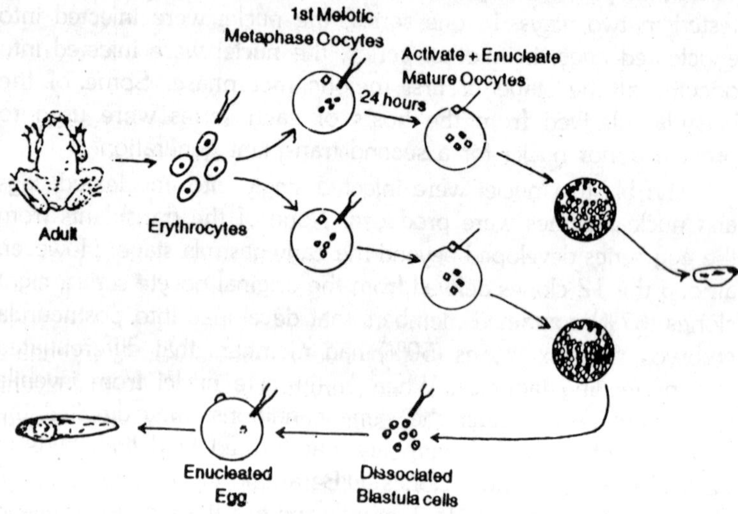

Fig. 6.1. Erythrocytes obtained by intracardiac puncture of adult Rana pipiens frogs were broken in distilled water by osmotic shock and microinjected into oocytes near the equator at first meiotic metaphase. Approximately 24 hrs later (at 18℃), when the oocyte matured, the matured oocyte (egg) was activated by pricking with a glass needle, and the egg nucleus was removed microsurgically. (Top) Original transplant generation; prehatching tadpoles resulted. In some cases, nuclear transplant blastulae were dissected and their animal hemisphere nuclei were transplanted singly into activated-enucleated eggs. (Bottom) First retransfer generation; swimming tadpoles resulted.

In addition, there was no evidence that the mitotic progeny of the erythrocyte nuclei lost there ability to replicate their genomes and continue cell cycling. Fourth, some tadpoles fed and therefore functioned as independent organisms. Thus, the developmental block previously exhibited in prefeeding tadpoles from differentiated somatic nuclei was surmounted. The progression of tadpoles to the feeding stage is a critical developmental event, because at this stage maternal (egg) components are largely used. Thereafter, continued development, further differentiation, and survival of the tadpoles are dependent on an external source of nourishment. An important factor that must be considered in evaluating the formation of nuclear transplants is the contribution of maternal RNAs and proteins to their development.

It is certain that the transplanted nucleus contributes a significant role to the development of nuclear transplants, because eggs lacking a functional nucleus but containing a functional centriole develop at best only into partially and abnormally cleaved blastulae. Nevertheless, there is not enough information to evaluate exactly the relative contribution of maternal RNAs and proteins to the development of nuclear transplants. For this reason, the survival of the nuclear transplants as independent organisms is crucial, because most maternal products are presumably used and degraded within a restricted early period. Therefore, the most critical evaluation of the genomic potential of differentiated somatic nuclei should be made on those nuclear transplants that function as independent organisms.

Mechanisms of Genomic Activation

The primary mechanism(s) in nuclear transplants that is responsible for the activation of the genome of differentiated cells is not known. However, it is quite clear that the molecular components in the cytoplasm control nuclear and genomic function. Some key results are cited from studies performed in oocytes, eggs, and cell cultures that characterize some steps in the activations process. A series of studies has shown that the state of the cytoplasm controls chromosome condensation, DNA synthesis, and mitosis.

The classic studies conducted by Brachet (1922) in sea urchins and by Bataillon and Tchou-Su (1934) in amphibians showed that when sperm prematurely entered oocytes, which were still undergoing maturation divisions, the sperm chromatin rapidly condensed into chromosomes similar to those of the oocyte. In later years the cytoplasmic control over nuclear and chromosomal activities was studied extensively in amphibian eggs and oocytes by nuclear transplantation. Nuclei injected into activated eggs behave like pronuclei; they enlarge, undergo chromatin decondensation and synthesize DNA. During nuclear activation, some nonhistone proteins leave the donor nuclei, while both histone and nonhistone proteins migrate from the recipient cytoplasm into the injected nuclei.

At the diplotene stage of the oocyte, when its nucleus (germinal vesicle) is greatly enlarged and is engaged in RNA but not DNA synthesis, transplanted nuclei also enlarge and synthesize new RNA products directed by the regulatory mechanisms of the oocyte. Here, too, an exchange of nuclear proteins between nucleus and cytoplasm is observed. When the nuclear envelope of the germinal vesicle breaks down, the oocyte chromosomes condense and become aligned on the spindle of the first meiotic metaphase. Similarly, injected nuclei also transform into metaphase chromosomes aligned on spindles. The cytoplasmic control of nuclear activity also operates in the mouse oocyte as well as in oocytes of other species. The same phenomena were reported in cultured mammalian cells when cells from different phases of the cell cycle were fused together. Cells in S phase induced cells in G1 phase to undergo DNA synthesis, whereas mitotic cells fused with G1,G2, or S cells caused premature chromosome condensation.

The main conclusion drawn from the above studies is that a new program of genomic expression is induced in test nuclei by a specific set of molecular factors residing in the oocyte, egg and different phases of the cell cycle, respectively. The availability of cytoplasmic extracts from activated eggs that can induce sperm to transform into pronuclei and mitotic chromosomes *in vitro* will ultimately permit the biochemical analysis of the factors in eggs that induce DNA synthesis and chromosomal condensation. With respect to extracts from oocytes, Korn *et. al*. (1982) identified a positive regulator that activates dormant 5S RNA genes of erythrocytes. The activating factor is sensitive to protease and heat treatment suggesting that protein(s) are involved. Heterokaryons have also provided clues into the process of gene activation. An outstanding example is the activation of certain dormant genes in the hen erythrocyte. The classic experiment by H. Harris (1974) involved the fusion of hen erythrocytes with HeLa cells.

In the heterokaryons, the nuclei of the parent cells remain separate, and each nucleus can be identified and monitored. The erythrocyte nuclei enlarged, their highly condensed chromatin dispersed, they resumed synthesis of RNA and DNA, and later, a series of chick proteins were synthesized. During the activation process, there is an uptake of mammalian nuclear proteins by the chick erythrocyte nucleus and a loss of histone 5 (H5) from the chromatin of the erythrocyte. Blau *et al.* (1985) fused mouse muscle cells with a variety of nonmuscle cell types of human origin. Such heterokaryons containing muscle cells do not undergo cell division, and therefore, the heterokaryons are stable and retain the complement of chromosomes. In these heterokaryons, various specific mRNAs and proteins, normally made by muscle cells were also made by the nonmuscle genomes.

In addition to extending previous studies on heterokaryons, the authors suggest that their system will be amenable to isolating the transacting regulators that induce muscle gene activation. In the case of *Rana* erythrocyte nuclei, we have shown that exposure of erythrocyte nuclei to the components in the cytoplasm from the stage of first meiotic metaphase and subsequent maturation stages leads to widespread activation of the genome. This activation involves in some manner the ability of noncycling and nonreplicating erythrocyte nuclei to undergo significant DNA replication in activated eggs.

Autoradiographic studies by Leonard *et al.* (1982) demonstrated that only 24% of adult erythrocyte nuclei transplanted to eggs synthesized DNA and, in these cases, only a small portion of the genome engaged in DNA synthesis. If, however, adult erythrocyte nuclei were first injected into oocytes that later matured and were activated, more than 75% synthesized DNA, and more than one half of these nuclei synthesized DNA in amounts similar to those of the egg pronucleus. Thus, exposure of erythrocyte nuclei to oocyte cytoplasm leads not only to induction of DNA synthesis in more nuclei, but also significant enhancement of the extent of replication of the genome. We have proposed that a remodeling of the proteins in the erythrocyte chromatic occurs in the oocyte cytoplasm that renders the chromatic capable of responding more normally to the signals in activated egg cytoplasm that induce the synthesis of DNA. Such remodeling of chromosomal proteins could lead to changes in chromatin structure and in patterns of DNA methylation that eventually permit the genome to respond later to the appropriate transcription factors and synthesize the appropriate RNAs during embryogenesis.

Other auto radiographic studies examined the nucleo-cytoplasmic exchanges of nonhistone proteins when endodermal nuclei were transplanted to eggs. These studies showed that during the first cell cycle there is an accumulation of cytoplasmic egg nonhistone proteins by transplanted nuclei and also a major loss of nonhistone proteins from the transplanted nuclei into the cytoplasm. Perhaps a similar but more effective nucleocytoplasmic exchange occurs in erythrocyte nuclei placed in the oocyte, not only because of a longer exposure to the cytoplasm, but also because the nuclei are exposed to additional factors not present in the egg. With respect to chromosomal proteins there is evidence that changes occur in the proteins of chromatin in germ cells during spermatogenesis, oogenesis, and after fertilization. Thus, there are precedents that substantiate our hypothesis that remodeling of chromosomal proteins may have occurred in erythrocyte nuclei exposed to oocyte cytoplasm. This hypothesis can be tested by examining appropriate chromosomal proteins, as well as the state of DNA methylation and the chromatin structure of specific genes in test nuclei before and after exposure to oocyte cytoplasm.

Such studies, together with the identification of specific extracts of oocyte cytoplasm that control molecular changes in the chromatin of test nuclei, might ultimately indicate the primary mechanisms involved in the activation of an entire genome of differentiated somatic cells. This information might also indicate how the germ cell chromatin of oocytes is prepared for participation in fertilization and embryogenesis. In addition to the remodeling of chromosomal proteins, an essential replication factor may be required. For example, Blow and Laskey (1988) have shown that sperm chromatin can re-replicate in a cell-free system extracted from *Xenopus* eggs, if the nuclear membrane is permeabilized or breaks down. They suggest that an essential replication factor gains access to DNA when the nuclear envelope breaks down during mitosis and that one role of the nuclear envelope is to limit DNA to one replication per cycle.

In our transfers of nuclei into oocytes a first meiotic metaphase, the nuclear membrane breaks down and the interphase chromatic is transformed into metaphase chromosomes aligned on a newly induced spindle. These events occur by 1 hr or earlier after transfer to the oocytes. It is likely that noncycling erythrocyte nuclei lack such an essential replication factor or have negligible amount, but acquire it from the oocyte cytoplasm after nuclear membrane breakdown. Nuclei injected initially into eggs retain their membranes until after the period of DNA synthesis and would therefore be deficient in the replication factor. The above speculation is consistent with our findings that adult erythrocyte nuclei incubated in oocyte cytoplasm undergo significant DNA synthesis in activated eggs and support tadpole development, whereas those injected in eggs undergo little DNA synthesis and support development only to the early gastrula stage.

7

GENES DURING DEVELOPMENT IN SEA URCHIN

The development of a complex organism from a fertilized egg most certainly requires elaborate temporal and spatial manifestations of differential gene activity. Nevertheless, the extent to which this is so has remained vague in spite of recent advances in methodology which provide elegantly detailed information about the structure and function of particular genes. In this report we summarize a variety of recent investigations of sea urchin embryos and make some generalizations about the extent to which gene expression changes during embryonic development, the timing and tissue specificity of these changes,and the possible levels of regulation of these changes. We also consider the respective roles of the maternal and embryonic genomes which function during embryonic development. These investigations have allowed us to identify and, in some cases, clone the DNA of genes exhibiting interesting activity during embryogenesis. Following fertilization, the sea urchin egg undergoes a period of very rapid cleavage. The fourth cleavage gives rise to blastomeres of three different sizes and developmental fates: macromeres, mesomeres and micromeres.

Micromeres cultured in isolation differentiate *in situ* into primary mesenchyme cells which ultimately secrete skeletal calcitic spicules they are thus determined and require no further interactions with other blastomeres to express this state of determination. Subsequent cleavages give rise to a blastula which ultimately hatches from the fertilization envelope. Primary mesenchyme cells begin to appear in the blastocoel a few hours after hatching, and gastrulation by invagination begins a few hours later. Following completion of gastrulation, skeletal formation is rapid, radial symmetry is lost, and the bilaterally symmetric pluteus larva forms. The pluteus is highly complex, consisting of a variety of differentiated tissues. Embryonic development to the feeding pluteus stage takes a few days, depending on the temperature and species.

Activation of Translation of Stored Maternal mRNA

The sea urchin egg contains proteins, RNA, and other materials required for embryonic development, there being no net change in mass until the pluteus begins to feed. In particular, sea urchin eggs contain a store of translationally inactive maternal mRNA, much of which quickly enters polysomes following fertilization. While this maternal mRNA accounts for most of the protein synthesis during early development mRNA is also actively synthesized within a few hours after fertilization; ultimately, mRNA transcribed from the embryonic genome replaces the maternal mRNA. The timing of these events is considered later in this review. We have been interested in defining the respective roles of these two classes of mRNA functioning in the embryo. Maternal mRNA might be expected to code for a special set of proteins required for cleavage or other events in early embryonic development.

To test this hypothesis we compared the proteins synthesized by the unfertilized egg with those of zygotes or early embryos. Cultures were incubated with ^{35}S-methionine, lysates were prepared, and proteins were resolved by two-dimensional electrophoresis according to O'Farrell (1975). Comparison of approximately 10,000 individual polypeptides showed very few differences in extent of labeling before and after fertilization in spite of the greater than 20-fold increase in amount of mRNA being translated. As applied, this technique detects only those polypeptides having isoelectric points between about 4.8 and 7.1 and molecular weights between about 10 and 125 k daltons; most of the mass of newly synthesized protein of eggs and zygotes falls within this range though some proteins, such as histones, are notably excluded. The sequence complexity of RNA in polysomes of embryos suggest that we detect only about one-tenth of the translation products expected.

Presumably, we detect only the product of the more abundant and/or efficiently translated mRNAs. In spite of these limitations it is clear that the activation of stored maternal mRNA involves, for the most part, a quantitative rather than qualitative change in the population of mRNA being translated; specific mRNA sequences are not being selected for translation. Most, and probably all, of the small amount of RNA being translated in the unfertilized egg is newly synthesized and unstable. The RNA which is recruited into polysomes following fertilization was synthesized during oogenesis and stored in a stable, untranslatable state in the egg. Thus the same mRNA sequence can exist simultaneously in same cell in two different states: translatable and unstable or untranslatable and stable.

While these two classes of RNA are synthesized at different times, it is presently not known how they are distinguished. The most favored hypothesis is that proteins associated with the stored maternal mRNA prevent its translation in the egg. Not all maternal mRNA sequences are recruited into polysomes shortly after fertilization. Some stored histone mRNAs do not begin to enter polysomes until just before the first cleavage. These appear to be sequestered in or near the pronucleus. One polypeptide not detected in eggs, but coded by maternal RNA, accumulates only during part of the cell cycle and is probably rapidly degraded at other times. When oocytes of another echinoderm, the starfish, are induced by 1-methyladenine to undergo meiotic maturation, many changes in pattern of protein synthesis occur; this requires changes in the population of maternal mRNA available for translation.

Specific maternal mRNAs present but not translated in oocytes appear in polysomes about the time of germinal vesicle breakdown. These stored maternal mRNAs are not sequestered in the germinal vesicle, nor is the release of its contents required for their translational activation. As in sea urchins, fertilization of the starfish egg does not result in a further change in the pattern of protein synthesis. Changes in patterns of protein synthesis are commonly associated with meiotic maturation of oocytes in a wide variety of organisms. We predict that meiotic maturation of sea urchin eggs is also associated with a change from a pattern of protein synthesis characteristic of oocytes to the pattern characteristic of egg and early embryos.

Patterns of Protein Synthesis During Embryonic Development

Substantial changes in the patterns of protein synthesis occur during later embryonic development in sea urchins. For example, a comparison of proteins synthesized in eggs and plutei of *Lytechinus pictus*. Polypeptide spots which have undergone changes in relative rates of synthesis are identified. Triangles pointing up indicate spots having increased intensity compared to the other stage; triangles pointing down indicate spots having decreased intensity compared to the other stage. Arrows point to spots detectible at that stage but absent at the other stage, for which the corresponding positions are shown by open circles. The definition of these "qualitative" changes is arbitrary. To normalize for the increasing absolute rate of protein synthesis, autoradiographic exposure was for the same total disintegration of radioactive protein loaded onto each gel. Since longer exposure times increase the number of spots which are detectible in some areas of the gel (but decrease resolution in other areas), some apparently qualitative changes must be result of limited sensitivity.

The patterns of proteins synthesized by *S. purpuratus* embryos have been subjected to quantitative analyses. Only about 20% of the polypeptides undergo any detectible changes in relative rates of synthesis during embryonic development. Of these, about half-change by at least 10-fold, but only about 1% of the spots change by 100-fold or more, for spots undergoing "qualitative" changes (About half of the total), these values are minimal estimates of the extent of change. No spot ever accounts for more than a few percent of the total protein synthesis of the embryo. The timing of these changes has been analyzed by labeling embryos with 35S-methionine for 1-2 hr at intervals throughout embryogenesis. The frequency with which spots have appeared or disappeared since the preceding labeling period ("qualitative" changes) have been plotted for each interval.

Few changes in pattern of protein synthesis occur in the first few hours of development, but changes become much more frequent about the time of hatching. In fact, nearly all the polypeptides synthesized in eggs and early embryos which become undetectible during embryogenesis disappear between hatching and the completion of gastrulation. Many of the spots which become undetectible are barely detectible at any stage and their disappearance is sometimes transient; these apparently qualitative

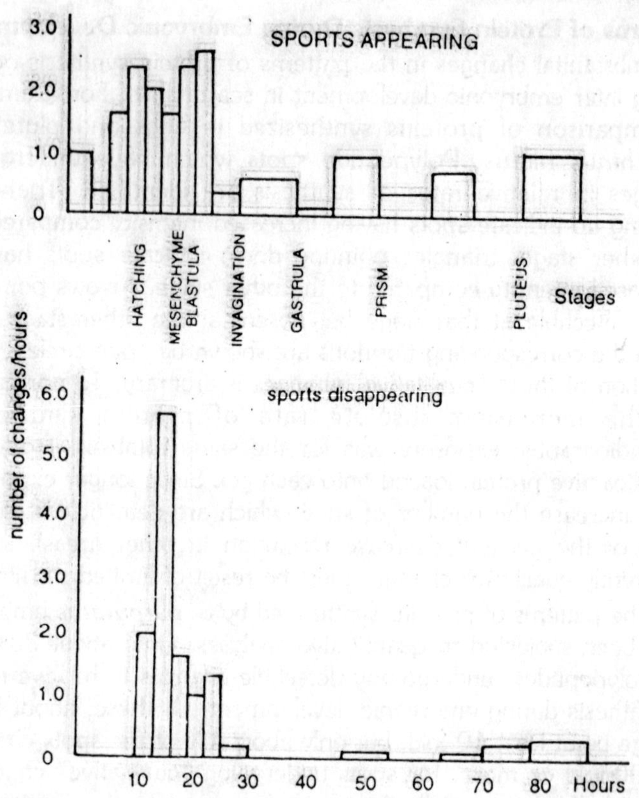

Fig. 7.1. Timing of "qualitative" changes in protein synthesis during development of S. purpuratus embryos.

changes are thus probably the result of minor variations in relative rates of synthesis or autoradiographic sensitivity. Appearances of new spots are most frequent in the interval between hatching and gastrulation as well.

The most striking quantitative increases, both the extend and number, also occur during the period between a few hours before hatching and the beginning of gastrulation. Many of the spots which appear or increase in intensity for the first time in this period continue to increase for many hours thereafter. By pluteus stage nearly all intensely labeled spots have undergone large increases in relative rates of synthesis. Of the polypeptides showing changes of more than 100-fold, all but one increase during development. Overall there is a modest increase in the number of detectible, newly synthesized polypeptides. Most of the polypeptides whose relative rates of synthesis change are never among the prominent translation products. Thus the overall pattern of protein synthesis is remarkably constant during embryonic development. These investigations of protein synthesis indicate that at about the time of hatching, extensive, rapid changes in gene expression begin. Hatching marks the beginning of a very interesting period in development in other ways as well.

Embryos cultured in the presence of the inhibitor of RNA synthesis, Actinomycin D, arrest at the time of hatching. An abrupt change in synthesis of histone variants occurs around the time of hatching; it is mediated by the replacement of early histone mRNAs by recently synthesized late histone mRNAs. The rate of cell division slows considerably at the time of hatching such that there is no longer an exponential increase in cell number. By mesenchyme blastula stage, a few hours after hatching, a large fraction of the RNA being translated is transcribed from the embryonic genome. Cultured primary mesenchyme cells begin to undergo extensive, rapid changes in patterns of proteins synthesis at a time corresponding to hatching.

If N-linked glycoprotein synthesis is inhibited by tunicamycin, gastrulation does not occur. Synthesis of the intermediates in N-linked protein glycosylation, dolichol and dolichyl phosphate, reaches a maximal rate a few hours after hatching. Changes in patterns of protein synthesis as a result of vegetalization induced by LiC1 first appear after hatching. Some or all of these development changes which are observed between hatching and gastrulation might be triggered by a common event or set of related events. Taken together, these observations suggest that a rather abrupt developmental transition from early to late embryonic development begins at the time of hatching. It is possible that many of the changes of protein synthesis observed at this time are co-ordinately regulated.

A variety of changes in cellular behaviour and transcription also abruptly occur in *Xenopus* embryos after the twelfth cleavage, the timing of these events appear to be related to achievement of a critical ratio of nucleus to cytoplasm. We have investigated the possibility that changes in patterns of protein synthesis observed during embryonic development might be mediated by regulation of the availability for translation of specific stored mRNA sequences. We compared the products of *cell-free* translation of RNA extracted from embryos of various stages to proteins synthesized *in vivo* using two-dimensional electrophoresis (Bedard and Brandhorst, in preparation). In general, an mRNA sequence translatable in wheat germ or reticulocyte lysates is detectible only at stages when it is being translated *in vivo*; this observation applies to nearly all of those cell-free translation products (about half) which co-migrate with in vivo products.

It remains possible that some mRNAs are stored in a form untranslatable in cell-free systems, as well as *in vivo*, but are ultimately processed to a translatable form (see below for more discussion of this possibility). The arrest of embryos cultured in Actinomycin D at the hatching blastula stage suggests that active transcription is required for the developmental changes which begin at hatching. Until then, translation of maternal mRNA is apparently sufficient for normal, though retarded, development (Gross et al., 1964). The simplest interpretation of the observations presently available is that the rapid changes in patterns of protein synthesis is occurring around the time of hatching are due to replacement of maternal mRNAs with the newly transcribed mRNA representing a distinct but overlapping set of genes.

At the time of hatching there are several hundred blastomeres, so it is likely that newly synthesized mRNA accumulates rapidly enough to account for the observed increases in synthesis of polypeptides. The rapid decline in synthesis of many polypeptides following hatching may be due to exhaustion of the maternal mRNAs coding for them through normal decay.

Patterns of Protein Metabolism During Embryonic Development

Changes in mass of individual polypeptides were analyzed separations of proteins of eggs and plutei. These patterns are even more similar than the patterns of newly synthesized proteins, but some important changes are obvious. Several very prevalent egg proteins decline greatly or are undectible by pluteus stage. These proteins decline gradually in mass during embryonic development and most are not detectibly synthesized during embryogenesis. They are synthesized and stored during oogenesis and processed or degraded during embryonic development and presumably include the yolk proteins. A few proteins accumulate substantially during development. These include a group of small acidic proteins, enriched in the ectoderm, whose relative rate of synthesis increases by about 200-fold.

A variety of patterns of protein metabolism can be discerned by comparing stained gels with autoradiographs of gels loaded with newly synthesized radioactive proteins. Some spots are never detectibly labeled during embryonic development, but no decline in their mass is detected; while some of these may lack methionine, most are probably quite stable. These include many proteins which associate with chromatin during embryogenesis and are stored in the egg. Many proteins are synthesized throughout embryonic development at a constant relative rate of synthesis but do not change in mass; presumably, their mass is maintained by turnover. The relative rates of synthesis of other polypeptides increase or decrease considerably during development, but they remain almost constant in mass. For example, the major actin variant (β-actin;) increases is labeling intensity by about 100-fold during development but does not change in mass. These observations indicate that different polypeptides have quite different stabilities and that these stabilities can change during development.

Post-translational events are probably critical components of the processes regulating gene activity during development. The most striking feature of these comparisons is the great similarity of proteins present in eggs and plutei. This observation demonstrates the extent to which oogenesis prepares the egg for embryonic development. This preprogramming presumably minimizes the extend of differential gene regulation required during development and allows for an accelerated rate of embryogenesis.

Tissue Specificity of Protein Synthesis in Sea Urchin Embryos

At the 16-cell stage the micromeres are determined to form primary mesenchyme even in isolation, but the patterns of proteins synthesized by micromeres are indistinguishable from those of mesomeres and macromeres. On the other hand, mesomeres/macromeres have a distinctive population of complex class (rare) maternal mRNA sequences compared to micromeres, but these are not represented on polysomes. Primary mesenchyme cells isolated at the beginning of gastrulation, which are derived from micromeres, have a pattern of protein synthesis quite distinct from epithelial cells. Isolated micromeres in culture undergo abrupt, extensive changes in protein synthesis at a time corresponding to the period between hatching and invagination (Harkey and Whiteley, 1982a); this is the period in the intact embryo when the primary mesenchyme cells appear in the blastocoel by ingression from the vegetal plate. Expression of the differentiated state of these cells, the formation of skeletal spicules, does not begin until several hours after these change in protein synthesis, toward the end of gastrulation. There are very few changes in their patterns of protein synthesis during the period when cultured primary mesenchyme cells secrete and elaborate skeletal elements.

The timing of changes in patterns of proteins synthesis during embryonic development suggest that critical changes in gene expression occur substantially earlier than overt expression of specialized morphology and function. On the other hand, the determined state of micromeres is not reflected in a detectibly unique pattern of protein synthesis. The silver-stained pattern of proteins extracted from micromeres is identical to that of mesomeres/macromeres as well. Primary mesenchyme cells constitute only a small fraction of the mass of cytoplasm in the embryo, and they may not be very accessible to radioactive amino acid precursors of protein synthesis.

Consequently, the synthesis of many of the mesenchyme-enriched proteins was probably not detectible in our investigation of protein synthesis in intact embryos. Thus, our observations on the timing of changes in protein synthesis probably underestimate the extent of change, particularly in the period following hatching. Unfortunately, the pH gradients of gels used in our investigations and those of Harkey and Whiteley (1982a, b) are quite different making direct comparisons difficult and tentative. By the completion of gastrulation, ectoderm and endoderm mesoderm fractions can be conveniently separate.

As discussed above, most of the labeled proteins in the endoderm/mesoderm fraction are probably synthesized by the endoderm. Comparisons of protein synthesis in these two fractions by two-dimensional electrophoresis indicate that the vast majority of polypeptides are shared in *S. purpuratus*. A few newly synthesized polypeptides are greatly enriched in the ectoderm or endoderm/mesoderm fractions of *L. pictus*. These proteins were labeled in intact embryos before separation of the tissues. The separation techniques employed can induce a stress response which leads to alterations in protein synthesis; some of the induced polypeptides are of the same sizes as those induced in heat shocked embryos. The relative rates of synthesis of all of these tissue-enriched polypeptides increases substantially during development, though some are detectibly synthesized in eggs. Some of these tissue specific proteins are well represented by mass in the egg; it is not yet known whether they are non-randomly distributed in the egg cytoplasm or segregated during early cleavage. The relative rates of synthesis of some ectoderm specific polypeptides begin to increase during cleavage while the synthesis of endoderm specific proteins generally increases after the beginning of invagination.

Cloned cDNAs corresponding to a family of ectoderm specific proteins have been identified and characterized. Recently, cloned cDNAs for tissue-enriched mRNAs have been identified for *Tripneustes gratilla* and *L. pictus*. With the exception of the tubulins which are enriched in ectodermal cilia, the function of these tissue-specific proteins is unknown. We have compared the newly synthesized polypeptides greatly enriched in ectoderm or endoderm/mesoderm of three widely divergent species: *S. purpuratus*, *L. pictus* and *Arbacia punctulata*. Except for the tubulins, the tissue-specific polypeptides are not shared by all three species, though there are some clusters of similar spots. Half of the newly synthesized polypeptides detected on two-dimensional gels are shared between *S. purpuratus* and *L. pictus*; thus the poor correspondence between tissue-enriched polypeptides is very surprisingly and suggests that there are weak constraints on changes in their structure during evolution.

Table 7.1. Detection of Synthesis of Paternal Proteins in Hybrid Embryos

Hybrid cross			*Number of Paternal Spots*					
		of common	*Cleavage*		*Hatching*		*Pluteus*	
Egg	*sperm*	*Spots*	*Obs.*	*Exp.*	*Obs.*	*Exp.*	*Obs.*	*Exp*
Sd	Sp	500	0	60	0	60	3	70
Sp	Sd	500	0	80	5	90	5	90
Sp	Lp	275	0	120	1	120	2	130

We are currently attempting to prepare monospecific antibodies to these tissue-enriched polypeptides in order to investigate their localization within the egg, embryo and blastomeres. It is likely that many of the other changes in protein synthesis observed during development are related to cellular differentiation, but these will probably only be localized by more sensitive technique such as *in situ* hybridization.

Change in RNA Population During Embryonic Development

Comparisons of rare polysomal mRNA prepared from seas urchin embryos of various stages demonstrate that most mRNA sequence are shared but that some are stage specific. Virtually all mRNA sequences associated with polysomes of embryos are represented in the stored maternal RNA, and there is a decline in the complexity of polysomal and the total RNA during development. Surveys of recombinant cDNA libraries indicate that most prevalent and moderately prevalent transcripts are present in similar abundance in eggs and plutei, though there appears to be a decline in abundance at intervening stages. Cloned cDNAs complementary to transcripts which accumulate extensively during development have been identified. Our observations on the extent of change in patterns of proteins synthesis during development suggested that a greater fraction of the clones complementary to mRNA in cDNA libraries should correspond to transcripts which change in prevalence. While there is a variety of possible explanations for this apparent discrepancy, we favour the following. Most of the polypeptides undergoing changes in relative rates of synthesis during development are minor, never being intensely labeled.

The most highly labeled proteins at gastrula stage of *S. purpuratus* are the major actin variant and two members of the family of small, acidic, ectodermal proteins. The moderately prevalent transcripts for these account for only about 0.05-0.1% of the mass of polyadenylated RNA in the cytoplasm, corresponding to about 100-200 copies per cell. The relative rates of synthesis of most polypeptides, including those undergoing changes in synthesis, are at least 10-100 times lower. If the mRNAs corresponding to these polypeptides are translated with efficiencies similar to those for actin and the ectoderm proteins, then they must be present in 1-20 copies per cell or fewer. Such rare transcripts are not detectible by the method usually employed to screen recombinant DNA libraries. On the other hand, if they are represented in only a few blastomeres, they could be relatively prevalent in these cells and account for the local accumulation of large amounts of specialized proteins.

Since there are many other rare polysomal RNA sequence having abundances similar to these estimates, one might expect a much larger number of spots detectible by two-dimensional electrophoresis. Increasing the pH or size range of the gels does not greatly increase the number of spots. We tentatively conclude that the mRNA sequences corresponding to the missing spots are translated less efficiently or might even be non-functional. The relative constancy of the RNA population during embryonic development and formation of distinct tissue layers suggests that many of these sequences may be the transcription products of genes expressed in all cells, coding for proteins having a house-keeping role. Contrary to this interpretation are the observations that most rare and moderately prevalent transcripts present in embryos are absent in adult tissues. The patterns of protein synthesis in several adult tissues are quite distinct from those in embryos as well (our unpublished observations). Similar to highly differentiated cells of other organisms, cells of specific adult sea urchin tissues show greatly enhanced synthesis and accumulation of a few polypeptides.

Expression of the Paternal Genome in Interspecies Hybrid Embryos

While it is well-established that translation of maternal mRNA accounts for most of the protein synthesis during the first few hours of development, the contribution of transcription of the embryonic genome to the mRNA population during early embryonic development has been vague. One approach to this problem is to analyze the expression of the paternal genome in interspecies hybrid embryos. This approach, for instance, has demonstrated that H1 histone genes are transcribed before the first cleavage and account for translatable mRNA in early cleavage-stage embryos. We have analyzed the synthesis of paternal proteins in three interspecies crosses. No paternal proteins were detected in early cleaving embryos; it should be noted here that unlike the genes coding for the early histone variants which are highly repeated, the genes coding for most of these proteins are present in one or a few copies per haploid genome.

Surprisingly, as late as pluteus stage only a few specifically paternal proteins were detected. This observation was confirmed for the *S. purpuratus* x *L. pictus* cross using a radioactive complementary DNA probe transcribed from polyadenylated, polysome-enriched gastrula RNA. The probe shares few sequences with RNA of *S. purpuratus*. Comparison of the kinetics of hybridization of this probe to excess RNA of *L. Pictus* or hybrid gastrulae indicates that most prevalent and moderately prevalent paternal sequences are reduced or absent in hybrid embryos. Many rare paternal transcripts are represented in hybrids. DNA-driven hybridization with this probe indicates that the missing paternal RNA sequences are maintained in the hybrid genome in normal amount. There are two types of explanation for these observations. The first is that expression of the paternal genome in hybrid embryos is defective. For example, cytoplasmic factors of maternal origin required for transcription or post-transcriptional processing may exist and act in a species-species manner.

The observation that the synthesis of paternal proteins is restricted in both of the reciprocal crosses between *S. purpuratus* and *S. droebachiensis* indicates that there is not a species-specific genomic or cytoplasmic dominance. It also suggests that if cytoplasmic factors are involved they have diverged rapidly in these two closely related species. It is also possible that the maternal cytoplasm or genome might repress the transcription of paternal genes as in the case of ribosomal genes in hybrids of two species of *Xenopus*. It is also possible that the paternal genome is rearranged in hybrid embryos. The other type of explanation is that most prevalent mRNAs translated in embryos are persistent maternal mRNAs synthesized during oogenesis and not replaced by newly transcribed mRNA during embryonic development. In this case these paternal transcripts would not be expected to appear in normal amounts in the hybrid embryos.

Analyses of the kinetics of accumulation of several transcripts which are prevalent in the egg using cloned DNA probes have indicated that these transcripts are rather stable and enter the cytoplasm slowly during embryonic development. The slow accumulation of new representatives of these sequences cannot account for their mass even late in development, indicating that these maternal transcripts are quite persistent. On the other hand, rare transcripts turn over rapidly, and newly transcribed rare mRNAs of the complex (rare) class would thus be expected to be represented in hybrid embryos even if some prevalent maternal mRNAs are not normally replaced. If some maternal mRNAs are persistent, some changes in gene expression during development may be the result of selective recruitment of these transcripts into polysomes.

Much of the maternal RNA stored in the egg has a peculiar structure similar to that of unprocessed or incompletely processed nuclear precursors of polyadenylated RNA includes interspersed self-annealing, repeated sequences reminiscent of some introns. Much of the egg RNA lacing poly(A) contains oligo (A) tracts-characteristic of nuclear RNA but absent in polysomal mRNA. It is possible that these maternal RNAs give rise to actively translated mRNA after completion of post-transcriptional processing or modification. Selective translational regulation of protein synthesis during embryonic development appears to operate in *Ilyanassa*. If maternal mRNA are persistent and selectively modified for translation, they might be non-randomly localized in the egg and/or segregated during cleavage into particular lineages. We are presently attempting to resolve these two possible explanation for the restricted expression of the paternal genome in hybrid embryos; both may apply to some extent.

We have constructed a recombinant cDNA library complementary to mRNA prepared from *L. pictus* gastrulae and have identified several DNA clones corresponding to transcripts specific to *L. pictus* embryos which are considerably less prevalent (or absent) in hybrid embryos derived from *S. purpuratus* eggs fertilized with *L. pictus* sperm. These cloned cDNAs were used to characterized their respective transcripts in *L. pictus* embryos. Several of these transcripts are present in almost constant abundance in eggs and throughout embryonic development. These transcripts might correspond to persistent maternal RNAs, but alternatively, they might be replenished by new transcription. In several cases investigated thus far by hybridization of cloned DNA probes to blots of RNA separated electrophoretically by size, there is no detectible change in size of these transcripts during development. That is, there is no evidence that these transcripts are processed during development. Other cDNA clones have been identified corresponding to transcripts which are barely detectible in hybrid embryos but accumulate during development of *L. pictus* due to new synthesis. The existence of this case of transcript indicates that at least some paternal genes normally expressed during embryonic development are hardly expressed in hybrid embryos. We are attempting to establish which events in gene expression account for the failure of these mRNAs to accumulate in hybrid embryos. The first step is to determine whether they are transcribed in hybrid embryos. Understanding the defect in expression of these paternal genes in hybrid embryos should provide valuable information about the interaction between nucleus and cytoplasm and the regulation of differential gene activity during embryonic development.

8

Gene Expression in Early Development

One of the most important problems pertaining to the mechanisms of development is to find out which events take place between the time a gene comes into operation and its phenotypic expression. In bacteria, the entire sequence of genetically regulated events is chiefly controlled at the transcriptional level. In eukaryotes, other points of control are the transfer of mRNA from the nucleus to the cytoplasm and the events preceding translation. These problems acquire special importance in embryonic development. Indeed, it is evident that the regulation of the rate of each one of these processes may play an important role in differentiation.

Although the mechanisms of regulation are genetically controlled, the regulation of many developmental process is controlled by the genotype via some cytoplasmic organization. The true picture seems to be still more complicated. The mRNA molecules which have been transcribed in the oocyte at the lampbrush stage are partially translated during oogenesis, partially during cleavage, and partially during later stages of development. This may be true of similar mRNA molecules, i.e., synthesized on one and the same gene, as well as of different mRNA molecules that are translated at different stages of development. And vice versa, at any one moment of development. mRNA molecules transcribed during oogenesis, as well as those synthesized in the nuclei of the embryo at a recent previous stage of the embryogenesis, are translated at the same time. At attempt is made herein to analyze these problems during early stages of development. But similar situations may be observed at later differentiation stages, when the function of the nuclei has terminated earlier than morphogenesis. Erythropoiesis may be a good example of such differentiation. One may suggest that cytoplasmic program of realization of gene expression is a general property of all the processes of differentiation.

Transcription

Changes in the Quantity of Templates

In the course of oogenesis and early development, the quantity of DNA which may serve as templates for transcription increases. In the early oogenesis of amphibians and many other animals, amplification of ribosomal genes occurs, which results in a some 1000-fold increase of the templates for rRNA synthesis. After this increase, the quantity of ribosomal non-chromosomal DNA in the oocytes nuclei is twice as great as of chromosomal DNA. In the course of development of the oocyte, the quantity of mitochondria containing DNA also increase. If in the small eggs of sea urchins the amount of this DNA is as low as 6 μg, in amphibians eggs it is 1000-fold greater and in birds it is a million times greater, than the amount of DNA in the haploid set of enchromosomes. When development starts, chromosomal DNA doubles at each division and increases in proportion with the increase in the number of cells.

At the same time, the number of mitochondria and, respectively, the mitochondrial DNA, does not practically change. Therefore, the proportion of cytoplasmic DNA diminishes, and when, for example, an amphibian embryo approaches the gastrula stage (40,000 cells), the mitochondrial DNA in every cell and, correspondingly, in the whole embryo does not exceed 1-2%, i.e., the value for the adult tissues. As the mass of the embryo practically does not change in early development, the volume of cytoplasm per nucleus rapidly decreases as cleavage proceeds. This means that, assuming a constant rate of transcription, the capacity to provide the protein-synthesizing machinery with new mRNA molecules increases. Hence, the choice of units for evaluating the intensity of the RNA synthesis depends on whether one is interested in the absolute rate in which case DNA should be taken as a reference; on the other hand, if the nuclear-cytoplasmic interrelations are the tissue in question (i.e., to what degree synthesis of RNA provides for synthesis of new proteins), then the mass of cytoplasm, or the whole embryo, is a more appropriate reference.

Onset of RNA Synthesis in Nuclei

As early as 1959 a method was worked out whereby the onset of nuclear activity controlling morphogenesis (i.e., the so-called morphogenetic function of the nuclei) could be demonstrated. At the time the experiments were started, available data on the arrest of development in lethal hybrids, in mutants, and in enucleated embryos suggested that in the amphibian and fish embryo nuclear functions begins to manifest itself at the time of gastrulation, and in the sea urchin at the mild-blastula (before hatching) stage. The method employed in our work was to inactivate the nuclei, either chemically or by means of heavy X-irradiation at successive stages of development, and to see to what stage such functionally enucleated embryos were able to develop. If the embryos irradiated at two successive steps of development ceased to develop at the same stage, one could conclude that between the two steps nuclei were exerting no morphogenetic function. On the other hand, if the development of the embryo irradiated later was arrested at a later stage, then this was taken as evidence of nuclear morphogenetic function.

The function is the stronger, the higher the ratio of time-gap between arrested stages to that between irradiated stages. As a result of the application of this method, it was shown that the morphogenetic function of the nucleus begins in the fish embryo at the mid-blastula stage (6 hours at 21°C), in amphibians at the late mid-blastula (stage 8-9) and in the sea urchin at early blastula (approximately 128 cells). An obvious suggestion is that the onset the nuclear morphogenetic function might coincide with the initiation of mRNA synthesis. This assumption was admittedly not entirely justified, as we know very little about the functional role of different mRNA's or, specifically, whether RNA's synthesized at any one stage serve a morphogenetic, or any other, role in the embryonic cells.

Yet, when methods of determining the mRNA synthesis became available, it was shown that in the loach (*Misgurnus fossilis*) and also in the frog there is good correlation between initiation of RNA synthesis and the stages at which morphogenetic function of the nuclei had been predicted to begin. Later, a similar coincidence was revealed in the sea urchin embryo. In recent years, however, evidence has been obtained in the trout and in the axolotl that synthesis of mRNA begins sometime before the morphogenetic function of the nuclei. One possibility is that these mRNA's are meant not for a morphogenetic function, but for the synthesis of, the example, nuclear proteins. In fact, at all the stages of development, including the ones before cleavage, a low level of RNA synthesis can be detected. It might be argued that at the early stages of development the very low level of synthesis is due to the small number of nuclei, and that the increasing synthesis at the mid-blastula stage (in loach) is related to the progressively increasing number of nuclei. However, there are facts that contradict this suggestion:

1. The kinetics of the increase of incorporation of radioactive precursors does not correlate with the increase in the number of nuclei. Between hour 2 and hour 5 of development of the loach, when the number of cells increases about 200-fold, the rate of incorporation increases not more than 2-3 times. On the contrary, between hour 6.5 and hour 8.5 the number of cells in the embryo increases less than 4 times, but the rate of synthesis increases more than 10-fold. Thus, if in the early stage of development of the loach the rate of incorporation into RNA is calculated per nucleus, the erroneous conclusion may be drawn that within the first 6 hours the rate of synthesis drops drastically.
2. Autoradiographic studies of amphibians, fish, and echinoderms do not reveal any significant uridine-3H incorporation during the early stages of cleavage. In the sea urchin, RNA synthesis reveals itself only at the 16-cell stage in the loach, at the mid-blastula stage; in amphibians, at the late blastula stage. At these stages incorporation increases rapidly from very low to rather high values.
3. Incorporation of the precursors into RNA of the embryos functionally enucleated by irradiation of the gametes prior to fertilization or right afterward, or incorporation in the cytoplasm of the egg, makes up a considerable proportion of the total embryonic incorporation.

The above observations suggest that, at least in fish and amphibians at the cleavage stages, nuclear RNA synthesis, if it exists, is not very pronounced. This agrees well with the fact that at these stages the G_1 and G_2 phases are practically absent and the S phase is extremely short, i.e., DNA synthesis proceeds at a high rate. But the most recent data by Timofeeva et al. (1972) show that in loach at the early blastula stage (4.5-5 hours of development) there is a relatively weak synthesis of tRNA precursors. The question of the beginning of synthesis of RNA in the sea urchin embryo is more controversial. For example, it was shown by us that in *Strongylocentrotus nudus* intensification of RNA synthesis occurs after the onset of the fifth cleavage, i.e., approximately at the onset of morphogenetic function of the nucleus. Our autoradiographic investigations did not reveal considerable RNA synthesis in the nucleus of the early embryo, although one would have expected very intensive incorporation in the small number of nuclei that the embryos possess at these early stages. Glisin and Glisin (1964) that during the early cleavage stages only terminal tRNA exchange occurs whereas at the 32-cell stage true RNA synthesis may be observed. Czihak (1965) by an autoradiographic method was able to reveal RNA synthesis in the micromeres of the 16-cell embryo.

Later there were reports of RNA synthesis at still earlier stages, at the four-blastomere stage; at the two-blastomere stage; and even prior to the first cleavage. Recently, synthesis of heterogeneous, non-ribosomal RNA was detected in anucleated fragments of activated eggs. A study of the hybridization properties of this RNA revealed it to be complementary to the mitochondrial, but not the nuclear, DNA. The rate of synthesis in these fragments proved to be similar to that in the intact early embryos, suggesting that the early RNA synthesis may be cytoplasmic. Hybridization experiments also indicate that at later stages, when the nuclei are known definitely to synthesize RNA, mitochondrial synthesis makes up a major part of the total RNA synthesis. According to Wilt (1970), in the sea urchin synthesis of nuclear RNA becomes detectable at the 16-blastornere stage. In conclusion, much has still to be learned about the timing of the early RNA synthesis in sea urchin nuclei, which in fact, may differ by several cleavages in various species. The possibility that there may be differences among the blastomeres in the different territories of the cleaving embryo has been suggested by Czihak (1965); this, however, has not been confirmed.

Rate of RNA Synthesis in Early Development

The increase in the rate of RNA synthesis in the early development of the embryo is a sum of three components: change in the rate of synthesis in the nuclei (this may result from an increase in the rate of transcription and/or of the number of the genes transcribed : increase in the total number of nuclei in the embryo; and finally, an increase of the proportion of RNA-synthesizing nuclei. It was shown by autoradiography that, if the isolated loach blastoderm at the mid-blastula stage is incubated with uridine-^{3}H, incorporation can be detected only in the cells of the basal layer, which is normally adjacent to the yolk.

At the later stage, more distant layer become involved in the synthesis; and at the ninth hour (late blastula), many more blastoderm cells are actively incorporating. The increase in the percentage of the synthesizing nuclei is shown in table 8.1. That these regional differences do not depend on the rate of penetration of the precursor is demonstrated by experiments in which the embryos were cut tangentially and the animal and the vegetal zone were incubated separately. Under these conditions, the earlier beginning of the RNA synthesis in the vegetal fragment was still retained. It was shown by biochemical methods that at the seventh hour the rate of RNA synthesis in the vegetal fragments was several fold higher than in the animal fragments (per milligram of protein). Finally, if the cells were dissociated and incubated with uridine-^{3}H, the percentage of incorporating cells increased with the age of the blastoderm at the time of dissociation. However, in this case the percentage of synthesizing cells is larger than when the whole-blastoderm is incubated, hence it is likely that dissociation removes some of the regional differences.

Table 8.1. Percent of Synthesizing cells in Early Development in Intact Blastoderm and in Dissociated Cells

Stages of development (hours)		*Intact blastoderm*	*After dissociation of blastoderm in separate cells*
Mid-blastula	7	14	39
Late blastula	9	21	48
Early gastrula	19	29	57

Besides the differences in RNA synthesis, it has been shown that the basal cells contain more ribosomes and display a higher rate of protein synthesis. Regional differences in the RNA synthesis were also revealed in amphibians and in sea urchin embryos. They are likely to reflect the ooplasmic segregation which occurs prior to the first cleavage. In teloblastic eggs, such as those of loach, the differences may depend also on the relationships with the yolk which lies underneath. Knowing the number of cells at different stages, the change in the number of the synthesizing cells, and the rate of synthesis in the whole embryo, one can evaluate the change of genome activity in development. For example, within the short period of time between the seventh and ninth hour of development, a 6- to 9-fold increase in RNA synthesis was observed in the loach. Within the same time, the number of cells in the embryo increases from 1700 to 5900 i.e., 3.5-fold, and the number of synthesizing cells increases from 14% to 21% i.e., 1.5-fold. This means that the synthesizing activity per nucleus increases only 1.4- to 1.7-fold. Similar values are obtained by the autoradiographic method, which is more direct but less accurate. After the tenth hour almost all the blastoderm cells are involved

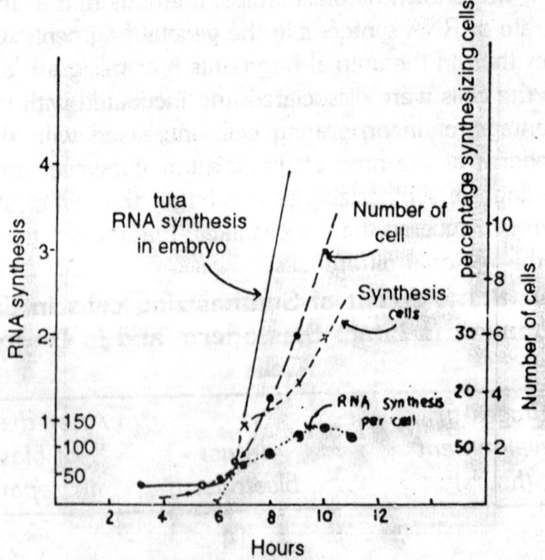

Fig. 8.1. Constituents of total RNA synthesis. O–O, Total synthesis per embryo, --•-•--, Cell number: X–X, percentage of RNA synthesizing cells; •...•. RNA synthesis per cell.

in RNA synthesis and the increase in total synthesis becomes proportional to the increase in the number of cells in the embryo. This means that the synthesizing activity per genome does not increase.

In amphibians RNA synthesis sharply increases between stages 8 and 9 ; in this time interval, the rate of increase in the number of cells is lower than that of the RNA synthetic activity. On the other hand, between stages 9 and 10 the increase in the rate of synthesis becomes equal to the rate of the increase in the number of cell: i.e., the rate of transcription per nucleus remains constant. We have shown that in *Strongylocentrotus nudus* between the early blastula (128 cells) and the mid-blastula (about 500 cells) stage the total synthesis per nucleus increases by 1.5 fold RNA increases by 5-to 6-fold; hence the rate of synthesis per nucleus increases by 1.5 fold are consistent also with an increase of the synthetic activity per nucleus. In the following hours, the increase of RNA synthesis in the embryo can be accounted for by the increase in the number of nuclei; in fact, at this time the synthesis of RNA per nucleus actually decrease.

The experiments carried out in our laboratory showed that in fish at the gastrula stage, and in the sea urchin at the stages following hatching, the overall RNA synthesis temporarily decreases. Since the number of cells is known to increase at this time, the observation implies quite a sharp decrease in transcriptional activity in nuclei. No such decrease for the same species was reported by other authors; this may be due to the fact that the stages compared were quite far apart. However, no final conclusion as to whether or not the observed decrease is a real one can be drawn, short of information about changes of the pool of uridine triphosphate and/or permeability changes. Thus, in fish and amphibians, and probably in sea urchin, the period of very low RNA synthesis during cleavage is followed by a relatively short period when the intensity of synthesis per nucleus somewhat increases (nor more than 2-fold); then the rate of synthesis per nucleus either remains constant or decreases. However, since the cell size decreases, if reference is made to unit of cytoplasm mass, then the RNA is found actually to increase; i.e., there is an increase of the transcriptional efficiency.

Regulation

We shall consider only one example of regulation of gene activity in development, i.e., the onset of the RNA synthesis at the blastula stage in loach embryos. The specific question concerns the factor(s) responsible for the stimultaneous switching-on the RNA synthesis on many genes with in the very short time of 0.5 hour. At the onset of synthesis is strictly related to a certain stage of development, the search should be restricted to the change in the course of development and may serve as a "clock" for the activation of the genes.

First of all, it is interesting to find out whether the clock operates in the embryo as a whole or whether it functions independently in every cell. To answer this question RNA synthesis in the whole embryo and in the dissociated cells was compared. In sea urchin embryos it was shown that dissociation at the late blastula stage does not prevent the synthesis of rRNA from being switched on. We have shown that, in early embryos of sea urchin and loach, intensification of mRNA synthesis may occur in the dissociated cells as well. Kafiani et al. (1971) compared the rate of RNA synthesis in isolated loach nuclei at the early blastula stage (prior to the intensification of the synthesis in the embryo) and at the late blastula stage (8 hours) when the synthesis is intensive enough.

The intensity of RNA synthesis is the isolated nuclei was found to display the same differences as the RNA synthesis by the whole cells. These differences could be due either to changes of the chromatin itself or of the enzyme. In the presence of an excess of exogenous RNA polymerase from *Escherichia coli*, RNA synthesis sharply increases and reaches the same level in the nuclei obtained from the early and late blastula stage. From this result the authors concluded that the intensification of the RNA synthesis in the late blastula depends on the greater activity of the RNA-polymerase in the nuclei. The degree of specificity of *E. coli* RNA-polymerase for the loach chromatin should, however, be verified. It should be mentioned in this context that the injection of the s-subunit of the *E. coli* polymerase into oocytes or early embryos of *Xenopus* drastically changes the pattern of RNA synthesis. Kafiani has also attempted to show that the major gene-activating factor is the change in the ionic composition of the loach embryos and, in particular, the change in the K:Na ratio.

It is known that, by changing this ratio, it is possible to cause the appearance of new puffs in the polytene chromosomes. Indeed, as the development of the loach embryos proceeds, the Na^+

concentration somewhat decreases while that of K^+ considerably increases. There is no evidence so far that alterations in the experimental conditions, resulting in anomalous values of ion ratio, can change the time of the onset of RNA synthesis. Shiokawa and Yamana (1967) reported having found a low molecular factor in amphibian eggs, which suppresses the rRNA synthesis from the early stages. These data were not confirmed in later investigations, although it is quite evident (and is also clear from the experiments on transplantation of nuclei) that the mechanism of the switch-on of RNA synthesis should involve some cytoplasmic factor. For example, Crippa (1970) has succeeded in isolating from *Xenopus* oocytes a protein factor which specifically suppresses rRNA synthesis by interacting with the rRNA cistrons.

Furthermore, from the data cited above it follows that the regulation of the RNA synthesis depends on the RNA-polymerase. However, it is unlikely that this the clock mechanism on which the onset of RNA synthesis is development depends. In the course of cleavage, the size of cells progressively decreases while the amount of DNA in every nucleus remains constant. This means that the ratio between the quantity of DNA and the mass of cytoplasm in the cells changes in the course of the early development: could this be a signal for switching-on the genes at a certain stage of development? This question has received a positive answer in experiments carried out in our laboratory in which the onset of the morphogenetic function of the nucleus was compared with DNA synthesis in haploid and diploid loach embryos. The experiments showed that in the haploid embryos these processes begin one division later than in diploids. An additional division of haploid having only half of the DNA causes a 2-fold decrease in cell sizes; as a result, the nucleo-cytoplasmic ratio becomes equal to that of diploids at the time of the onset of RNA synthesis. This correlation may be considered to the indirect experimental evidence in favour of the above hypothesis.

Transport of RNA to the Cytoplasm

The transfer of RNA to the cytoplasm has two most important peculiarities: it is preceded by a partial degradation of large RNA molecules in the nuclei and it proceeds in time. The known example of the changes RNA undergoes in the nuclei is the processing of the high-molecular precursors of ribosomal RNA and precursors of mRNA carrying information about hemoglobin in erythroblasts. That only the RNA molecules which have undergone processing pass to the cytoplasm and that the time of departure is comparable to, or longer than that of the processing, indicates that these two phenomena are related; i.e., the rate of the RNA transfer from the nucleus seem to be determined by the rate of its maturation in the nucleus. It is still a difficult task to give an accurate kinetic description of the process of RNA transfer from the nucleus to the cytoplasm.

Four processes which are very difficult to differentiate occur simultaneously in the cell: synthesis of RNA, degradation of its major part in the nucleus, transfer of the remaining RNA to the cytoplasm and subsequent degradation of this portion in the cytoplasm. The rates of all four processes are different for different kinds of RNA. For example, in the case of rRNA the rate of 18S rRNA transfer to the cytoplasm is higher than that of 28S rRNA. It is highly probable that for different kinds of mRNA these rates are also different. The process of RNA transfer from the nucleus still cannot be adequately described by a system of differential equations, as we do not know what model to choose to describe this mechanism.

A concentration dependence, when the rate of RNA transport to the cytoplasm is proportional to its concentration in the nucleus would be the simplest decision. At the other extreme would be the assumption that the time of transfer of every kind of molecule from the nucleus is predestined. Differences observed in the processing of RNA molecules of different kinds (rRNA and mRNA of hemoglobin) and in the fraction of RNA released to the cytoplasm in different tissues may be interpreted to mean that, in the course of embryonic development, the character of RNA transfer should be changing. Experimental evidence supporting this suggestion is not plentiful. On the one hand, some authors showed that in sea urchin embryos some portion of the label is found on the polysomes shortly after the beginning of synthesis. At the same time a substantial portion of the newly synthesized RNA remains in the nucleus for quite a long time (at least 2 hours): After gastrulation, the labeled RNA leaves the nucleus much faster.

Direct measurement of the RNA in the cytoplasm show that in the early stages of development of sea urchin this process is rather slow, but as development proceeds the rate of release increases 10- to 15- fold. Thus, in the course of development, the fraction of RNA undergoing degradation, the time the departing molecules are held in the nuclei, and their rate of transfer change. We studied RNA transfer from the nucleus to the cytoplasm in the loach embryo by autoradiographic and biochemical methods. However, an estimate of the rate of RNA transfer to the cytoplasm proved to be difficult, as chase experiments could not be done owing to the large pool of radioactive precursors that accumulated in the embryonic cells with in a short time.

Incorporation of the label in RNA gradually decreases, the deceleration being difficult to estimate accurately because the rate of RNA synthesis rapidly changes in the course of development. RNA synthesis in the nuclei of the loach embryo begins at the mid-blastula stage (6 hours at 21°C). Autoradiographic data show that no transfer of this RNA to cytoplasm occurs during the first hours. It is likely that part of this RNA undergoes degradation within the nuclei while part accumulates, to be transferred to the cytoplasm in the course of the following hours of development. At the later stages (early gastrula, 10 hours) when transfer has started, only light RNA (2-10 S) leaves the nuclei while the heavier fractions (more than 30 S) remain in the nuclei.

Table 8.2. Transition of RNA from Nuclei to Cytoplasm at different stages of Early Development

Stages of (hours)		*Number of grains* *Nucleus*	*Cytoplasm*	*Whole cell*	*percent of transition*
Mid-blastula	7	71 ± 3	1	-72 ± 3	1
Late blastula	8	86 ± 2	2	88 ± 2	2
Late blastula	9	108 ± 4	19 ± 1	127 ± 4	15
Early gastrula	10	115 ± 4	20 ± 3	135 ± 5	15

Another difficulty encountered in the quantitative autoradiographic estimation of the rate of RNA transfer from the nuclei to the cytoplasm in early development is that in the course of the experiment the cells divide one or several times. The resulting distribution of the label between the daughter cells, and the decrease in the size of the cells and their nuclei, require the introduction of some corrections that diminish the accuracy of

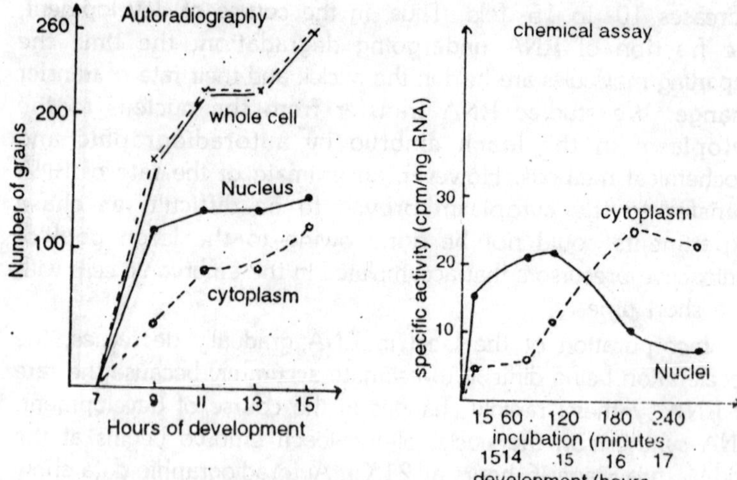

Fig. 8.2. Transition of RNA from nuclei to cytoplasm. Autoradiographic data (left) and biochemical assay (right). Blastoderms of loach embryos were incubated with uridine-^{3}H: for autoradiography, 2 hour at mid-late blastula; for biochemistry, 15 minutes at early gastrula. Then blastoderms were washed with an excess of cold uridine and chased with it.

measurement. Nevertheless, some quantitative conclusions can be drawn. Fig. 8.2 shows the results of an experiment in which after a 2- hour incubation with uridine ^{3}H, the embryos were transferred to the "cold" medium; the synthesis of the labeled RNA continued although its rate decreased 4- to 5-fold. Two hours after the beginning of the "chase", the rate of RNA synthesis and the rate of RNA transport from the nuclei become similar, and this is reflected in the fact that the quantity of labeled RNA in the nucleus is almost equal throughout the several hours that follow. At the same time, the cytoplasmic label rapidly grows.

Assuming that RNA in the cytoplasm does not undergo degradation during the experiment, one can calculate that within the first 2 hours of synthesis about 30% of the total labeled RNA goes to the cytoplasm although some part of it (about 5-10%) does so as late as 6 hours after being synthesized. Without going into the details of this calculation, it is clear that the labeled RNA synthesized from the remnants of the radioactive pool of the precursors during the later hours is insufficient to account for the increase in the quantity of radioactive RNA in the cytoplasm if the major portion of labeled RNA synthesized earlier has not contributed to it. If part of RNA in the cytoplasm undergoes degradation, the part of RNA which leaves the nuclei late will be still greater. The later transport of some part of RNA may be explained by two models. The first, a stochastic model, suggests that the transport of RNA depends only on its concentration within the nucleus.

In this case, the rate of transport of labeled RNA after pulse labeling should decrease asymptotically. If some portion of RNA does not go out of the nucleus or does so very slowly, one has a situation resembling the experimental evidence: some decrease in the rate of transfer and some decrease in the content of the labeled RNA in the nucleus. An alternative model, a regulated one, suggest that the RNA synthesized is gradually released to the cytoplasm. If many kinds of RNA have been synthesized in the nucleus, and they have different rates of transfer, the release of the label will be asymptotic, or close to it. A more elaborate control may be that different kinds of RNA go to the cytoplasm one after another, thereby determining the sequence of translation of various proteins. Thus, according to the data cited above, in loach the RNA synthesized in the mid-late blastula goes to the cytoplasm at various stages of gastrulation and the initial stages of formation of the axial organs. There are a few facts supporting the second model: for example, absence of the RNA transfer at the early stages of synthesis, transfer only of the RNA of the 2 S to 10 S class at the early stages of development, the slow dynamics of transition, and other indirect data.

Evidence that different RNA's synthesized simultaneously, but transferred early and late, carry different information would be unequivocal proof of this model. Passive transfer of RNA to the cytoplasm requires only restricted permeability of the nuclear membrane. For active, controlled transfer, a mechanism of higher complexity is needed, which should particularly include a binding, chemical affinity of nuclear RNA to the nuclear structures—chromatin or the membrane. The existence of the binding is confirmed by the possibility of thermal fractionation of RNA in the course of phenol extraction. The RNA fractions obtained in accordance with Georgiev's method at low and high temperatures agree with the concept of the nuclear and cytoplasmic RNA.

In the nuclei there is a small amount of readily extracted RNA that might correspond to the molecules going out of the nucleus. Our investigations on RNA migration during mitosis also testify to the existence of binding between RNA and the nuclear structures. In these experiments loach blastoderms were pulsed with uridine-^{3}H and then chased with "cold" uridine in the presence of high concentrations of actinomycin D. Within an hour almost all the cells underwent mitotic division. In the dividing cells the label was dispersed over the whole cell whereas in the interphase cells it was concentrated in the nuclei. Hence, the conclusion is that during mitosis the newly synthesized RNA goes out of the nucleus and afterward it returns to the nuclei of the daughter cells. Some experiments were made with the culture of fibroblasts front Chinease hamster, in which RNA transfer from the nuclei occurs relatively quickly.

The monolayers of fibroblasts growing on the glass were incubated with uridine-^{3}H for 40 minutes, the last 30 minutes in

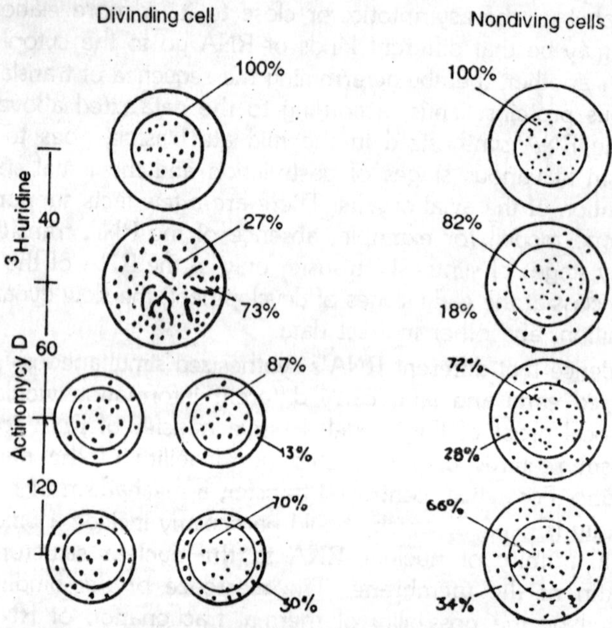

Fig. 8.3. Distribution of labeled RNA between nuclei and cytoplasm in dividing and non-dividing cells. Fibroblasts of Chinese hamster in monolayer culture were incubated with uridine-^{3}H (40 minutes) and Colcemid (the last 30 minutes). Then the metaphase and interphase cells were shaken from the glass in succession and incubated separately without Colcemid but with actinomycin (2 μg/ml). Silver grains over nuclei and cytoplasm were counted.

the presence of colcemid. Then the round metaphase cells were treated with trypsin for a short time and then shaken off the glass. In the cells synchronized in this way, the metaphase cells amounted to 90%. A portion of the metaphase cells was fixed immediately, and another portion was placed in the medium without colcemid but with actinomycin. Under such conditions cell division could continue, but *de novo* RNA synthesis was inhibited. For control, the interphase cells were used which were taken from the glass after the metaphase cells had been removed and transferred to the medium with actinomycin. During mitosis, almost all the label is evenly distributed in the cell; its concentration on the chromosomes is somewhat greater than in the cytoplasm.

On the other hand, in the interphase cells only 18% of the labeled RNA is found in the cytoplasm, although those cells that have not undergone mitosis have more time for synthesis (40 minutes). In 1 hour, when all the cells isolated in the metaphase are about to end mitosis, most of the labeled RNA is again found in the nuclei of the daughter cells, and only 13% of the grains are to be found in the cytoplasm. Most of this is most likely RNA which had gone out of the interphase nuclei either before mitosis or immediately after it. By this time, in the interphase cells which have not undergone division about 28% of the labeled RNA has moved to the cytoplasm. Thus, in this case too, most of the newly synthesized RNA, if not all of it, goes out of the nuclei for the duration of mitosis and then return, almost completely, to the nuclei of the daughter cells. Hence, cellular division is rather a retardation than an enhancement of the normal transfer of RNA to the cytoplasm. The migration of RNA from the cytoplasm to the nucleus at the end of mitosis requires some explanation.

In the work of Goldstein et al. (1969) it was shown that the nucleus of an amoeba labeled in its RNA, when transplanted to a non-labeled amoeba, loses part of its RNA, which moves to the nucleus of the host cell. Thus, RNA can migrate through two nuclear membranes; since about half of all RNA is found in the nuclei and half in the cytoplasm, the nuclear RNA must reversibly bind to the nuclear structures. If these data are valid also for higher animals, the behaviour of RNA in mitosis and interphase may be explained in terms of affinity to the nuclear structures (chromatin seems to be the most likely candidate, although the inner surface of the nuclear membrane cannot be ruled out).

During cell division the structure of chromatin changes sharply; it becomes spiralized into chromosomes. It affinity to the nuclear RNA decreases so much that at last RNA disperses in the cell. When mitosis is over and the chromosomes become despiralized, their ability to bind RNA is restored and the nuclear RNA concentrates in the nucleus. This takes place either before the nuclear membrane is formed after it, but then it is no obstacle to the nuclear RNA, as follows from the experiments with amoeba. In the interphase nucleus the absorption properties of chromatin do not change, but the properties of the RNA molecules changes as a result of processing and the monocistonic fragments of mRNA go to the cytoplasm. Thus, both in mitosis and in interphase, the release of RNA from the nuclei is due to the decreased affinity to chromatin; in mitosis these changes occur in chromatin, and in the interphase in RNA. Making the above suggestion still more speculative, one may assume that the sites of RNA responsible for binding with chromatin are localized in the fragments of this molecule which undergo gradation; and it is the separation of mRNA from these fragment that is the mechanism of RNA departure from the nuclei.

We do not consider here the question of the role of proteins that are bound to RNA in the nuclei. These proteins can be responsible not only for RNA cleavage [they have been proved to possess endonuclease activity], but possibly for binding with chromatin. Thus, the study of the behaviour of RNA in mitosis may be a handy model for the elucidation of the mechanisms controlling the fate of RNA after it has been synthesized. There are still many things to be clarified about the step (realization of hereditary information) which follows transcription. We are quite sure now only about the existence of processing and also that RNA release from the nucleus could be relatively slow, especially in the embryonic cells. How the process is controlled is still obscure. However, regardless of whether there is an active control over the pattern of RNA release from the nuclei or whether this merely the outcome of the structure of the nuclei and the properties of RNA, the time of release is extremely important in embryonic development. The fact that mRNA synthesized at the blastula stage and responsible for gastrulation is released from the nuclei throughout this period of morphogenesis is a good illustration.

Translation

Informosomes

mRNA transferred from the nuclei to the cytoplasm of the embryonic cells become incorporated in the informosomes and polysomes; these differ chemically in the amount of protein they bind, and hence the difference in their buoyant densities. In CsCl density gradients, poly-ribosomes band as one peak with a density of 1.51 maximum; informosomes give a peak at 1.40. While ribosomes and poly-ribosomes make up a part of the cell mass and may be detected by optical density, no optically measurable amount of informosomes has been obtained so far, and informosomes are usually detected by the radioactivity of this RNA. Informosomes were found and described in detail first in loach embryos and then in sea urchin embryos.

There has been a few reports about informosomes in tissues of adult organisms. Some properties make informosomes similar to the ribonucleoprotein complexes found in the nuclei, which gives grounds for believing that the two groups of particles are identical. Even in the early reports it was suggested that informosomes are in intermediate stage between RNA synthesis in the nucleus and its translation, i.e., that mRNA, either before or upon being released from the nucleus, associates with proteins giving rise to the informosome particles, which then to polyribosomes where the RNA is translated. Thus, these particles can be instrumental in regulation at the translational level. It is surprising, therefore, that descriptions of informosomes arouse great interest. The existence of informosomes has been questioned by several authors, the main objection being the possibility of formation of RNA-protein artificial aggregates during homogenization.

However, recent work from Spirin's laboratory gives new data supporting the concept of informosomes as real subcellular particles. The question of the role of informosomes is the transfer of mRNA to polyribosomes is still an open one. About 80% of the labeled RNA transferred from the nuclei to the cytoplasm of loach is associated with informosomes, i.e., it is in the area of the 1.4 g/cm^3 peak, with only 20% in the polysome peak (1.51 g/cm^3). This ratio does not change after prolonged incubation. One may assume that at each moment only 20% of mRNA is translated, and 80% remains in a "masked" form and informosomes continuously exchange mRNA; other explanations are also possible.

Therefore, until this question is resolved, the role of informosomes remains obscure, although it is evident that the fact that a considerable part of the mRNA is associated with informosomes should be taken to indicate their great importance in the regulation of translation in the cell.

The Intensity of Protein Synthesis in Early Development

It the course of early development, the rate of protein synthesis increases in all animals to reach a maximum at a stage which is different in different organisms—e.g., at the blastula stage in the sea urchin, in the loach at the end of gastrulation. The rate of synthesis decreases in the sea urchin during hatching and in the loach at the beginning of organogenesis. Later, another increase in the rate of protein synthesis takes place. Incorporation of labeled amino acids in protein depends not only on the rate of synthesis, but also on the rate of penetration of amino acids into the cell and on the size of the amino acid pool. Therefore, for an accurate determination of the rate of synthesis, the specific activity of the amino acid pool should be estimated at every stage investigated, and this makes the study much more complicated. In the preliminary experiments with loach it was shown that, at the stages in which an increased incorporation of arginine into protein was observed (at the end of gastrula), both permeability to arginine and the size of the arginine pool increase.

No appreciable change in the specific activity of arginine was observed, hence the increased incorporation of labeled arginine in the protein may be interpreted as being due to an increase in the rate of protein synthesis. It is likely that the greater permeability to amino acids of the cells of the loach blastoderm ensures an intensified transport of amino acids from the yolk to the blastoderm, which may result in a larger amino acid pool, which in turn provides the conditions for a higher rate of protein synthesis. A comparison of the curves of protein synthesis at various stages of development prompts some speculations—e.g., that in the sea urchin an increased protein synthesis at the blastula stage is necessary for hatching and gastrulation; or that a slowing down of the rate of synthesis prior to gastrulation (in sea urchin) or organogenesis (in loach) is associated with the switching of synthesis from "old" to "new" protein.

In fact, there are very few data on such "switchover" and, more generally, on the role of the proteins synthesized in the early development. Fractionation of proteins or immunological methods have shown rather minor changes in the patterns of the proteins synthesized. This, however, is not surprising if one considers the very small amounts of proteins that are likely to be synthesized that are beyond the sensitivity of the methods at present available. These minor proteins may be responsible for morphogenesis. Besides, in the majority of experiments the analyses deal with soluble proteins, whereas differentiation is most likely to be associated with proteins of the cell surface and other structural, rather insoluble, proteins. In the early development, nuclear proteins make up a considerable portion of the newly synthesized proteins.

The mass of the nuclei increases many times more rapidly than that of the embryo, hence it is not surprising that in sea urchin and loach nuclear proteins amount to one-fourth to one-half of all the proteins synthesized. This is why increased protein synthesis at the early stages is to some extent indicative of an increase in the mass of the nuclei. However, this phenomenon cannot account for all the fluctuations in the rate of synthesis. It may seem surprising that the templates for histones, which are not highly specific nuclear proteins, are not entirely provided for during oogenesis, but, rather are partially transcribed during early development. The data of these authors suggest that RNA synthesis and nuclear control of morphogenesis are not necessarily synonymous.

Dependence of Protein Synthesis on RNA

In the first experiments of Gross on protein synthesis in actinomycin treated sea urchin eggs, it was shown that early total protein synthesis does not depend on that of RNA. However, the recent papers of that laboratory report that at the early stages actinomycin changes the share of synthesis on small polyribosomes, where histones are synthesized on newly formed templates. Evidently, a decreased level of synthesis on the new templates, entailed by the suppressed RNA synthesis, is compensated for by the higher level of synthesis on mRNA produced during oogenesis. We studied protein synthesis in sea urchin and loach embryos either treated with actinomycin or submitted to heavy X-irradiation at various stages of early development. Both treatments produce a similar effect on protein synthesis, although the mechanisms of action are very different. Actinomycin interieres with RNA synthesis, whereas heavy doses of X-rays partially decrease the level of RNA synthesis but inhibit an increase in the synthesis during development. Heavy doses of radiation seem to impair transcription in such a way that the RNA synthesis after irradiation cannot serve as a template for protein synthesis.

It is essential that in both cases the does used ensured complete inactivation of the genetic function of nuclei. Actinomycin-(or X-ray) induced inactivation of sea urchin nuclei during the earliest post-fertilization stages of development does not change the pattern of protein synthesis. Thus, the eggs were treated with actinomycin prior to fertilization, and in the first hours after fertilization protein synthesis went on increasing, achieving the early blastula level. This increase means that protein synthesis at these stages occurs chiefly on the pre-existing templates stored during oogenesis. Thereafter, however, the rate of protein synthesis markedly decreases; this means that at these stages in normal development the molecules of mRNA being translated are those synthesized in the nuclei of the embryo Inactivation of nuclei at all following stages also causes a rapid decrease in protein synthesis which is obviously due to the decay of the short-lived mRNA. Nevertheless, the level of protein synthesis, always lower than that of control, is maintained for a long time, until the death of the embryos. Protein synthesis during oogenesis or in the nuclei of the embryo. Thus, protein synthesis in early development may occur on three kinds of templates: the ones stored during oogenesis, and the short and long-lived ones synthesized in the embryo.

The data on the dynamics of protein synthesis after nuclear inactivation at different stages has enabled us to estimate the quantity of mRNA and the time of its appearance and disappearance—i.e., the character of genetic control over protein synthesis. Of course, one should not except complete correlation between the rate of synthesis and the quantity of templates. It is possible, for example, that the translation on the old templates (when the synthesis of the new ones is blocked) will be unnaturally long, and their role in normal development will be erroneously overestimated. Fig. 8.4 shows similar data obtained with early (left) and later (right) loach embryos inactivated with heavy doses of X-rays. One can seen that the inactivation of nuclei at any moment during the first 6 hours of development (21°C) causes the same effect; i.e., protein synthesis slowly increases, achieving

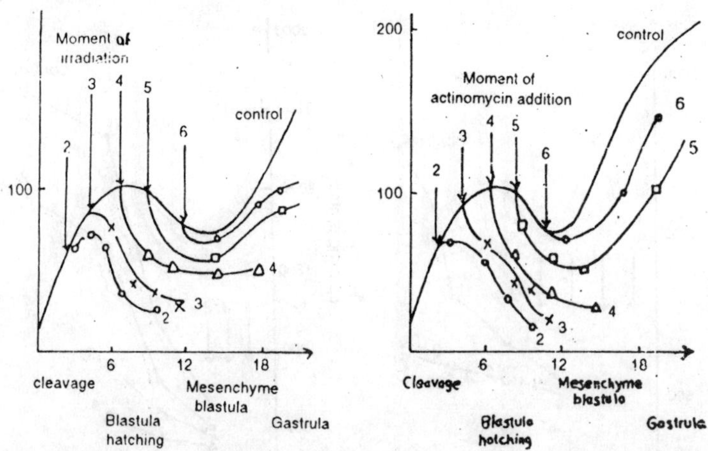

Fig. 8.4. Protein synthesis in sea urchin eggs after X-irradiation (left) or actinomycin treatment (right). The embryos were X-irradiated (10 krad) or actinomycin was added (25µg/ml) at moments of development indicated by arrows. Aliquots of embryos at different stages of subsequent development were taken and incubated with ^{4}C-labeled amino acids for 1 hour; the specific activity of protein was determined.

the level of the ninth hour or normal development. Morphologically, the irradiated or actinomycin-treated embryos also stop developing at the ninth-hour stage and then die; apparently in the loach the templates providing protein synthesis and development until late blastula are stored during oogenesis, as in the sea urchin.

The effects of nuclear inactivation at subsequent times during development strongly depend on time of irradiation—the later the embryos were irradiated, the more protein did they have time to synthesize on the templates produced before the moment of inactivation. For instance, the difference in protein synthesis between curves 2 and 3 may be due to the appearance of the templates synthesized between the stages indicated by arrows 3 and 2, i.e., within 1 hour between hours 7.5 and 8.5 of development. A comparison of the right and left diagrams makes it clear that genetic control over protein synthesis in the loach is not realized until the sixth-hour stage, begins at this stage, and coincides with the onset of RNA synthesis. It is clear that mRNA synthesized within a very short period of time is being translated for a long time during subsequent stages of development. And, vice versa, at any moment during early development, translation

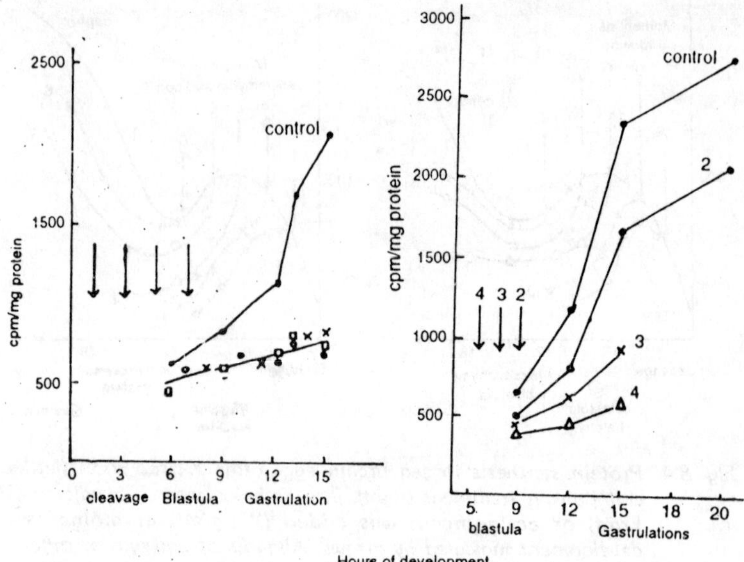

Fig. 8.5. Protein synthesis in loach embryo irradiated at different developmental stages. The embryos were X-irradiated (20 krad) at moments indicated by arrows. The blastoderms were isolated from the yolk at different stages of subsequent development and incubated for 1 hour with ^{34}C-labeled amino acids; the specific activity of protein was determined. Rates of protein synthesis after irradiation at different moments during early development (left) fit one curve.

occurs both on newly synthesized templates and on templates formed during oogenesis. Unfortunately, not only is our knowledge about the changing types of proteins synthesized in development confined to a small number of soluble proteins present in high concentrations, but it now appears that the data on mRNA mostly refer to the RNA's transcribed from repeated sequences. Therefore, we cannot say to what extent the pattern of mRNA synthesis and the pattern of translation predetermine the precise time table or programme of appearance of proteins responsible for differentiation.

RNA synthesis and its release into the cytoplasm is the necessary prerequisite for the synthesis of the proteins; but this does not mean that the programme for their synthesis depends entirely on the synthesis of RNA. For example, we don not know to what degree to lifetime of mRNA of different types is predetermined by their structure and to what degree by the ability of cytoplasmic structures to distinguish between different types of mRNA in such ways to accomplish their mRNA and the cytoplasmic components in the control of translation, one should consider the facts demonstrating independence of some parameters of translation from nuclear control.

Nonnuclear Control at the Translational Level

A classical example of the independent character of protein synthesis and the changes in undergoes in the course of development is the activation of synthesis after fertilization of the sea urchin egg. Without going into the details of the rich literature dealing with this problem, the mechanism of this process can be described as the activation of ribosomes and "unmasking" of mRNA stored during oogenesis. This is, however, not a unique example as at the subsequent stages of rate of synthesis has been also proved to be controlled by cytoplasmic mechanisms. This can be seen in the data illustrated.

After late nuclear inactivation, the level of protein synthesis is lower as compared to controls, but the patterns of the curves tend to be similar to those of the control. For example, in sea urchin embryos, irradiated or treated with actinomycin before hatching, protein synthesis continues to decrease, as in the control, but the rate of decrease is more pronounced. However, at the stages corresponding to the mesenchyme blastula and the onset of gastrulation, the rate of protein synthesis increases just as in the control, although the level of the control is not reached. This is due entirely to the cytoplasm mechanism, as the nuclei in these embryos are completely inactivated. In the case of loach a similar situation is observed.

Protein synthesis increase in the first hours of development up to mid-blastula, although the nuclei have not been inactivated at the later stage, protein synthesis still increases although no new templates capable of being translated are formed. This phenomenon is still more significant if irradiation has been carried out at the end of gastrulation (16 hours). Protein synthesis in the control decreases temporarily and then goes up again. In irradiated embryos with inactivated nuclei, the same effect is observed; i.e., protein synthesis goes down (the process is more rapid than in the control at the expense of degrading short-lived mRNA) and rises, as in the control. Being at all times lower than in the control, protein synthesis achieves the level of the moment of irradiation.

In other words, in loach and sea urchin, intensity of protein synthesis depends not only on the presence of the RNA templates but also on the cytoplasmic regulatory mechanisms which tend to maintain synthesis at the level characteristic for the given stage of development. Some information about this process may be obtained from the evidence about the maintenance of the normal level of total protein synthesis of sea urchin embryos treated with actinomycin prior to fertilization.

At the stages when part of the synthesized protein is being normally translated on the new templates, the synthesis of these proteins, which occurs mainly in the light polyribosome, is inhibited in the actinomycin-treated embryos; in a compensatory way there is an increase in protein synthesis in large polyribosomes, on the templates stored during oogenesis. One may believe that embryonic cells process a specific regulatory feedback mechanism which maintains in a non-specific manner the level of protein regardless of what proteins are being produced. It would be simpler to assume that protein is limited by some substance; then the absence of some templates would stimulate translation on others. However, if the concentration of this limiting substance changes regularly in the course of development, it means that we are dealing with a cytoplasmic regulatory mechanisms at the translational level. Two more examples will demonstrate that in embryonic development the quantitative relationship between RNA and protein synthesis is not very strict.

When loach eggs or spermatozoids are irradiated with heavy does of X-rays before fertilization, one obtains haploid embryos which seem to develop normally until late organogenesis and often from abnormal but moving larvae. At the mid-late blastula stage, the haploid embryos synthesize half as much RNA per nucleus) as the diploid embryos. At later stages in haploid embryos, a compensatory increase of the number of cells occurs and RNA synthesis in the embryos approaches that of the diploid controls. Yet in these embryos protein synthesis is at all times maintained at a normal level; i.e., it does not differ from the diploid controls. Apparently, the 2-fold decrease in the quantity of the templates formed at the blastula stage is compensated for by their more efficient translation. A different situation is observed in nucleocytoplasmic haploid loach X goldfish (*Carassius auratus*) hybrids. If irradiated loach eggs are fertilized with goldfish sperm,

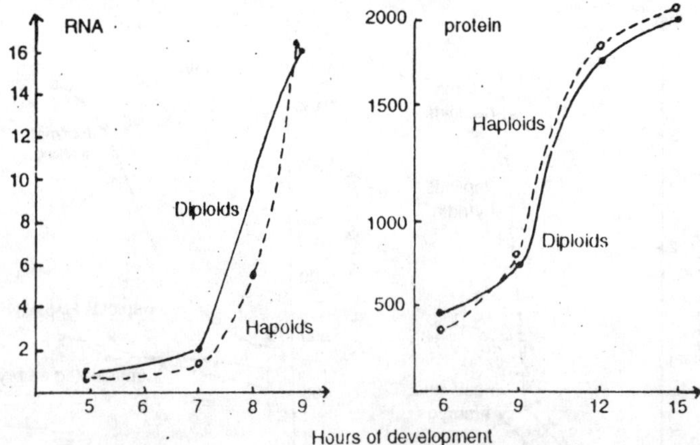

Fig. 8.6. RNA and protein synthesis in diploid and haploid loach embryos. Halpoid embryos were obtained by fertilizing the eggs with sperm irradiated with heavy doses of X-rays (50 krad). RNA and protein synthesis was determined after 1 hour of incubation with uridine-^{3}H or ^{34}C-labeled amino acids and express as specific activity.

they develop only to the late blastula stage, i.e., in the same way as the anucleated embryos obtained by irradiation of both gametes. On the other hand, if non-irradiated loach eggs are fertilized with goldfish sperm, the result will be well-developed hybrids possessing the traits of both parents. Hence, the goldfish nucleus in the irradiated loach cytoplasm cannot support development. And yet these embryos have a normal level of RNA synthesis which is just as intensive as in loach haploids. Unlike the case of loach haploids, protein synthesis in nucleocytoplasmic hybrids proceeds in a way similar to that anucleated embryos, i.e., slowly reaches the 9-hour stage (late blastula) and then ceases to grow. Thus, it seems that, in the absence of new loach RNA, goldfish templates cannot be translated in the loach cytoplasm. This experiments demonstrates that the *presence* of mRNA is not sufficient condition to allow normal translation to take place. If this is so translation control is, in turn, under some kind of nuclear control.

Thus, in embryogenesis translation is also a point of regulation in the pathway of realization of genetic information. Nuclear control might be exerted via the composition and quantity of mRNA on both rate and types of proteins synthesized. However, the fact that half the amount of mRNA of a haploid embryo may be

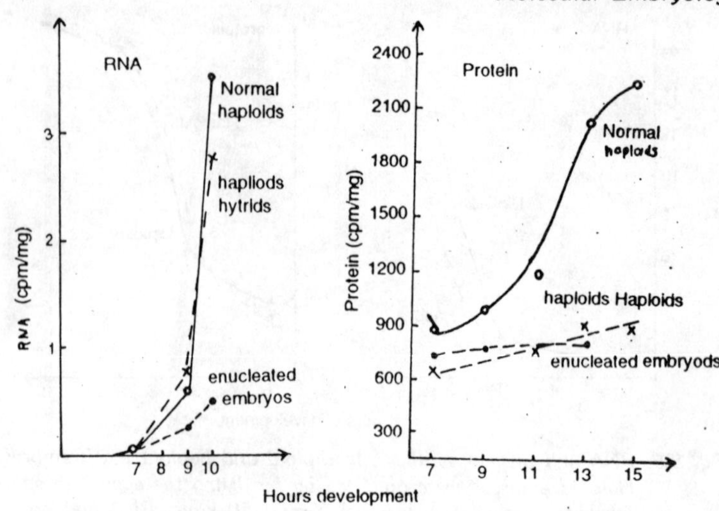

Fig. 8.7. RNA and protein synthesis in haploid loach embryos, haploid androgenetic (nucleocytoplasmic) hybrids of loach X goldfish, "anucleated" loach embryos. Irradiated loach eggs (20 krad) were fertilized with normal sperm of loach, with sperm of goldfish, or with irradiated (50 krad) loach sperm. Assau of RNA and protein synthesis.

translated at the same rate as in a diploid one, should be taken to mean that synthesis normally occurs not at its highest possible rate. Only degradation of a major fraction of short-lived mRNA, as the case after actinomycin treatment, may cause a decrease in the rate of protein synthesis. The remaining long-lived mRNA continues to be translated at a submaximum rate, and a certain stage of development synthesis on the same templates may increase considerably. One of the mechanisms involved in the regulation of the rate of protein synthesis is the change in the amount of ribosomes taking part in the translation, i.e., the quantity of polyribosomes. This is the case in the early development of sea urchin, but it is quite possible that there are other mechanisms and that the rate of translation itself may be regulated.

The lifetime of mRNA is of importance for the character of protein synthesis. All we know is that some mRNA's function in the cell for a short time, and others much longer. The difference between long-lived and short-lived mRNA have not been elucidated. But they seem to be inherent in the structure of RNA, i.e. determined by nucleus. Finally, direct control over the composition of the proteins synthesized may consist in the selection of the types of mRNA being translated. There is at least one example of such regulation.

In the course of spermatogenesis, RNA synthesis is arrested until meiotic divisions (in *Drosophila*) or right after them (mouse). However, it is only after this that differentiation of the spermatozoid begins. Evidently, it occurs without direct nuclear control, at the expense of mRNA synthesized before meiosis. One possibility is that during the entire sequence of all these complex events is determined by the same proteins. However, it was shown by cytochemical and autoradiographic investigations that in the maturing spermatozoids of *Drosophila* the lysine-rich histones are replaced by the arginine-rich histones. This process is sensitive to puromycin; i.e., it is a true *de novo* protein synthesis. No such synthesis was observed throughout spermiogenesis; consequently, the translation of the mRNA coding for the specific proteins of the spermatozoid head begins as late as several days after it has been synthesized.

Regulation of Enzymatic Activity

In this section we consider two examples demonstrating the stage of realization of genetic information following translation; i.e., the enzymatic function of the protein synthesized may also involve regulatory mechanisms. It should be noted that by regulation we mean, not the maintenance of a constant level of enzymatic activity (of the type of allosteric inhibition), but rather the control over the changes in the enzymatic activity in the course of the embryonic development which is realized independently of genomic control. In the early development of loach embryos. The rate of glycolysis increases to a maximum at the end of gastrulation. An analysis of the activity of all glycolytic enzymes led Milman and Yruowitzki (1966, 1967) to the conclusion that the increase in rate of glycolysis cannot depend on an increase enzyme activity. It turned out that, in fact, in the course of development the activity of the enzymes of gluconeogenesis, i.e., glucose-6-phosphate dehydrogenase, fructose diphosphatase, and phosphoenolpyruvate carboxylase, gradually decreases. The decrease in the activity of these enzymes is not associated with the formation of some inhibitor, but rather it depends on the degradation of the enzymes—alternatively, on the rate of degradation prevailing over that of synthesis.

Such a simple mechanisms makes it possible to intensify the respiration of the embryo as developed proceeds. Whether or not this mechanism is genetic is just a question of terminology. On the one hand, the intensification of glycolysis occurs within the functioning enzymatic system, and the changes in its functions do not require additional nuclear control or macromolecular synthesis. On the other hand, the short lifetime of the enzymes of gluconeogenesis is compared to the enzymes of glycolysis is likely to be genetically programmed. However, we suggest that the processes that can take place autonomously in the cytoplasm should be considered as "non-genetic" or "cytoplasmic." We have also revealed some changes in the activity of asparate aminotransferase in the sea urchin that is also an example of non-genetically

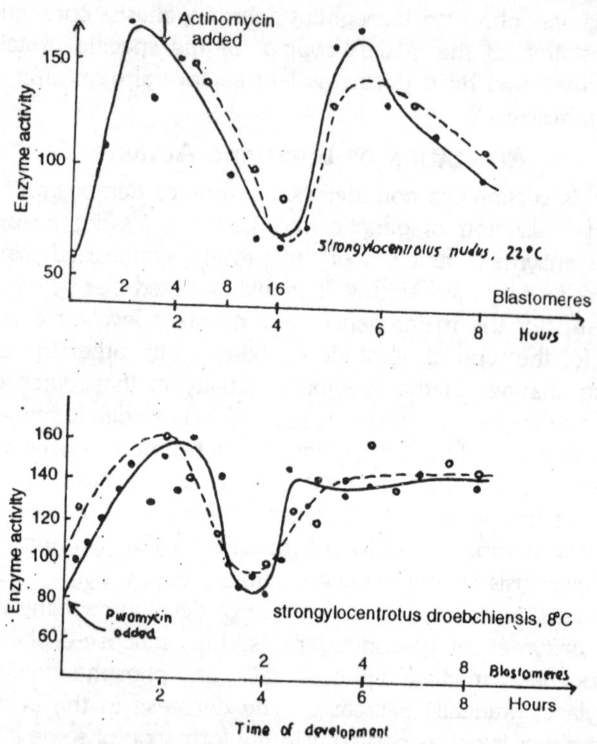

Fig. 8.8. Changes in asparate aminotransferase activity in early development in the sea urchins Strongylocentrotus nudus (top) and S. droebachiensis (bottom). Enzyme activity is expressed in arbitrary units, •—• Control; •-•, in the presence (top) of actinomycin (25 μg/μl) or bottom of puromycin (100 μg/ml).

regulated alteration of enzymatic activity in early development. This enzyme changes its activity in a rather characteristic way which is difficult to associate with the morphological or biochemical events of early development.

It is interesting that its activity is not affected by either actinomycin or puromycin. A sufficiently high concentration of puromycin is known to inhibit completely even cleavage; nevertheless, the behaviour of asparate aminotransferase in such an uncleaving, but fertilized, egg is absolutely similar to that in the normally developing embryos of the control. We had only limited success in elucidating the mechanism of the above activity changes. They may be partially accounted for by the concentration of the cofactor, pyridoxal 5-phosphate, in the egg, although some of the changes seem to be connected with the enzyme protein itself. If this is the case, the significance of the changes in enzymatic activity is difficult to evaluate. But it is quite apparent that the changes do not depend on direct genetic control, synthesis of macromolecules in general, or even morphological manifestations of development.

Genetic Aspects of Early Mammalian Development

Knowledge of the genetics of pre-implantation development in mammals and man is increasing rapidly with our improving control over these stages. It would be impossible to do full justice to the entire subject is one lecture. Fortunately, Dr. Ralph Brister has covered major areas of this topic in his description of the synthesis of macromolecules during preimplantation development. I will concentrate largely, although not entirely, on our own work. Most of my remarks will also be devoted to the oocyte and the embryo, but first a few comments are called for on the genetics of spermatozoa, especially with reference to attempts at the identification of different types of spermatozoa from their haploid phenotype.

Is there Haploid Expression in Mammalian Spermatozoa?

This question has been asked repeatedly, with divergent answers. Differences have been reported to exist between X and Y-bearing spermatozoa for example, in response to changes in pH or time of insemination in relation to ovulation. This contention implies that phenotypic differences exist between these two types of spermatozoa, i.e., in their charge or metabolic pathways. Yet when we examine the biochemistry of spermatogenesis the evidence tends to suggest that such differences do not exist. RNA synthesis by the XY bivalent ends before meiosis and does not resume in the spermatid. This bivalent also has the typical Beterochromatic appearance of chromosomal regions inactive in RNA synthesis. Phenotypic characteristics determined by sex-linked genes would thus display 'diploid' inheritance from the spermatogonium rather than haploid inheritance from the spermatid.

Moreover, large areas of the cell wall disappear or are not formed between neighbouring spermatids. This situation could result in considerable mixing of the products of adjacent spermatids; 'metabolic compensation' occurs between some adjacent cells in culture when cell contacts are present. If therefore appears doubtful that only gene products typical of X or Y-linked genes are present in spermatozoa. The situation with autosomal genes could be different, because there is evidently weak synthesis of RNA by autosomes in spermatids. Again, however, the dominant contribution to the differentiating spermatozoa would appear to be provided from the earlier diploid stages. When we actually look for firm evidence of expression in mammalian spermatozoa, the only clear cut example is that in Tt^{12} heterozygous male mice. Differences between spermatozoa carrying T and t^{12} can be illustrated merely by altering the time of mating with respect to the time of ovulation.

If the mating and ovulation coincide, each class of spermatozoa fertilizes one-half of the eggs; however, if the spermatozoa have to wait during a period of delayed ovulation then the great majority of eggs are fertilized by t^{12} spermatozoa. This is clear evidence of a 'haploid' effect in spermatozoa. Other examples quoted of haploid expression are very doubtful. Blood-group antigens provide a classic example : after many claims that these antigens could be detected on spermatozoa and showed "haploid" expression, opinion has now changed and there is a major probability that the earlier workers merely detected "sector" antigens adhering to spermatozoa. The suggestion that anomalies in the inheritance of a human translocation were due to distortion in fertilization ratios has been withdrawn. Another example may be the melanising activity of rabbit spermatozoa, heterozygotes for pigment genes producing two types of spermatozoa—one with the other without melanising activity.

Recently fresh data has been brought forward revealing the existence of transplantation antigens in humans and mouse spermatozoa, those on human spermatozoa showing haploid expression. It will be of interest to follow subsequent developments in this field. The foregoing examples show that haploid expression might exist, but the evidence is largely unconvincing except in one case. There is clear evidence that the diploid soma of the male can exert profound influence on the inheritance of characteristic in spermatozoa. These data have been extensively analyzed by Beatty and his collaborators. Before leaving this topic, I must stress that techniques designed to detect the X or Y chromosome in the sperm-head methods to separate X or Y-bearing spermatozoa by sedimentation, or other methods relying on physical differences between sperm due to the X-Y difference, could be of potential value. But repeated claims to have used these method of separation are consistently denied : a recent paper provides details.

Gametogenesis and the Origin of Chromosomal Anomalies in Animals and Human Embryos

A considerable number of humans abortuses are non-diploid. Of the total 20% are triploid, and the origin of these is probably through delayed fertilization resulting in polygyny or, more likely, polyandry. But the majority of non-diploids are trisomics. How do these arise? The possession or lack of one chromosome could obviously be due to non-disjunction, and this could occur during meiosis or mitosis. Errors in mitosis after the early cleavage divisions could result in mosaicism; mosaics occur but most trisomics are uniform. If the origin of trisomy is due to the formation of univalents, it can be detected in diakinesis; if failure of separation of chromosomes is the reason, then anaphase is the optimal stage for examination.

Studies in meiosis in human spermatocytes has shown that univalents are exceedingly rare among the autosomes, but that the XY often shows end-to-end association. In contrast, examination of mouse oocytes revealed that with increasing maternal age, the number of univalents increased markedly. Similar anomalies were seen in human oocytes but much more data on frequency is needed, Univalents were fund among all the sizes of chromosomes, perhaps more frequently among the smaller sizes. Based on these observations, autosomal trisomics would arise through errors in oocytes, but sex chromosome trisomics could arise in either spermatocyte or oocyte. There is strong supporting evidence for this suggestion. Using the Xg^2 marker on the sex chromosomes, Race and Sanger (1969) found that sex chromosome trisomy could arise in spermatogenesis or oogenesis. Different methods are needed for the analysis of autosomal trisomy, and one important piece of evidence is that the incidence of such human trisomies is correlated with maternal age and not with paternal age. Clearly these data support the suggestion that autosomal trisomies arise in oocytes, and that sex-chromosome trisomies arise in both spermatocyte or oocyte. Other suggestions have been made for the origin of trisomy in oocytes, e.g., that decreased frequency of intercourse in older couples leads to delayed fertilization. The experimental data from animals supports neither this contention, nor various others.

Genetic Aspects of Pre-implantation Development

Studies on invertebrates have shown that the early period of development after fertilization is largely controlled by maternal templates inherited from the oocyte. The situation in mammalian embryos appears to the different. Transcription evidently begins in the 4-celled stage or earlier Dr. Brinster has reviewed other pertinent evidence in the symposium. Various other data can be added in support of this suggestion, e.g., the death of YO mouse embryos during early cleavage. Moreover, genes are also active in the late morula, e.g., those causing the death of homozygous $t^{12}t^{12}$ mouse embryos. By the blastocyst stage there is considerable evidence of genetic activity, as shown by the death of some species hybrids at this stage, and by the presence of sex chromatin in rabbits and other blastocysts.

Some claims have been made that the transplantation antigens of the foetus are expressed in these stages. If true, these reports could obviously be important in our understanding of maternal/foetal interactions. In our own work we could not detect these antigens using mixed cell antiglobulin tests. Despite the initiation of genetic activity in these early stages, it is clear that major alterations in chromosome number are complete with early development. Triploids can survive to post-implantation stages, but die shortly after implantation. A curious fact about triploids is that the number of sex chromatin bodies in the nuclei is determined by different rules to those effective in the sex chromosomes polysomies. Recently, haploid and diploid parthenogenetic mouse embryos have been shown to survive to implantation or beyond. Haploid was induced by electrical stimulation of oocytes or by exposure of the oocytes to enzymes. It is surprising that earlier attempts to induce gynogenesis (i.e. the initiation of the development of the egg by a spermatozoon with its chromosomes previously inactivated by irradiation, dyes etc.) merely resulted in abnormal early cleavage by the haploid embryos.

Manipulation of Embryos

Earlier work has shown that the blastomeres resulting from the early cleavage divisions of mammalian embryos are very liable. Some of the blastomeres can be destroyed without preventing embryonic development and fusion of cleaving eggs and morulae can result in chimaeric embryos. The blastocyst can now be manipulated in various ways, and still give rise to full-term foetuses. Pieces of trophoblast can be dissected from sheep or rabbit blastocysts. The trophoblast excised from rabbit factuses served to sex the blastocyst by means of the sex chromatin body in females.

A recent advance promises to extend his work, for the long arm of the Y chromosome distal to the centromere can now be stained selectively in interphase or mitosis by means of the dyes quinacrine mustard or quinacrine hydrochloride. These dyes should serve as excellent markers for male embryos at all stages of development. Equally challenging is the recent work showing that entire inner cell masses or individual cells masses or individual cells can be inserted into blastocysts, and can give rise to cells colonising large areas of the resulting foetus. The inner cells mass can be withdrawn entire from one blastocyst, and placed into a recipient blastocyst. The inner cell mass of the recipient and donor can intermingle, and so lead to the formation of chimaeric foetuses. Alternatively, the inner cell mass of the host can be excised before the donor inner cell mass is inserted.

Even more astonishing is the insertion of a single cell into a recipient blastocyst, the donated cell producing cell lines that colonise vast areas of the developing foetus. Sex-linked genes are proving of great value in plotting events during cellular differentiation in the early mammalian embryo. According to the Lyon hypothesis, one X chromosome is inactivated during early development. Inactivation is usually random, so that approximately one half of the cells in the embryo express genes on one X chromosome, and the remaining half express genes on the other. Deviations from equality could obviously occur in particular tissues where there are only one or a few stem cell precursors in the embryo. Gandini et al., (1968) analyzed tissues of humans heterozygous for mutants at the glucose-6-phosphate dehydrogenase locus for the ratio of the two types of cell, and indicated that a limited number of stem cells (8 or less) are precursors of the haemopoietic tissue in the human foetus. This form of "clonal" development could obviously apply to other tissues.

The conclusions of Gandini et al. are negated if non-random inactivation of the X's, of it cell selection for one cell type occurs extensively. These phenomena were evidently absent from their material and did not invalidate their conclusions. But non-random inactivation or cell selection apparently does occur in the erythropoietic lines of cells in the Lesch-Nyhan syndrome. One consequence of clonal differentiation from a few precursor cells could be to provide an opportunity for selectively colonising particular tissues of a host embryo. Placing two or three precursor cells of a tissue in a blastocyst could lead to the extensive colonisation of that tissue in the foetus. Cells specific for particular tissues might be obtained by using the embryonic inducers to control the differentiation of cells in the donor blastocyst. Outgrowths of cells from rabbit blastocysts show considerable differentiation in culture in the presence or absence of inducers.

9

Genes Regulating Amphibian Metamorphosis

Metamorphosis is one of the most dramatic development changes in late embryonic life. Some of the classical notion of functional maturation, as in the case of acquisition of urea formation in Amphibia, have emerged from a biochemical study of amphibian metamorphosis. The initiation and completion of metamorphosis in both amphibians and insects is under obligatory hormonal control which distinguishes it from other hormone-dependent late developmental changes. It is well known that an anuran tadpole or an insect larva will never turn into its adult form if the respective endogenous metamorphic hormones, thyroxine and ecdysone, were withdrawn or prevented from reaching the target tissues. Another feature that distinguishes both amphibian and insect metamorphosis from ordinary adaptational responses is that the process is begun and completed in anticipation of a change in environment.

The hormone serves to trigger a predetermined programme of developmental changes in order to prepare the embryo for an environment suitable for adult life. Since this chapter will deal with induction of new function and proteins, it is important to note that the hormone in no way directs cell to acquire differentiative characters but allows already differentiated but immature cells to acquire adult functions and structures. Ever since the discovery by Gudernatsch (1912) that exogenous thyroid hormone will cause the precocious onset of metamorphosis in tadpoles, hormonal induction of the process is a well-established procedure of studying sequential biochemical changes in a variety of amphibians and insects. In most studies little has been learned about the mechanism of action of the hormone, later usually serving as convenient tool to induce or delay developmental process in free-living embryos. It is in this context of a developmental tool that thyroid hormones will be treated in this article which is restricted to amphibian metamorphosis.

There are great similarities in the types of phenomena involved in insect metamorphosis which has been reviewed by many authors. The following two main facets of the regulation of protein synthesis leading to a well-ordered and sequential acquisition of new functions or structures during metamorphosis will be considered separately: (1) the regulation of RNA synthesis, formation of enzymes, and alterations in cellular structures in tissue, such as the hepatocyte, that undergo further development or functional maturation; (2) the importance of new or additional protein synthesis in tissues, such as the tail, that are programmed for death or regression. In accordance with the aims of this publication the account given below is based on personal interests and ideas and is not meant to be an exhaustive review of the subject. For the latter, the reader is referred to several reviews that have appeared in the last few years.

The Role of Hormones in Amphibian Metamorphosis

Net long after the discovery by Gudernatsch (1912) that feeding tadpoles on mammalian thyroid tissue induced precocious metamorphosis, it was found that thyroidectomy of larvae (or feeding larvae on antithyroid drugs) prevented their development into adult frogs or toads. The dormant thyroid tissue of the developing larvae is activated into producing and secreting the two thyroid hormones, L-thyroxine and, 3, 3, 5-triiodi-L-thyronine (T_3), by thyrotropic hormone (TSH) produced by the anterior pituitary. The larval pituitary itself is under neural control. Thus eventually it is some environmental factor, such as illumination, temperature, salinity, that acts as the initial trigger for metamorphosis.

Although a hypothalamic tripeptide, thyrotropin-releasing factor (TRF), of the kind recently described in mammals has no yet been discovered in amphibian larvae, there is much indirect evidence to suggest that a similar neurochemical link may exist between external signals and the production of endogenous thyroid hormones. Of great interest in considering hormonal control of metamorphosis is the relatively recent discovery in amphibians of "*juvenilizing*" factors of the type known for insects. The level of insect juvenile hormone is well known to determine whether or not or how the larval target tissues will respond to the metamorphic stimulus of ecdysone. It now seems that prolactin obtained from mammalian pituitaries exerts a similar juvenilizing effect on tadpoles in that its administration at the same time of spontaneous or thyroid hormone-induced metamorphosis arrests, delays, or modifies the process. In salamanders and newts (urodeles), which do not undergo the same type of metamorphosis as in frogs and toads (anurans), prolactin causes the terrestrial forms to return to water, a phenomenon known as "*water drive*" or "*second metamorphosis*."

Proteins Involved in Metamorphosis

Accompanying the dramatic visible morphological changes during metamorphosis tail (regression, limb emergence, positioning of eyes, etc.) are profound functional and biochemical changes in most tissues. Many of the changes or acquisition of new functions necessary for the organism for a terrestrial life are a consequence of the induction or preferential synthesis of proteins. The induction of the enzymes of urea cycle leading to the switch from ammonotelism to ureotelism during amphibian metamorphosis is new indeed a classic of the biochemical basis of development. First, virtually every type of cell in the embryonic tissue is subjected to the action of thyroid hormones, and none fails to respond in some important way. (This is not too surprising if one is declining with a rapid transition from an aquatic to a terrestrial life). Second, the hormone acts locally and directly on different cells, as for examples Kollros (1942) has shown that only the part of that tail to which thyroxine was applied underwent lysis, leaving the rest of the tail intact. In a series of elegant experiments, Wilt (1959; Ohtsu et al., 1964) had shown that only that eye to which thyroxine was applied developed rhodopsin, leaving the other with the larval visual pigment, porphyropsin.

There is thus no indirect systemic action of the hormone although an alteration in the metabolic pattern of one tissue is eventually bound to affect the activity of another via secondary adaptational routes. Third, the types of proteins whose synthesis is induced is determined by the nature of the cell, that is to say that there is qualitatively no hormone-determined protein synthesis.

In some cases, the positioning of some types of cells determines the response. For example, epithelial cells of the tail synthesize collagenase while those on the skin accumulate collagen under the influence of thyroid hormones or that thyroxine may provoke regression as well as growth in adjacent cells of Mauthner's neurons. The last points is of considerable importance in the definition of the role of developmental hormones in the genetic control of protein synthesis. The multiplicity of responses to thyroxine in the same organisms makes it highly unlikely that the hormones has any inherent informational content to determine, say, via an interaction with genes or repressors, to induce the synthesis of a fixed number of proteins. When one extends the above list to be very different actions of thyroid hormones in fish, bird, and mammals, it is only reasonable to conclusion that once a cell has the receptor to recognize the hormone then it makes use of the hormone as a trigger mechanism to initiate processes determined, but not expressed, during early differentiation.

Regulation of Protein Synthesis During Metamorphosis

Current Concepts of Regulation of Protein Synthesis in Animal Cells

It is now widely accepted that, although some fundamental concepts of DNA transcription and messenger RNA translation are universal, regulation of protein synthesis in nucleated cells of higher organisms involves additional control steps not described in microorganisms. Several reviews have been devoted to this topic, and the reader's attention is drawn to two publications edited by San Pietro et al. (1968) and Wolstenholme and Knight (1970). The features mentioned briefly below are important for the interpretation of the work to be described later.

Translational Control

The contrast between long-lived proteins synthesized on unstable bacterial messengers and some animal proteins of short half-life synthesized on relatively stable messengers has prompted many investigators to propose an extranuclear regulatory mechanism in higher organisms. Much of the earlier evidence for a translational control of protein synthesis was based on indirect manifestations of inhibitors of RNA and protein synthesis and did not establish the exact level at which such a control could be affected. The explanation of Wool et al. (1968) for the anabolic actions of insulin and of Korner (1970) for that of growth hormone are based on an almost direct modification of the functioning of ribosomal subunits. But the currently most favourable hypothesis of translational control is that put forward by Tomkins to account for the induction of tyrosine aminotransferase in rate hepatoma cells. Tomkins has proposed the existence of a cytoplasmic repressor whose function is to control stability or availability of messenger RNA, for translation. It is the synthesis of such a repressor that is thought to be under hormonal control. No such evidence is yet available for metamorphosis or similar late embryonic developmental systems, and direct translational control important as it may eventually turn out to be, will not be discussed in this article.

Nuclear Restriction and Breakdown of RNA

A substantial amount of RNA synthesized in the nucleus is rapidly turning over, heterodisperse and of high molecular weight. This RNA is not a precursor of rRNA, mRNA, or tRNA and DNA-RNA hybridization studies have shown that many of the species of RNA within the nucleus do not appear in the cytoplasm. Britten and Davidson (1969) have suggested that the rapidly turning over intranuclear RNA hybridizes very rapidly turning over intranuclear RNA hybridizes very rapidly with DNA and may be a product of repeating DNA sequences or "redundant" DNA. Although a precise role for this RNA is far from established, they and others have thought it may be somehow important in differentiation and growth.

Selective Transfer of RNA from Nucleus to Cytoplasm

An intranuclear restriction of one whole class of RNA poses the question of what mechanisms control a selective transfer of RNA from the nucleus to the cytoplasm. The transfer of both ribosomal and messenger RNA seems to require the formation within the nucleus of ribonucleoprotein particles which are then incorporated into cytoplasmic polysomes. Several workers have now detected informosome-like particles both in the nucleus and the cytoplasm. The role of the protein associated with such informational particles in the polysomes may be an important factor in the availability or rate of mRNA translation. Very little is yet known about the nature of these proteins, their site of synthesis, how they react selectively with mRNA, etc. but any developmental process must depend on their ready availability.

Structural Requirement for Protein Synthesis

This article deals in some detail with an important question which has not received much attention concerning regulation of protein synthesis in animal cells. It relates to the association between cellular structure and protein synthesis, particularly the attachment of ribosomes to membranes of the endoplasmic reticulum. Such an attachment is known to affect protein synthesis, but so far the main role for the structural association is thought to be that of secretion of proteins. However, the special role of membrane-bound ribosomes in bacteria and the presence of highly active membrane-bound ribosomes in predominantly non-protein secreting tissues suggests some other function for the membrane-ribosome attachment. It is of particular interest to note that the ribosomes and membranes to which they are attached are turning over continuously and that there exists a tight co-ordination between the proliferation of membranes and ribosomes when additional demands are made for protein synthesis during growth and development.

Regulation of Protein Synthesis in the Developing Tadpole Hepatocyte During Metamorphosis

The liver of the tadpole has been intensively studied in the laboratories of Frieden (1967; Friedsen and Just, 1970) and Cohen (1966, 1970) with respect to the synthesis of proteins that characterize amphibian metamorphosis, such as urea cycle enzymes, serum albumin, and adult haemoglobin. These and other workers had firmly established, especially in the bullfrog, *Rana catebeiana*, that administration of thyroid hormones to pre-metamorphic tadpoles causes a *de novo* synthesis of these proteins. Because of this firm biochemical background, the author's laboratory has over the last seven years investigated the formation and turnover of nuclear and cytoplasmic RNA in the pre-metamorphic bull frog hepatocyte at different stages after hormonal induction of metamorphosis in order to understand the nature of the process of induction. Cohen's group have also been extensively investigating the metabolism of RNA *in vivo* and in isolated preparations of tadpole liver.

Thyroid Hormone Induced Formation of New or Additional Proteins

In bullfrog tadpoles (*Rana catesbeiana*) a rather long lag period of about 6 days elapses after exogenous thyroid hormone administration before new or additional proteins, characteristic of the metamorphic change in the function of the liver, could be detected. The increase of urea cycle enzymes or the appearance of serum albumin in the blood is preceded by a day or two by a rather abrupt increase in the rate of amino acid incorporation into protein *in vivo* per unit of ribosomal RNA. The decline in the rate of incorporation is only an apparent one caused by the progressive changes in levels of free amino acid as regression of organs like tail, intestine, and gills gets under way.

To some extent the lag period preceding the increase in protein synthetic rate represents the time for additional RNA to be synthesized and processed in the nucleus. Some of the RNA synthesized during the lag period is certainly important for the *de novo* synthesis of hepatic metamorphic proteins since actinomycin D, administered with or soon after thyroxine will prevent the rise in carbamyl phosphate synthetase. However, hormonal control of transcription is not the exclusive mechanism for the induction of these proteins, and it is thought that thyroid hormones are also required for a sustained translation of messenger RNA or for some process of maturation of the protein molecules. Cohen and his

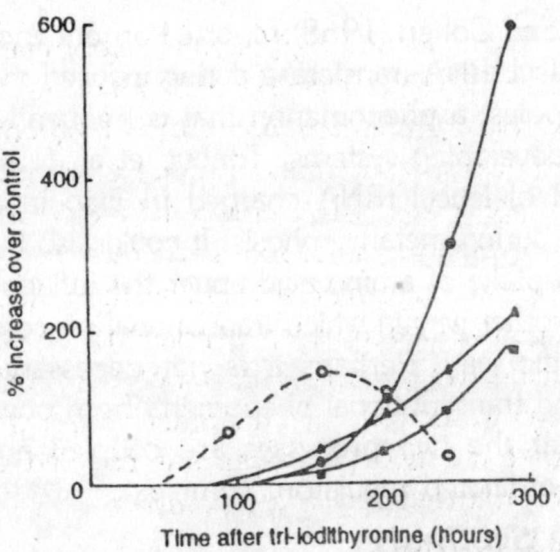

Fig. 9.1. Schematic representation of the lag period for the induction by triiodothyronine of de novo protein synthesis and the stimulation of amino acid incorporation into hepatic protein of Rana catesbeiana tadpole. o--o, specific radioactivity of protein recovered in the liver microsomal fraction 40 minutes after the administration of a mixture of 14C-labeled amino acids; •–•, carbamyl phosphae synthetase; Δ–Δ, specific activity of cytochrome oxidase in mitochondrial fraction; ■–■, serum albumin accumulation in blood.

colleagues have concluded that thyroxine exerts a dual action in regulating the rate of formation of the key urea cycle enzyme, carbamyl phosphate synthetase as well as for glutamate dehydrogenase.

It seems that part of the induction is due to control of transcription which involves all species of RNA and that almost simultaneously the hormone controls the activation of an inactive form of the enzyme. These results were based on the discrepancy between the detection of the enzyme by immunochemical methods and measurement of enzyme activity. Part of the inactive enzyme was preformed and part synthesized following hormone administration. Shambaugh et al. (1969) have further found that thyroxine also stimulates the synthesis of the inactive form of the messenger for carbamyl phosphate synthetase, but it is not certain whether or not it involves mechanism based on the inactivation by the hormone of a cytoplasmic translational repressor of the type described by Tomkins for explaining the induction of tyrosine amino transferase in cultured rat hepatoma cells by cortisol. How or at what rate-limiting step of translation the hormone exerts an effect on protein synthesis is not clear although a few observations may be relevant.

Unsworth and Cohen (1968) reported an enhanced activity of hepatic aminoacyl tRNA transferase during induced metamorphosis of bullfrog tadpoles, a phenomenon that is frequently observed in many rapidly developing systems. Tonoue et al. (1969) found an altered pattern of leucyl tRNA charged *in vivo* in a number of tadpole tissues during metamorphosis. It could also be argued that the enhanced uptake of amino acid under the influence of thyroid hormones is another way in which translational process is facilitated in a non-specific way. Perhaps it is not necessary to separate translational and transcriptional phenomena from one another, but to consider that the two processes are coupled and integrated into a well co-ordinated regulatory complex.

Nuclear RNA Synthesis

It can be that there occurred, well within the latent period of 5-6 days for new proteins to be detected an acceleration of the rate of nuclear and cytoplasmic RNA synthesis in bullfrog tadpole liver following induction of metamorphosis with triidothyronine. Somewhat similar results in the same species of tadpole induced with thyroxin have also been observed in Conen's laboratory. The curves for rates of RNA synthesis have been derived from values of specific activity of RNA obtained after correction for changes in uptake of the radioactive precursor which occur earlier after triiodothyronine, as has also been bound by Eaton and Frieden.

An early perturbation in the pool size or uptake of precursors of RNA and protein in target tissues has been observed with hormones known to affect protein synthesis. The question of pool sizes in radioactive labeling of constituents of non-regressing tissue is even more complicated when regression occurs in other tissues in the same organism during metamorphosis. Thus, the abrupt downward trend in the specific activity of nuclear and cytoplasmic RNA as was also noticed for amino acid incorporation, is not due to a sudden reversal of the accelerated rate of synthesis but merely reflects a dilution of radioactive precursors caused by the autolysis in regressing tissues, such as the tail, gut and gills. It would be tempting to suggest that the additional RNA made following hormone administration includes messengers for proteins like urea

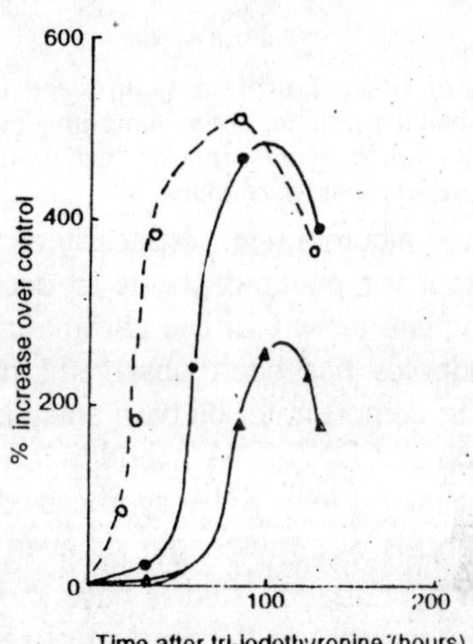

Fig. 9.2. Schematic representation of the stimulation of nuclear RNA synthesis and its turnover into the cytoplasm of liver of Rana catesbeiana tadpoles after induction of metamorphosis with triiodothyronine, o–o. Specific activity of nuclear RNA labeled with uridine-^{3}H not corrected for changes in distribution of radioactivity in the acid-soluble fraction: •–• specific activity of nuclear RNA after correction for changes in acid-soluble radioactivity, Δ–Δ specific activity of RNA recovered in cytoplasmic ribosomes and polyribosomes.

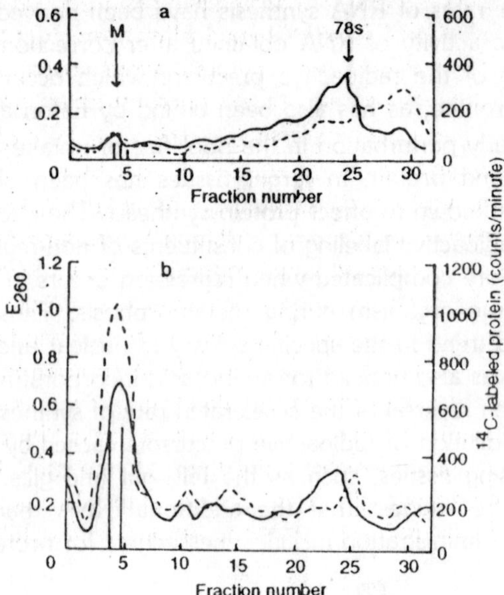

Fig. 9.3. Distribution of nascent protein synthesized in vivo on hepatic polysomes obtained from (a) pre-metamorphic bullfrog tadpoles and (b) animals in which metamorphosis had been induced 7.0 days before the preparations were made.

cycle enzymes, serum albumin, etc., especially as a sustained RNA synthesis is important for metamorphosis to occur. Furthermore, an alteration in template activity of live chromatin from thyroxine-treated bullfrog tadpoles has been observed. In our laboratory, however, we failed to demonstrate, by base analysis, sucrose density gradient fractionation and DNA-RNA hybridization that a significant part of the additional nuclear RNA synthesized *in vivo* at the onset of metamorphosis was messenger or even DNA-like RNA.

Our DNA-RNA hybridization studies have shown that whereas there did occur a net increase in the amount of readily hybridizable nuclear RNA, there was also a drop in the overall "hybridization efficiency" of the RNA formed during metamorphosis. This paradoxical finding really suggest that very small increases in DNA-like RNA following hormone administration are accompanied by relatively massive increases in the rate of synthesis of ribosomal RNA. This is not to say that no changes occurred in the nature of genes transcribed under hormonal influence but that one does not yet have sensitive enough methods to detect a small change in a wide spectrum of nuclear RNA molecules with eventual messenger function. Much of the DNA-like RNA in the nucleus, which is also rapidly hybridizable, is not found in the cytoplasm, and although it may have a function in differentiation it is not thought to be direct precursor of cytoplasmic messenger RNA.

Cytoplasmic RNA

A consequence of the burst of nuclear RNA synthesis is the appearance of additional cytoplasmic RNA, mainly in the heavier polyribosomal aggregates with a concomitant decrease in the relative amount of monomeric ribosomes or ribosomal subunits. This build-up of polyribosomes occurs at a times just preceding the appearance of new proteins, as can be judged from a comparison. An increased rate of accumulation of newly synthesized ribosomes and polyribosomes coinciding with new or additional protein synthesis seems to be a common feature of regulation of protein synthesis during growth and development. However, unlike the situation of hormone-induced growth in many mammalian tissues, there is no appreciable accumulation of ribosomes per cell during the initial phase of induction of metamorphosis (6-10 days).

Table 9.1. Double-labelling or Ribosomal RNA Demonstrating that Both Breakdown and Synthesis of RNA are Accelerated at the onset of Metamorphosis

Day on which orotic acid¹⁴C administered	*Metamorphosis induced*	*specific activity (cpm/mg of RNA)* ^{3}H	^{14}H	$3C^{14}C$ *raton*
0	–	17,350	4,600	3.77
	+	14,900	5,230	2.85
2	–	15,765	5,680	2.78
	+	8,010	7,825	1.02
4	–	15,500	3,650	4.24
	+	6.360	11,300	0.56
6	–	12,870	4,100	3.13
	+	5,085	9,420	0.54

This conclusion was borne out by the double isotope labeling studies shown in table 9.1, which suggest that there is an accelerated turnover of cytoplasmic RNA and ribosomes as additional ribosomes appear after hormone administration. It seems that there is some mechanism within the developing cell that selectively breaks down "old" ribosomes and allows the preferential accumulation of "new" ribosomes formed after the metamorphic stimulus was applied.

Distribution of Ribosomes in the Cytoplasm and Proliferation of Endoplasmic Reticulum Membranes

What is perhaps of utmost importance in studying the sequential events that occur between the initial burst of RNA synthesis following triiodothyronine administration and the appearance of newly synthesized proteins in a redistribution of ribosomes attached to membranes of the endoplasmic reticulum. That such a redistribution may have occurred was first suggested in experiments in which liver mitochondria free supernatant fractions from premetamorphic and induced tadpoles were titrated against Na deoxycholate. When the fraction of ribosomes remaining attaching to microsomal membranes was measured as more deoxycholate was added, those from metamorphosing tadpoles were found to be more tenaciously bound to endoplasmic reticulum. The next question was to find out whether there was a redistribution of all "old" and "new" ribosomes on pre-existing and stable membranes or whether there was also some alteration in the formation or proliferation of membranes.

It is known that cellular membranes, even in resting cells, are not metabolically stable but turning over quite rapidly. To study this problem we compared accumulation of newly formed ribosomes, their distribution on endoplasmic reticulum and the labeling of membrane phospholipids as an index of membrane proliferation. As shown in Figure there occurred an almost simultaneous increase in the synthesis of ribosomes and membranes of the endoplasmic reticulum, especially those to which ribosomes are found (rough endoplasmic reticulum). There were no obvious qualitative changes in the types of membrane phospholipids synthesized, but it is interesting to not that the simultaneous increase in the formation of the two rough endoplasmic reticulum components is accompanied by an enhanced rate of protein synthesis *in vivo*. Titration of mitochondria-free extracts with sodium deoxycholate showed that the newly formed ribosomes in induced animals were more tightly bound to membranes of the rough endoplasmic reticulum, i.e, they required a higher amount of detergent for their release. It can be seen that the "heavy rough membrane" fraction, i.e., the one in which ribosomes are more

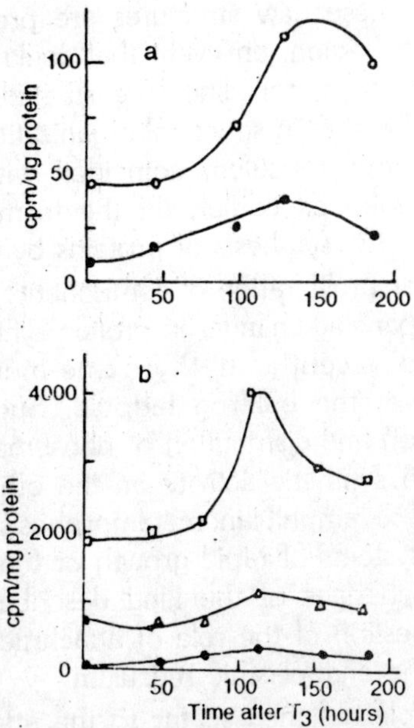

Fig. 9.4. Co-ordinated proliferation of endoplasmic reticulum and the increase in hepatic protein synthesis in vivo in the heavy rough membranes during triiodothyronine-induced metamorphosis of young bullfrog tadpoles. (a) The appearance of newly synthesized (^{32}P-labeled) heavy rough membrane phospholipids (o) and RNA; (•) (b) the recovery of labeled nascent proteins formed in vivo after a short pulse of radioactive amino acids in heavy rough membranes 0, light rough membrane (Δ), and free polysomes (•).

densely packed on membranes, was more active in protein synthesis *in vivo* than were free ribosomes or the "light rough membrane" fraction.

Although the evidence is too tenuous to warrant associating a distribution pattern in broken cell preparations with differential protein synthetic rates in the intact cell, it is tempting to suggest that the protein biosynthetic response to the hormone may be contained in the newly formed ribosomes. The above ribosomal redistribution changes are so marked that it is easy to corroborate them by electron microscopy. These studies revealed that induction of metamorphosis caused a shift in the distribution of ribosomes from around the simple vesicular membrane structures of the immature larvae to the more complex double lamellar structures more commonly seen in mature tissues. At the initial stage of metamorphosis, these new structures are predominantly located in the perinuclear region, an event that could by easily discerned by light microscopy of toluidine blue stained preparations. It is interesting that this shift in structural organization of polyribosomes on the endoplasmic reticulum coincided with the biochemical observations, mentioned earlier, on the formation of RNA and phospholipid and the synthesis of proteins by polyribosomes.

Rather similar proliferation of components of the endoplasmic reticulum accompanying changes in protein synthesis has also been described more recently in thyroxine-induced and natural metamorphosis of the bullfrog tadpole. Such a co-ordination between formation and distribution of ribosomes on the one hand and their protein synthetic activity on the other is not a feature that is exclusive to amphibian metamorphosis but is found in a wide variety of systems of rapid growth or functional maturation. Ultrastructural changes of the kind described above raise the more general question of the role of attachment of ribosomes to membranes of the endoplasmic reticulum.

The major role assigned so far to the attachment is that of facilitation of secretion of proteins for export. However, the proliferation of endoplasmic reticulum or a preferential shift from free to membrane-bound ribosomes is a characteristic of growth and development. This is true for both non-secretary and secretory cells in which relatively more protein for intracellular use is made during rapid growth. It is likely that the attachment of ribosomes to membranes serves some other role during growth and development. The significance of the coupling of structural alterations and biosynthetic function that is noticed in tadpole liver during metamorphosis, mentioned above, may be that: (a) a response to growth and developmental stimulus is contained in newly formed ribosomes, and (b) that there is a topographical segregation on membranes of the endoplasmic reticulum of differently precoded polyribosomes carrying out the synthesis of different groups of proteins.

There is, of course, no direct experimental evidence to verify either of these suggestions, but some recent work from our laboratory on the additive effects of growth and developmental hormones in mammalian organs provides an indirect support. In these experiments, when growth hormone, triiodothyronine and testosterone were administered in different combinations, the increased rates of formation of ribosomes and the membranes of the rough endoplasmic reticulum were co-ordinated with those of synthesis of different classes of proteins in the liver and seminal vesicles. Since these hormones have quite different latent periods of action, the simultaneous administrated in different combinations, the increased rates of formation of ribosomes and the membranes of the rough endoplasmic reticulum were co-ordinated with those of synthesis of different classes of proteins in the liver and seminal vesicles. Since these hormones have quite different latent periods of action, the simultaneous administration of any two caused stepwise increases in rates of proliferation of ribosomes and rough endoplasmic membranes, each burst corresponding in its time-course and magnitude to that induced by each individual hormone. It is assumed in the above suggestions, that an exchange between polysomes and membranes proceeds at a low rate relative to the lifetime of these structures.

An indirect approach to testing the idea of a topographical segregation would be to identify induced proteins on proliferating rough endoplasmic reticulum during development. We are now attempting by histochemical methods of detection and localization of ornithine and asparate carbamyl transferases as they increase during induced metamorphosis. The recent work of Pitot et. al. (1969) on the immunochemical location of serine dehydratase has shown that this inducible enzyme is almost exclusively synthesized on membrane-bound and not free ribosomes of the liver. In a different system, Ledue et al. (1968) have shown that antibody synthesized at the initial stages of stimulation of the lymphocyte by a given antigen is localized in the perinuclear rough endoplasmic reticulum structures. Even in reticulocytes it seems that different classes of proteins are synthesized on membrane-bound and free ribosomes. The suggestion of a segregation by membrane attachment of different population of ribosomes carrying out the synthesis of different classes of proteins may not be as far-fetched as it may seem.

The Role of DNA Synthesis

The role of DNA synthesis is quite obvious in those tissues that are rapidly formed during metamorphosis, such as the limbs and lungs. However, the possible role of a restricted DNA synthesis is not clear in those tissues that do not grow rapidly but undergo a functional maturation, i.e., the tadpole liver and skin. That a relatively small burst of DNA synthesis during metamorphic maturation may be important and needs careful investigation is emphasized by the work of Topper's group on a limited DNA synthesis or cell division as a prerequisite for milk protein synthesis in prolactin-induced development of mammary gland in culture.

McGarry and Vanable (1969) have indeed found that cell division is a prerequisite for thyroxine-induced maturational changes in skin gland from *Xenopus* in organ culture. Ingram's work on the switch from fetal to adult haemoglobin during bullfrog metamorphosis suggests that DNA synthesis is important in that a different population of nucleated erythrocytes now takes over the synthesis of haemoglobin for terrestrial life. It is thought that the switching itself takes place in tadpole liver. A different kind of involvement of DNA synthesis may underlie the observations of A. M. Campbell et al. (1969) that triiodothyronine stimulated mitochondrial DNA polymerase in bullfrog tadpole liver. It is, however difficult to say whether an early alteration in cytoplasmic DNA metabolism is part of some general developmental process or they it merely is a first stage of an increase in mitochondrial size and number that is known to occur during metamorphosis.

Requirement of RNA and Protein Synthesis for Tissue Resorption

Tissue regression or cell death is an important and integral part of many embryonic developmental process. During amphibian metamorphosis, regression of organs like the tall gut, and gills accompany the maturation or formation of organs like the liver, limbs and eyes. A question that had not been answered until recently was whether or not regression or cell death was also based on a hormonal regulation of biosynthetic processes as we have seen above for maturation of the liver. We studied this question in our laboratory by using the technique of thyroid hormone-induced regression of the isolated tadpole tail maintained in organ culture.

The technique of tail organ culture was already successfully used by Weber (1963) and the induction of regression *in vitro* corresponds in magnitude and speed to that seen in the intact tadpole. Decrease in the size of the amputated tail *in vitro* was accompanied by an increase in the activity of enzymes involved in regression such as cathepsin, phosphatase, and deoxyribonuclease. Earlier work on the comparison of the properties of deoxyribonuclease and cathepsin in regressing and no-regressing tadpole tails had suggested that a part of the additional enzyme activity following hormone treatment may differ from the basal enzyme present in non-regressing tails. It is therefore interesting to note that there is a burst of both RNA and protein synthesis just when regression sets *in vitro*.

Recently, Tonoue and Frieden (1979) have found that *in vivo*, administration of triiodothyronine to bullfrog tadpoles rapidly (within 1-3 hours) lead to a decrease in the incorporation of radioactive leucine in proteins of the tail and other regressing tissues. This apparently opposite result from that found in organ cultures may be a manifestation of different properties of thyroid hormone. Besides the fact that, an enhanced rate of protein synthesis takes 1-2 days to be manifested, the rapid decrease in labeling of proteins may be due to changes in the precursor pool sizes although these authors seem to rule it out on the basis of indirect observations.

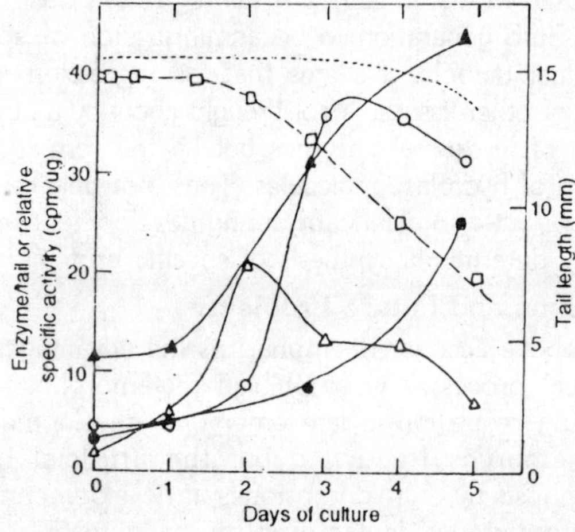

Fig. 9.5. Accumulation of cathepsin • and deoxyribonuclease Δ and burst of additional RNA and protein synthesis during regression of tadpole tails induced in organ culture with triiodothyronine added to the medium, (o–o). Incorporation of uridine-H into RNA; Δ–Δ incorporation of ^{14}C-labeled amino acids into protein of a mixture of labeled. Incorporation of day 0 is the average value for controls; all other points refer to samples to which T_3 was added... Tail length of controls over the duration of the experiment; —, length of triiodothyronine-treated tails showing a marked onset of regression between the second and third day after culture.

Not only thyroid hormones but all hormones, whether they affect growth and development or cause a rapid change in metabolic activity, are known to cause rapid alterations in pool size or uptake of sugars, amino acids and nucleotides in their target tissues. To determine whether any of the additional RNA and protein synthesized at the onset of regression induced in cultured tails was essential for the process itself, we turned to the use of inhibitors of RNA and protein synthesis. From these experiments, it was quite clear that inhibition of RNA synthesis with actinomycin D or of protein synthesis with puromycin or cycloheximide completely abolished the regression in tail organ cultures that was induced by triidothyronine. It was also found that inhibition of protein synthesis abolished the hormone induced increase in hydrolytic enzyme activities.

Eeckhout (1966) has also noted, in a different species of the tadpole, that protein synthesis inhibitors (such as puromycin and cycloheximide) arrested tail regression in organ cultures, and Weber (1965) observed that in *Xenopus* tail regression was more sensitive than was limb generation to be administration of actinomycin D to the intact tadpole. It seems that tail regression, and perhaps even that of other tissues, is not brought about by a direct hormonal activation of lysosomal enzymes but by the formation of a new population of hydrolase molecules. Thus, not only cell growth and maturation, but also cell death during development may require a genetically determined synthesis of specific proteins.

Conclusions and Future Problems

The above account re-emphasizes the advantages of studying biochemical processes in amphibian metamorphosis as a model for studying regulation in late embryonic development. Not only are the embryos free-living but the artificial induction of metamorphosis by thyroid hormones at developmental stages well before spontaneous metamorphosis have made it possible to establish a sequence of events preceding the acquisition of adult functions and structures. Among the earliest responses of target cells is a readjustment of permeability barriers to a variety of nutrients and precursors of macromolecules. However, the eventual specificity of developmental changes is most likely to reside in the nature of new or additional species of RNA and protein molecules formed. Due to a combination of the complexity of nuclear RNA, the high degree of nuclear restriction of the rapidly turning over nuclear RNA and the inadequacies of the currently available analytical techniques it has not been possible to relate the burst of RNA synthesis with selective gene activation at the onset of metamorphosis. In the liver of the bullfrog tadpole a relatively long period of time elapses between an accelerated synthesis of RNA in the nucleus and the appearance of new proteins that characterize metamorphic changes. During this period there occurs a substantial increase in the turnover of ribosomes in the cytoplasm accompanied by their redistribution on membranes of the endoplasmic reticulum.

At the same time, there is a tight coordination in the rate of formation of ribosomes and the proliferation of membranes of the endoplasmic reticulum to which they are bound. This phenomenon of structural reorganization of the protein synthesizing machinery is common to many late embryonic developmental systems, and its significance may reside in a topographical segregation of precoded polysomes engaged in the synthesis of different classes of proteins. A prominent features of amphibian metamorphosis is the convenience of studying tissues regression or cell death. It seems that regression is not merely due to activation of existing lysosomes but that an activation of RNA and protein synthesis underlies the process of regression.

Experiments with inhibitors of RNA and protein synthesis in the resorption of tadpole tail in culture have shown that cell death during metamorphosis requires the formation of new proteins just as it is necessary for those cells that are programmed for further growth and development. Thus it is now possible to describe sequential phenomena concerning the synthesis of RNA and protein at the onset of metamorphic maturation or growth. But there are several questions still to be resolved. For example, the nature of RNA made, its transfer into the cytoplasm, and its role as a messenger will have to be clarified—a problem now facing almost all work on differentiation. Are new genes really transcribed or is there a switch in the selection and transfer to the cytoplasm of the types of RNA molecules which are constantly made since an early stage in differentiation?

In the cytoplasm, it will be most worthwhile exploring the possibility, by a combination of biochemical, histochemical, and immunochemical techniques, of identifying a separate class of ribosomes preferentially satisfying the demand for metamorphic proteins. The fact that cell growth and cell death proceed simultaneously within the same organism creates problems of a flux of precursor pools which is impossible to control. It is obvious that more emphasis will now have to be devoted to studying the late embryonic developmental process in culture. Organ cultures of tadpole tail, skin, and liver have already been studied, and it would be useful to extend these studies to dispersed cell cultures, providing that cultured cells retain the full developmental competence to respond to the hormonal stimulus. Finally, the question of developmental competence is related to that of the initial site of thyroid hormone action, i.e., the hormone receptors. Some recent experiments from our laboratory have shown that in *Xenopus* larvae metamorphic competence was acquire very early in development. Metamorphic competence was assessed biochemically by a change in the overall rate of synthesis of DNA, RNA, phospholipids and protein, water loss and altered permeability to anions like phosphate. These changes indicate that the cells (undetermined) which were initially unresponsive have suddenly become sensitive to thyroid hormones.

It should be realized, however, that in normal development there is a variation in the magnitude of sensitivity of different

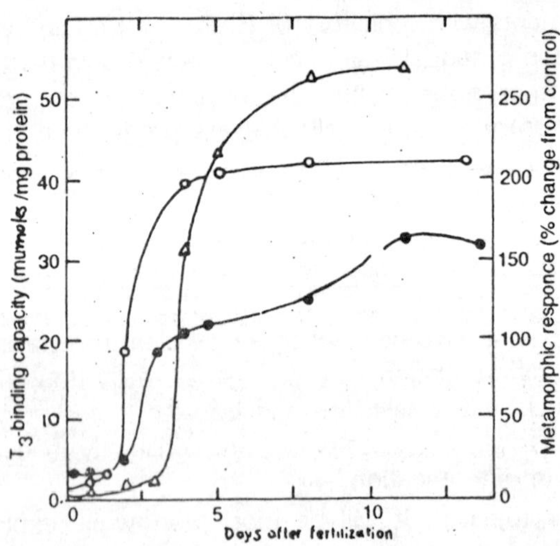

Fig. 9.6. Correlation between the appearance of temperature-sensitive binding capacity for triiodothyronine (•) and the acquisition Xenopus larvae.

tissues as a function of the age of the tadpole. One interpretation of such experiments was that the acquisition of sensitivity to the hormone was a consequence of the first appearance of hormone receptors. Thus when the capacity of *Xenopus* larvae to bind radioactive thyroid hormones was measured, there was an excellent correlation between high affinity (average Kd = 10^{-9} to 10^{-10}) thyroxine-binding sites and the rate at which the organism becomes hormone-sensitive. It has yet to be shown that one is dealing not with non-specific binding, but with true receptors whose interaction with the hormone would lead to the normal chain of events responsible for metamorphic changes. A selective rather than random distribution of hormone-binding sites has also to be determined. In the meantime, however, the study of co-ordinated acquisition of metamorphic competence and hormone-binding offers a developmental approach to the problem of hormone receptors which is now a key tissue in understanding the biochemical basis of hormonally regulated processes.

Metamorphosis and Gene Action

The improved resolution of starch gel electrophoresis and the use of enzyme histochemical techniques (3, 4) have greatly stimulated research on individual and developmental variation of proteins. Not surprisingly, this improved method of enzyme electrophoresis has been applied to amphibian metamorphosis. It was observed that appropriate one-third of the enzymes are specific to either the tadpole or the frog phase of the life cycle, whereas two-third of the major soluble proteins are specific to the different phases.

It was thus concluded that, although enzyme changes in the major soluble proteins, some of which may well be enzymes with unknown activities, are quantitatively more important. Thus many enzymes appear to be present in both phases of life cycle—e.g., are amphienzymes. In addition, a few proteins changes at times other than metamorphosis, and several proteins are present only during the metamorphic period. Two different approaches can be used to decide if certain structural genes are active in both phases of the life cycle, as well as to decide if others are expressed only in specific phases.

1. Isolation and characterization of enzymes with identical substrate specificities from tadpole and adult phases. In particular, if such enzymes are the products of different genetic loci, then

they should differ in at least part of their amino acid sequence, such a difference being most readily demonstrated by peptide mapping ("fingerprinting"), although preferably being carried to the ultimate point of complete primary structure determination.

2. If two proteins are coded from different genetic loci, then genetic variation of one should not be paralleled by genetic variation of the other.

Each approach has its problems. The first is laborious. For race enzymes, large amounts of tissues are needed in order to obtain sufficient pure protein. This means that individuals must be pooled. The second approach is biochemically less sophisticated, if considerably simpler. As small amounts of tissue can be extracted and submitted to electrophoresis, many studies can be done without combining organs or individuals.

It has recently been realized that the pattern of genetic variation of many enzymes and other proteins is very easily determined by starch gel electrophoresis. For a number of proteins, heterozygotes for a polymorphism have the two zones characteristic of the corresponding homozygotes—e.g., serum albumin. For other proteins they may be multiple banding in homozygotes but heterozygotes have simply the sun of zones present in both homozygotes—e.g., transferrin. For still other proteins, heterozygotes are easily detected by the possession not only of the two zones corresponding to those of appropriate homozygotes but also of one or more hybrid zones of intermediate electrophoretic mobility. For example, polymorphism of 6-phosphogluconate dehydrogenase in man, the quail (*Coturnix coturnix*), snails *Cepaea nemoralis* and *C. hortensis*), and a sea anemone (*Actinia equina*) involves a single hybrid zone in heterozygotes. For a number of vertebrate lactate dehydrogenases, the pattern of individual variation involves basically three hybrid zones in heterozygotes. There are several complications to the approach either through biochemical characterization or through genetic variation in studying the proteins of different phases in a life cycle:

1. Appearance of a new proteins does not necessarily mean gene activity in the sense of messenger RNA formation followed immediately by protein synthesis. There can be a long delay between gene transcription and activation of the messenger RNA-ribosome complex.

2. Some proteins are the product of more than one genetic locus, e.g., the distinct α and β chains of most vertebrate haemoglobins. For a β chain mutation in man the polymorphism is confined to the major adult haemoglobin Hb A. For an α chain mutant there will be parallel variation of Hb A, Hb A_2, and Hb F.
3. Proteins can have different properties and still have been the product of the same genetic locus if there has been a subsequent difference in chemical modification. For example, transferrin conalbumin in galliform birds differ electrophoretically and yet always show parallel genetic variation. It has recently been shown that these proteins differ not in amino acid sequence but in the carbohydrate moiety attached to sequence.

The present paper includes additional data on individual variation of certain proteins in amphioxus and amphibians. Amphioxus is in some reports the most primitive known living chordate. It metamorphoses from a planktonic larva to a benthic adult phase which, though still motile, spends much of its time partly or completely buried in coarse sand. Although the matter is by no means settled, it has been suggested that the metamorphosis of amphioxus is controlled by thyroxine or some related iodine compound.

Materials and Methods

Amphioxus, *Branchiostoma lanceolatum*, were obtained from the English Channel in the vicinity of Plymouth, England, Larvae were collected in plankton tows during the summer and adults were collected by bottom trawling with appropriate fine-meshed netting. The amphibians sampled were edible frogs, *Rana esculenta*, common frogs, *Rana temporaria*, and newts, *Triturus vulgaris*. Treatment of samples, electrophoresis in vertical starch gel, and histochemical localization of enzymes are detailed in other papers.

Amphioxus larvae are very small, as well as thin and elongate; thus it was necessary to pool 30 or 60 larvae at a time. Adult amphioxus were pooled for direct comparison with larvae. But, in addition, variation in a number of enzymes was studied in 40 individual-adult amphioxus. Separation of organs in adult amphioxus is difficult; the best that can be done with reasonable rapidly (so as to avoid autolytic modification) and with minimal contamination is the division of adults into body musculature (including notochord and nervous system), viscera (including the digestive system), and, depending on sex, ovary or testis.

RESULTS

Enzymes of Amphioxus, Branchiostoma Lanceolatum Esterases

Amphioxus esterases are resolved best in the pH Ferguson-Wallace discontinuous buffer, although good resolution of some esterases is obtained in pH 6.0 phosphate. The major amphioxus esterases are relatively non-specific in that almost identical patterns, differing only in relative activity in a few zones, are obtained whether the enzymes are localized by hydrolysis of a-naphthyl acetate, α-naphthyl butyrate, or 5-bromoindoxylacetate. None of these enzymes has activity on α-naphthyl phosphate, there being at least two distinct and active acid phosphatases in adult amphioxus, although no detectable phosphatase activity in extracts

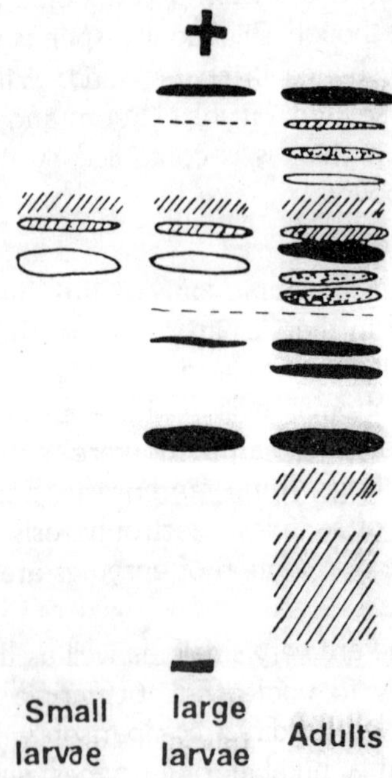

Fig. 9.7. Esterases of larval and adult amphioxus, Branchiostoma lanceolatum.

of larvae. The esterase pattern for small larvae (5 mm long), large larvae (8 mm long) and adults.

There is some individual variation in electrophoretic position of two of the more anodal esterases of adult amphioxus, though most of the esterase zones were identical in sample of 32 adults screened singly. It is clear that larval amphioxus lack many of the esterases present in adults. Only three zones of esterase activity are detectable in small larvae; the number increases to eight in large larvae. Seven of these eight zones are identical in position to zones found in extracts of adults. There is one conspicuous esterase of larval amphioxus which does not appear to be present, at least in anywhere near the same level of activity, in adults. In addition, adult amphioxus have at least seven esterases not present in larvae.

Dehydrogenases

No activity could be found in extracts of either larval or adult amphioxus to sliced starch gels at the completion of

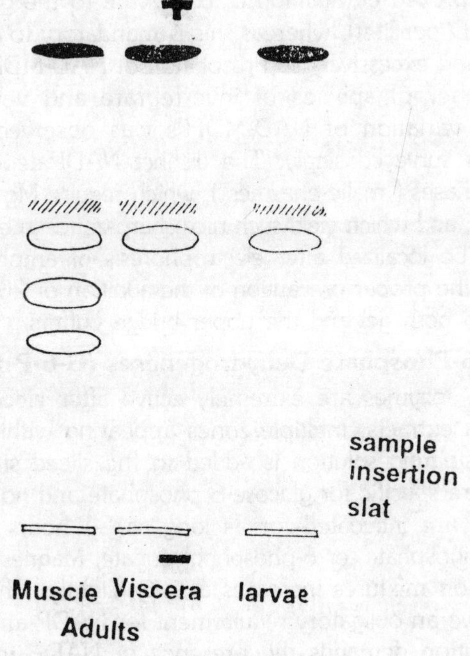

Fig. 9.8. NAD-dependent malate dehydrogenase isozymes of larval and adult amphioxus, Branchiostoma lanceolatum.

electrophoresis: *lactate*, *isocitrate* (neither NAD-nor NADP-dependent IDH's), glutamate, 6-phosphogluconate, ethanol, glyceraldehyde, α-glycerol-phosphate, or α-glyceraldehyde-phosphate. Very strong activity was observed for NAD-dependent malate dehydrogenase (abbreviated NAD-MDH) and glucose-6-phosphate dehydrogenase (abbreviated G-6-P DH), each of which yields a series of multiple zones (isozymes).

NAD-Dependent Malate Dehydrogenases (NAD-MDH's)

Extracts of larval amphioxus yield the same NAD-MDH pattern of two major and one trace isozymes as found in extracts of adult viscera. Adult muscle has these isozymes plus two more major NAD-MDH's which migrate somewhat more slowly. A curious observation is that these NAD-MDH's of amphioxus resolve so well in alkaline starch gels containing borate. In many extracts of other species the supernatant and mitochondrial NAD-MDH's resolve best in pH 6 to 7.5 phosphate gels, which give fairly poor resolution of the amphioxus NAD-MDH's.

In addition, resolution of the amphioxus NAD-MDH's is only slightly improved by addition of coenzyme to the gel (50 or 100 mg of NAD per liter), whereas this is mandatory to avoid spurious variation and excessively bad resolution of NAD-MDH's in extracts of a number of species of invertebrate and vertebrates. No individual variation of NAD-MDH's was observed in 40 adult amphioxus surveyed singly. The distinct NADP-dependent malate dehydrogenases ("malic enzymes"), which require Mg^{++} or MN^{++} for full activity, and which yield such nice sharp zones in electrophoresis, could not be localized after electrophoresis of amphioxus extracts even with the proper precaution of the addition of 20 mg of NADP per liter to both gel and the upper bridge buffers.

Glucose-6-Phosphate Dehydrogenases (G-6-P DH's)

These enzymes are extremely active after electrophoresis of amphioxus extracts, multiple zones appearing within 10 minutes after the staining solution is added to the sliced starch gel. The enzymes are specific for glucose-6 phosphate and no zones appear when gels are incubated for as long as 24 hours with glucose, glucose-1-phosphate, or 6-phosphogluconate. Magnesium ion added to incubation mixtures increases activity slightly. The multiple G-6-DH's have an obligatory requirement for NADP and, in addition, best resolution demands the presence of NADP in the gel and bridge buffers (20 mg NADP per liter). There are basically three sets of G-6-PDH isozymes. As variation at each set is distinct from variation at the others, the three sets are the product of distinct genetic loci. This is especially clear for the two rapidly anodal sets of G-6-P DH's. The most anodal set of zones is designated Locus I, the next set Locus II, and the slowly moving and poorly resorbing set, Locus III. Locus I G-6-P DH's are especially active in extracts of testes or ovary and Locus II G-6-P DH's are especially active in extracts of viscera; however, the specificity is relative and not absolute. Muscle contains primarily the Locus III G-6-P DH. Larvae have active Locus II and Locus III G-6 P DH's but no Locus I. Thus, metamorphosis leads ultimately to the activation of one of the three G-6-P- DH loci. The nature of individual variation in larvae was not studied; however, in adults many different electrophoretic patterns were observed.

With one exception the pattern of individual variation for both Locus I and Locus II G-6-P DH's can be explained by postulating five alleles at Locus I and four alleles at Locus II, heterozygotes having any two zones and homozygotes having one heavy zone. The exception is one individual with four zones at Locus I, although only a single zone at Locus II. A further problem is that there is often considerable difference in staining intensity between the two zones in postulated heterozygotes. Unfortunately, G-6-P DH, whether in sea anemones or in quail extracts, has a tendency to form satellite zones. Nevertheless, the pattern of variation of amphioxus G-6-P DH is unusual for a dehydrogenase in that there are no hybrid zones in heterozygotes. Except for sex-linked erythrocyte G-6-P DH in mammals, other G-6-P DH has been obtained; as for mammalian liver G-6-P DH and avian erythrocyte

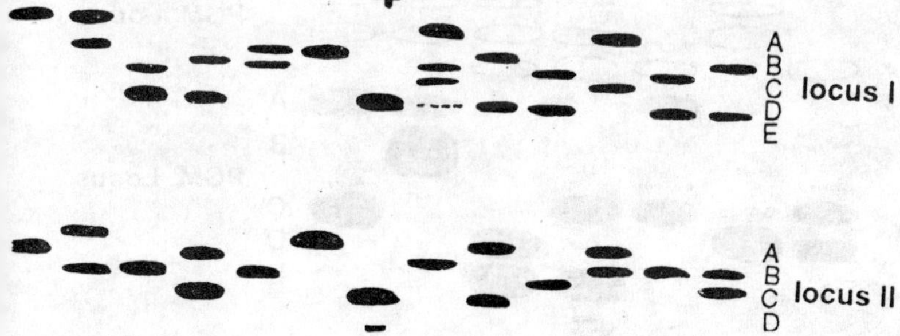

+
A
B
C
D
E
locus I

A
B
C
D
locus II
−

Fig. 9.9. Individual variation of Locus I and Locus II glucose-6-phosphate dehydrogenases of adult amphioxus, Branchiostoma lanceolatum.

G-6-P DH's in vertebrates have only hybrid zone in heterozygotes– and in sea anemones the pattern resembles lactate dehydrogenase in that heterozygotes have five zones, three of which are hybrid isozymes (unpublished studies by the senior author on *Actinia equina*). No evidence for sex linkage of amphioxus G-6-P DH, the amphioxus G-6-P DH loci are autosomal. The absent of hybrid zones in heterozygotes in the variation of amphioxus Locus I and Locus II G-6-P DH's, together with the extremely rapid electrophoretic mobility in starch gel, are suggestive of a smaller and simpler and structure than other dehydrogenases. It is possible that, in contrast to other G-6-P DH's those of amphioxus lack a stable quaternary structure, a point which is compatible with the observed data but which can only be proven by careful biochemical characterization of the isolated enzymes.

Phosphoglucomutase (PGM)

Extracts of adult amphioxus also have a very high activity of phosphoglucomutase. Initially it was not realized that amphioxus PGM is completely inactivated by even brief freezing and thawing. Thus, PGM was studied in only seven individuals. However, the range of individual variation of PGM among those seven adults is considerable. In addition, there is a surprising similarity in the patterns observed in amphioxus PGM and those observed in studies on variation of human PGM. *Amphioxus* and man both have two or three minor, rapidly moving zones of PGM, whose variation is independent of the variation of the more active and more slowly moving zones of PGM. The minor zones are determined by a locus, designated PGM_2 in man, which is distinct from that

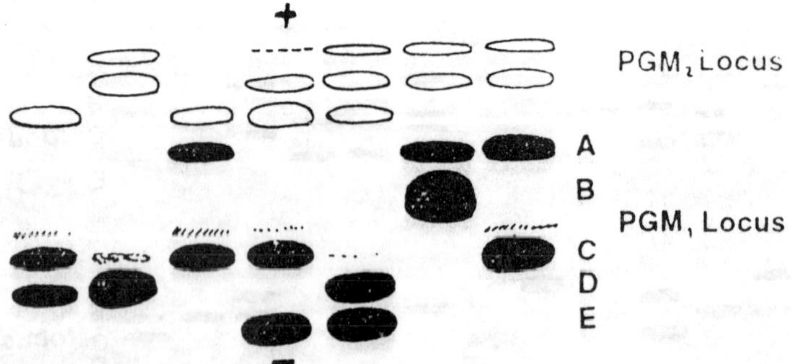

Fig. 9.10. Individual variation of seven adult amphioxus, Branchiostoma lanceolatum PGM_1 and PGM_2 phosphoglucomutases.

determining the major zones, designated PGM_1 in man. For man, the existence of two distinct PGM loci has been confirmed by the observation of independent segregation in appropriate pedigrees; whether or not the two PGM loci in amphioxus are linked can only be determined by appropriate mating experiments, which have not been done.

As with variants of human PGM, the heterozygote is a simple sum of the zones of the corresponding homozygotes. Whereas in man two alleles at PGM1 are relatively common and other variants are rare: in only seven adult amphioxus, five different positions for the major zone are observed, suggesting five different reasonably common PGM_1 alleles. The PGM_1 locus appears to be active in most tissues, whereas PGM_2 is most readily observed is gonad and, irregularly, in some viscera extracts.

Aldolase

A single zone of fructose-1, 6-diphosphate aldolase is observed in extracts of adults. It is probable that this necessary enzymes occurs in larvae but, if so, its presence could not be detected after starch gel electrophoresis of extracts.

Peroxidase

Adult amphioxus but not larvae have from one to four zones of peroxidase, migrating fairly rapidly anodally in alkaline buffer (pH 8.0 discontinuous and pH 8.5 borate). There is extreme variation in the number of and relative activity of the peroxidase isozymes. There is no obvious correlation in the number and activity of peroxidases with six or nutritional state. The variations defies rational genetic explanation, unless one postulates two different peroxidase loci, each with null alleles in significant frequency. The absence of these isozymes in pooled larval samples suggests that peroxidases becomes conspicuous only in the adult phase. Though some amphioxus adults have pink colour, which has been casually called myoglobin, no evidence for muscle haemoglobin could be found in these individuals. The peroxidase activity represents a true peroxidase and not haemoglobin.

Dehydrogenases of Triturus vulgaris, Rana esculenta, and R. temporaria

Newts, Triturus Vulgaris

In an initial survey of eight individual newts, four larvae and four adults, it was observed that both adults and larvae had only a single LDH isozymes in all tissues. One individual adult was observed with the typical five LDH isozymes expected of a heterozygote for a polymorphism. However, these studies were done utilizing only the pH 8 Ferguson-Wallace discontinuous buffer, which yields very sharp zones of LDH. When these samples were run in a pH 6.7 phosphate buffer (μ = 0.04), which also yields good resolution of LDH's, it was observed that all newts have distinct heart- and muscle-type lactate dehydrogenases. These are H_4 and M_4— and the intermediate isozymes, corresponding to H_3M, H_3M_2 and HM_3, typical of mammalian, avian, and some "lower' vertebrate LDH's, do not occur in certain amphibian, including the common newt. The individual newt-with five LDH zones in alkaline buffer turned out to have an M chain LDH polymorphism. Thus, in the newt hybrid isozymes do not form readily between H and M chains, although the heterozygote for the M chain polymorphism possessed the typical five zones as found in heterozygotes among birds and mammals.

No significant differences were observed between adults and larvae and, in the newt, it is clear that metamorphosis involves no changes in LDH gene expression. Similarly, larval and adult newts have similar glutamate dehydrogenases and the two zones of NAD-dependent malate dehydrogenase. Larvae have less G-6-P DH than adults but the position of the two zones is similar, even if larvae lack nearly all of the more anodal zone. No individual variation of these other dehydrogenases was observed among the small sample of eight efts.

Frog Lactate Dehydrogenases (LDH)

In contrast to the newt, the muscle and heart-type LDH's of *Rana esculenta* and *R. temporaria* are easily differentiated by electrophoresis in any of the conventional alkaline buffers. Tadpoles have primarily the H4, LDH in all tissues, whereas adults have M_4 as well, especially in muscle and liver. There are only traces of zones intermediate between H4 and M4. Thus, our results are identical to those of Sathe et al. studying *Rana pipens* and *R. palustris*.

Bullfrog *R. Catesbeiana* H chain has more of a tendency of combine with the M chain, although there is still only the H_2M_2 molecules formed and little, if any, of the H_3M and HM_3 tetramers. We have observed, though inconsistently, that refrozen and thawed samples containing H_4 and M_4 frog LDH's tend to increase in the amount of the trace intermediate isozymes although these never are as prominent as in extracts of appropriate avian or mammalian tissues. Electrophoresis of unfrozen frog tissues often fails to yield any intermediate LDH isozymes between H_4 and M_4, except for some H_2M_2 in bullfrogs. The bullfrog is also different from the other ranids in the more rapid electrophoretic mobility of the H_4 isozyme and that the M_4 isozyme occurs in appreciable amounts in tadpoles, especially in muscle and liver. Bullfrog adults also have a heterogeneity of the H_2M_2 and M_4 LDH's that is not paralleled in *Rana pipiens*, *R. esculenta* and *R. temporaria*.

Genetic Polymorphism at the H Chain LDH Locus in Rana esculenta

The hypothesis that the H chain is the same in both tadpole and adult phases is proven by finding a particularly clear polymorphism at the H chain locus in *Rana esculenta*. The Pattern of variation is exactly the same as observed in LDH polymorphism of several vertebrates, including pigeons, mice, and men, for which the genetic interpretation has been confirmed by pedigree studies or appropriate matings. In both tadpole and adults of *R. esculenta*, there are three different H chain alleles. Heterozygotes have five zones, including three hybrid isozymes not found in homozygotes. Five of the six predicted phenotypes for a triallelic variation have been observed in the small sample of frogs so far investigated. It

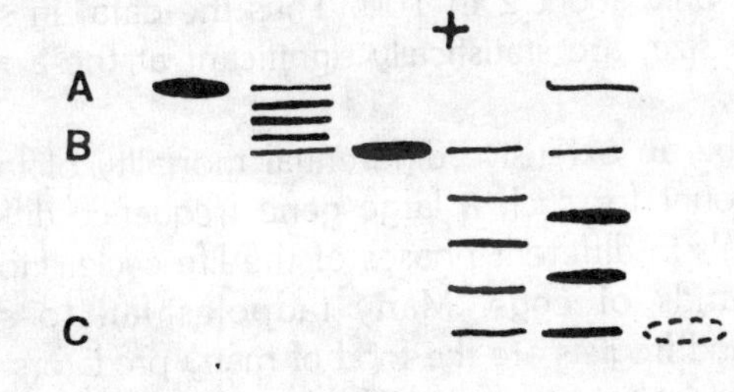

Fig. 9.11. Polymorphism of the H chain lactate dehydrogenase of tadpole or adult edible frog, Rana esculenta.

is of interest that the sample of eight adults yielded 1 AA, 1 AB and 6 AC, whereas the sample of eight tadpoles yielded 2 AB, 2 AC, 3 BB, and 1 BC. Thus, the estimated gene frequencies are in adults $p_A = 0.56$, $p_B = 0.06$, and $p_c = 0.38$; and in tadpoles $p_A = 0.2$, $p_B = -.59$, and $p_c = 0.19$. Because of the small sample size, the difference in p_A in the two phases, as well as in p, are not statistically significant. However, the difference in gene frequency of the B allele, 0.56 in tadpoles versus 0.06 in frogs, in the two phases is statistically significant.

The probability of such an unequal distribution occurring under the null hypothesis of no significant difference can be easily calculated – see Mather for discussion of the use of binomial and multinomial distributions for analysis of significance of differences in small samples. The total sample of 16 *R. esculenta* represents 32 alleles at the H chain locus. The probability of the 10 B alleles being distributed 9 in one phase and 1 in the other, or 10 in one phase and 0 in the other, is, from the appropriate binomial expansion:

$$p^{10} + 10\,p^9q + 10\,pq^9 + q^{10}$$

where, because of the equal number of tadpoles and frogs in the sample, $p = q = {}^1/_2$. The exact value for the above sum of four terms is $22 \times ({}^1/_2)^{10} = 0.0215$. The probability of observing such a difference in the frequency of the B allele in tadpoles and frogs due to chance is only about 2 in 100. Thus the data, in spite of the small sample size, are statistically significant at the a = 0.05 level.

There must be an extensive differential mortality of tadpoles and frogs to account for such a large gene frequency difference between individuals in different phases of the life cycle. However, frogs lay thousands of eggs. Many tadpoles fail to survive metamorphosis, and froglets are the food of many predators. There is plenty of numerical room for establishment of such different "genetic elites" in any species with a large number of zygotes. Survey of 6 adults and 44 tadpoles of *R. temporaria* failed to reveal any LDH variants at either the H chain or the M chain locus.

Frog Malate Dehydrogenases

There are two major NAD-dependent malate dehydrogenases in both tadpole and adult *R. esculenta* and *R. temporaria*. The tadpole enzymes migrate slightly more rapidly anodally than the frog enzymes in pH 8.0 discontinuous buffer. However, the resolution of these enzymes is poor and the electrophoretic mobility depends on the concentration of the extract and the organ extracted. Thus, the tadpole and frog differences are probably not significant. No clear individual variation was observed among 16 *R. esculenta* and 30 *R. temporaria*, although some type of individual variations exists in the position of the more slowly moving NAD-MDH isozyme in *R. esculenta*.

Frog Glutamate Dehydrogenases

In discontinuous buffer, pH 8.0 glutamate *dehydrogenase* of *R. temporaria* adults has two major zones. Tadpoles consistently have only the faster of these zones. However, in pH 6.7 phosphate buffer the tadpole and frog glutamate dehydrogenases present a different pattern of heterogeneity; thus it is possible that the tadpole and frog GDH's are completely different. No individual variation was observed among 30 *R. temporaria*. Thus, we cannot yet say whether the electrophoretic difference between the tadpole and frog GDH's is the result of different GDH genes being active in the tadpole and frog phases of the life cycle, or the result of some difference in chemical modification. It is well-known that vertebrate liver glutamate dehydrogenase is an allosteric enzyme.

Discussion

Enzyme Differentiation Versus Adaptation

This comparative approaches indicates that in both amphioxus and amphibians, where metamorphosis is triggered (at least in amphibians) by thyroxine or similar iodinated compounds, some enzymes are unique only to one phase or the other of the life cycle. In general, as differentiation proceeds there is an increase in the number of different enzymes detectable after electrophoresis of extracts, e.g., larval amphioxus have only one possibly unique larval esterase, whereas adult amphioxus have at least seven different unique adult esterases. Similarly, in *Rana temporaria* and *R. esculenta*, though not in *R. catesbeiana*, only traces of the M LDH chain are synthesized prior to metamorphosis.

In the newt, however, we have observed no differences between larvae and adults, whether the comparison involves H chain LDH, M chain LDH malate dehydrogenases, or glutamate dehydrogenase. Additional studies on the same eight individuals have revealed neither any significant haemoglobin differences, nor differences in phosphatases; adults have two sets of esterases present in significant levels of activity which are not present, or barley detectable, in larvae; however, at least six other esterases are identical in adults and larvae. Thus, most of the newt enzymes are amphienzymes and are unaffected by metamorphosis. The patterns of enzyme differentiation show some degree of variation between species of the same genus.

The best example in the present studies is the much greater activity of the M chain locus in bullfrog tadpoles as compared with-tadpoles of other related ranids. Similar differences suggesting species differences in "control genes" have been observed in haemoglobin differentiation—e.g., the persistence of larval haemoglobin in adult lampreys of the semineotenous species *Entosphenus lamottei*, the difference in the time of appearance of the major and minor adult haemoglobins in the embryos of galliform birds, the relative differentiation an absolute differentiation of coelomic and water vascular haemoglobins in sea cucumbers, and the differences in the presence or absence of haemoglobin components, as well as their number, in embryonic, fetal, and adult mammals. Although enzyme changes in different phases of a life cycle are often studied as examples of hormonal influence on the activation of genes—or at least the activation of messenger RNA-ribosome complexes—it must not be forgotten that some of these differences in protein phenotype reflect differences in the environment. To what extent are such examples of enzyme differentiation altered by differences in the environment?

It is known that the relative amounts of the H and M chains of lactate dehydrogenase can be altered by oxygen level independently of other embryological events. Accordingly, it is just possible that the tadpole-frog difference in the amount of the M_4 isozyme reflects differences in the availability of oxygen and would be reduced or eliminated if tadpoles and frogs were kept at identical oxygen levels. A further paradox is posed by the fact that tadpoles of some species have primarily the H LDH chain type, which is considered to be more suited in its kinetic properties to metabolism in obligatory aerobic tissues. However, tadpoles live in environments of variable, and often low, levels of oxygen—for which the absence of a Bohr effect in tadpole haemoglobin is ideally suited. Thus, at our present state of understanding—or misunderstanding—tadpoles are better adapted anaerobically than frogs in respect to haemoglobin, but better adapted aerobically than frogs in respect to lactate dehydrogenase. In general, the greater the morphological difference between phases of a life cycle, the greater the difference in the habitats of larvae and adults—although the correlation is not perfect, especially among the some insects and amphibian.

It is of interest that the highest frequency of amphienzymes is found in the newt, where adults spend considerable time in the same aquatic environment as the larvae. It may well be that some of the specific enzymes being studied as examples of differentiation are, in fact, really or more significance in terms of adaptation to the difference in larval and adult environment. As suggested earlier, from the stand point of changes in morphology, the major soluble and insoluble proteins are particular importance, even though *de novo* synthesis, or activation, of specific modifying enzymes is also necessary for those morphological changes—e.g., the collagenase of tadpole tails. It is worth bearing in mind that metamorphosis in some marine invertebrates, e.g., tunicates, barnacles, and polychaetes, can be initiated and finished within a few hours, in some cases as little as 15 minutes.

It is unlikely that all the necessary protein changes are accomplished by *de novo* protein synthesis. Rather, reliance must be placed more on synthesis or activation of a few key modifying enzymes, which in turn can alter certain major proteins, which might be additional enzymes or structural proteins. Very likely there is a series of such activation steps involving successive amplification, as in the activation of the "factors" in blood clotting. It is proposed to call such changes in development that involve the chemical modification of already synthesized proteins transubstantiation. Transubstantiative changes may involve proteolytic modification, as in the case of collagenase, addition or removal of non-protein moieties, such as the carbohydrate changes in transferrin, or modification of configuration by combination with a small molecule, e.g., an allosteric effect. The very unusual difference between adult and fetal shark *Squalus* haemoglobin may reflect a transubstantiative change which results in the loss of two tryptic or seven chymotrypitic peptides of what are otherwise identical proteins.

Consequences of Individual Variation

In biochemical and endocrinological studies it has often been the practice to combine a given organ or gland from many individuals in order to obtain sufficient protein. In relation to comparison of tadpoles and adult frogs, the electrophoretic studies warm that such a procedure may give rise to differences in enzyme activity or other properties even when the enzymes are fundamentally identical in both phases of the life cycle. This can result in three ways:

1. Pooling of a small number of individuals may result, purely by chance, in combining different numbers of various variants for a protein polymorphism common to both phases. This can be minimized by increasing the size of the pooled sample.
2. When there is a change in the gene frequencies of alleles for a polymorphic locus active in both phases of the life cycle, this difference in variants will manifest itself in a pooled sample no matter how large that sample is. The data on *R. esculenta* B allele of the H chain LDH locus indicates that considerable differences in frequency exist in a sample of tadpoles and frogs from the same locality. This could result from seasonal differences in selection pressure, for adults will have been exposed longer to any progressive changes than will tadpoles. O, this could result from differential selection arising from the difference in the tadpole and frog environments. The large number of eggs laid by frogs makes it perfectly possible to have very different "genetic elites" in the two phases, the only restrictions being those imposed by the fact that any unusual distribution in frogs must arise from what is already present in tadpoles and vice versa. Such a system could stabilize polymorphism which would otherwise be unstable. If the different H chain variants of *R. esculenta* differ in their enzyme kinetics, a point which has not been investigated, then pooled samples of tadpoles and frogs would show differences in kinetics because of the differences in the frequency of the variants, although fundamentally the same LDH genes were active in both phases of the life cycle. Similarly, a pooled sample of tadpoles or frogs would yield somewhat different electrophoretic patterns or chromatographic elution profiles.
3. Another observation is that protein polymorphism are by no means rare. In some species from 20 to 73 per cent of various proteins show genetically typeable variation. Furthermore, not all such polymorphisms involve only two or three alleles. Several protein polymorphism are known where from 6 to

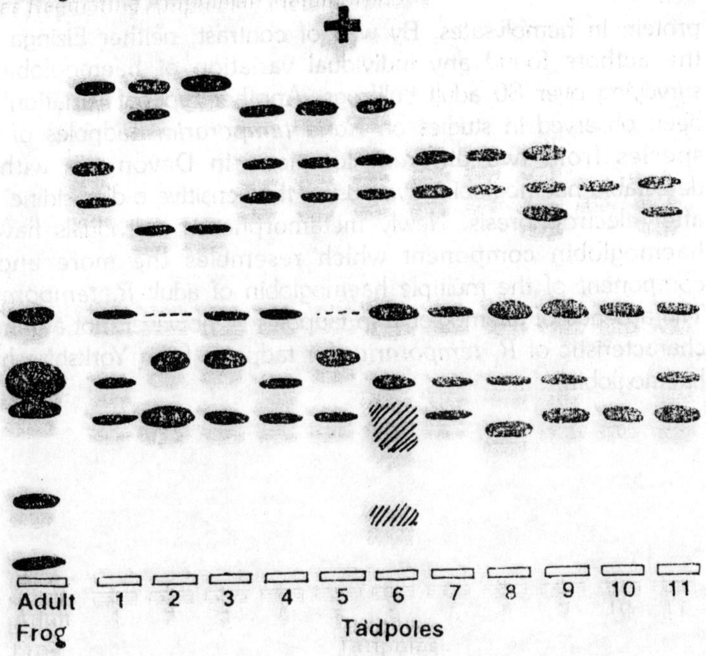

Fig. 9.12. Individual variation of the haemoglobins of 11 bullfrog Rana catesbeiana tadpoles out of a sample of 13.

13 different alleles all occur in relatively similar gene frequencies, probably representing natural selection for diversity. A small sample from such a population will invitably miss many of the possible phenotypes. What is worse is that two different samples will tend to contain quite different phenotypes, simply by probability theory and without any necessity for differential selection. Consider the amphioxus Locus I and Locus II-G-6-P DH's. With a total of five different alleles at one locus and four alleles at another, there are 15 × 10 = 150 different possible phenotypes of this one enzyme.

In view of the number of studies which have compared tadpole and adult frog haemoglobins, especially for determination of the number of polypeptide chains present, it is instructive to see the diversity of variation that occurs in one population of the bullfrog *Rana catesbeiana*. A diagram of the 11 different haemoglobin patterns observed among 13 tadpoles by *Elzinga*. In addition, many tadpoles have appreciable amounts of non-haemoglobin protein in hemolysates. By way of contrast, neither Elzinga nor the authors found any individual variation of haemoglobin in surveying over 80 adult bullfrogs. Another type of variation has been observed in studies on *Rana temporaria*. Tadpoles of this species from two different locations in Devon are without detectable haemoglobin, even using the sensitive o-dianisidine test after electrophoresis. Newly metamorphosed individuals have a haemoglobin component which resembles the more anodal component of the multiple haemoglobin of adult *R. temporaria*. The absence of haemoglobin in tadpoles is, however, not a general characteristic of *R. temporaria*, for tadpoles from Yorkshire have haemoglobin.

10

PRINCIPLES FOR REGENERATION

Many kinds of living organisms regrow appendages that are crushed or torn off in the mishaps of an active life. Human beings have scarcely any abilities of this sort, a fact which contributes to their curiosity about the mechanisms of regeneration in more resilient organisms. This curiosity runs deeper because regeneration in many ways resembles the initial normal development of an animal's structures. Normal development plus regeneration, collectively called *morphogenesis*, presumably operates by some general rules that we might at least elucidate empirically as a prelude to ferreting out deeper mechanisms. Yet for all the imaginative and neticulous efforts of at least four generations of developmental biologists, few general rules have stood the test of time.

If principles of widespread applicability exist they remain tantalizing obscure. In this situation, the whimsically dubbed *clockface* ("*clock-phase*") model of regeneration created a stir among developmentalists of both experimental and theoretical bents. Its most fundamental and radical tenet is that positional information is stored in cells as a "phase" or an "angle" analogous to the hour-hand position on a face of a clock, a quantity topologically unlike the conventional scalar variables of physiology. This has several startling implications which have been verified experimentally in several contexts, so it seems to call four some interpretation, perhaps involving a rhythmic or somehow periodic process. It also turns out that this model, unless adorned with special hypotheses to cover for exceptions, makes other predictions that are not observed. Moreover, a less original interpretation suffices to account for the remarkable regularities that gave the clockface model its first appeal. This relatively prosaic rendering assumes that positional information is encoded, not in a single angle or phase, but in a pair of transverse concentration gradients. This is a completely ordinary idea, essential qualitative facts about regeneration in epimorphic fields.

The Clockface Model

The clockface model was contrived as the vehicle for a brilliant resynthesis of observations by Vernon French, Susan Bryant, and Peter Bryant. They draw on experiments with the fruit-fly larva's "imaginal disks" (the precursors of the adult fly's various appendages), with the legs of cockroaches, and with amphibian limbs. Their original paper is a *tour de* force of focused experimentation and organization of data. They come up with two principles which together account for an extraordinary diversity of peculiar experimental results. Let me first remark that what one takes to be "the principles" is presently subject to so much subjective interpretation that there are as many versions as there are people retelling this story I will retell it based on four principles. The reader will want to examine the bibliography for other versions and for more detailed references to the original experimental papers.

It should also be noted that the experiments under consideration here do not span the whole diversity of regenerative processes. We focus here on situations in which cells seem to have no intrinsic directional polarity, seem to retain their differentiated identity, and seem to impose on adjacent cells (possibly newly created from dividing populations) to adopt a similar identity. The main point introduced by French et al. is that each tiny patch of tissue is somehow labeled permanently with an unchanging intensive quantity—a tissue specificity—that behaves like time in a cycle or like a hue or like an angle in that any one of them denotes a point on some abstract circle of states. French et al. (4) placed the digits 1-12 as indicators of local tissue specificity around the circumference of a limb or other developmental field. Many of their diagrams thus resemble the face of a clock. Although this notation gives the model its name, it unfortunately also requires a numerical discontinuity where none is intended biologically.

A more apt analogy is provided by the seasonal states of an ecosystem: spring grades into summer grades into fall grades into winter grades into spring with no discontinuities. The numerical analogy is unfortunate, but the principle is clear and tantalizingly suggests something oscillatory, somehow periodic, underlying. Without committing ourselves to any such interpretation, it will still be convenient to accept the "clockface" metaphor and refer to the quantity denoted around its rim as a "phase. Independent of its circlelike topology, the import of a cell's neighbourhood bearing this putative state label is that an appropriate stage in regeneration, tissue will develop structures (sensory hairs, muscle attachments, colour patches) corresponding to the local label. This is an application of Wolpert's principle that tissue specificity, encoded in a local quantity of unknown origin, is conceptually distinct from the cell's interpretation of that "positional" specificity and from other convert internal states such as "determination". The following simple rules make use of the putatively *circular* label of cell states, and suffice to systematize a lot of otherwise very perplexing experimental results.

Rule 1: Each little patch of cells is labeled with a phase which is part of a smooth phase gradient across the tissue. Thus there exists a smooth map from the tissue to an abstract ring of biochemical specificities. Once established, this map does not change. (Some independent second label, e.g., the proximo-distal level of cell specificity, is implicitly assumed, whereby to distinguish cell types on the two dimensions of an organism's surface.

Rule 2 introduces an exceptional point near which this map is not smooth.

Rule 2: In the normal limb, the phase values are supposed to run one full cycle around the limb axis (the azimuthal direction). If a limb were an open-ended cylindrical surface, this would pose no problem. But besides having an inside, here discretely ignored, a limb has a foot or a hand at the end: its surface resembles the glove with phase increasing full cycle around the open end. Herein lies the tantalizing nucleus of this whole subject. It phase increases smoothly through a cycle around the perimeter of a two-dimensional patch of tissue, let us say that the patch has "winding number" $W = 1$ (or 2, should phase increase through *two* full cycles, etc.). It is an unfortunate fact of geometry that only if the winding number is zero can a distinct phase value be attributed to every point inside. To convince yourself of this, try to sketch on the tissue a curve along which phase = 1, and an adjacent curve along which phase = 2, and so forth. All these curves start from the corresponding points along the tissue's boundary, and extend inward. If the winding number of phase along that border is zero, then each curve that enters the patch from one point on the border can again exit the patch at another point where the phase value is the same.

But if the winding number is not zero, then that integer number of full sets of phase contours enter and can't get back out. Wherever they converge, two points an arbitrarily small distance apart may have radically different assignments. This is a discontinuity. If W=1 there must be a discontinuity in the assignment of phase values somewhere; the tissue could evade this discontinuity only by placing it in a hole (a region devoid of living, labeled cells). One might suppose that the discontinuity would appear along some "seam" running the length of the leg (the proximo-distal direction) like an International Date Line, but none was found. The discontinuity is apparently more compactly localized than that. It might be an isolated phase "singularity". Alternatively, there could be several isolated phase singularities, e.g., three, of which two have opposite handedness and cancel out.

Rule 3 corresponds to the Rule of Intercalation.

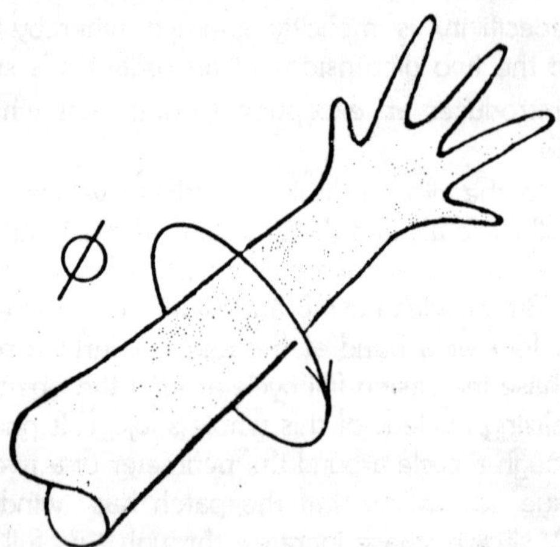

Fig. 10.1. The circular dimension around an appendage is putatively encoded in the cells by a time-independent quantity that increases smoothly from an initial value and returns to it without ever decreasing.

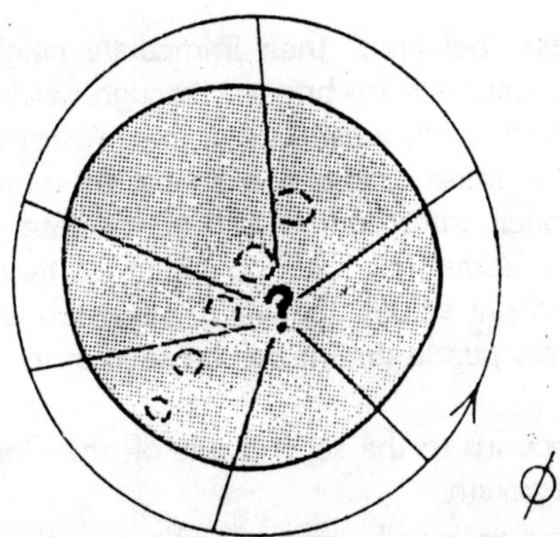

Fig. 10.2. The epidermis is stretched flat. The dashed rings indicate fingernails. From the five points on the abstract circle of phase values contour lines cross the epidermis, joining cells of like value. If the perimeter spans a full cycle, then these contour lines inevitably converge on the borders of an internal hole, or, if there is none, on a region of ambiguous phase (a "phase singularity").

Rule 3: So long as the phase gradient is shallow enough (if adjacent cells are sufficiently similar in phase), cells divide only to replace those that happen to perish. But if cells with normally non-adjacent phases are juxtaposed (e.g., by cutting off a leg, rotating it, and sticking it back on), then those cells begin to divide in earnest. As proliferation continues, the new cells take on phase

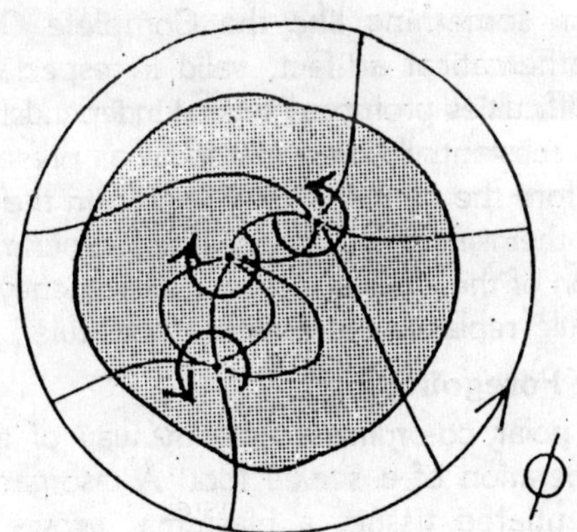

Fig. 10.3. As in but showing that any picture may be supplemented by any number of additional paired singularities of opposite handedness without violating the boundary conditions.

values intermediate "between" their immediate neighbours. Thus the phase discontinuity is soon bridged through newly regenerated tissue. Proliferation continues until it has restored the initial shallowness of the phase gradient in space. In some versions of the clockface model, intercalation of intermediate phase values might, without an additional rule, go either of two ways: there are two arcs of phase values "between" any pair of points of a circle. We defer this puzzle until it can be seen in its more general aspect.

Rule 4 corresponds to the second rule of, the Complete Circle Rule of distal outgrowth.

Rule 4: A new limb will grow out wherever there is a phase singularity. It is left- or right-handed according to whether the winding number of phase around the singularity is + 1 or − 1. Additional limbs of paired handedness are possible, and as we shall see, occur in fact. Within a patch of tissue whose border has winding number W around the phase ring, there will emerge R right limbs and L left limbs in such a way that R-L = W. For an exposition of this winding number formulation. In its original formulation, this rule was worked as though cells make a local decision based on fulfillment of a global criterion (non-zero integer winding number on a distant ring of tissue). This is implausible physiologically, and lends itself to the construction of paradoxes, besides.

In the approach taken here, emphasizing strictly local principles of cell interaction, something like the Complete Circle Rule emerges as a mathematical artifact, valid in especially simple situations. These difficulties prompted several independent analyses that converged on substantially the same ideas as presented next. It was not long before the contradictions implicit in the Complete Circle Rule made themselves felt in published experiments, too: an amended version of the clockface model subsequently appeared in which a local rule replaces the former global rule.

Examples of the Foregoing

Applying the polar co-ordinate rules by way of illustration, consider the regeneration of a served foot. A disorganized layer of vaguely differentiated tissue, a blastema, grows from the cylindrical edge to cover the stamp. By Rules 1 and 3, it must contain a phase singularity of the same handedness as the original foot. So by Rule 4 a new foot of the same sort replaces the old.

In terms of winding numbers, $W = 1$ (or -1), so in the simplest case $R = 1$ and $L = 0$ (or $L = 1$ and $R = 0$). Consider a second example in which a right foot is cut off and replaced by a left foot served from the other leg.

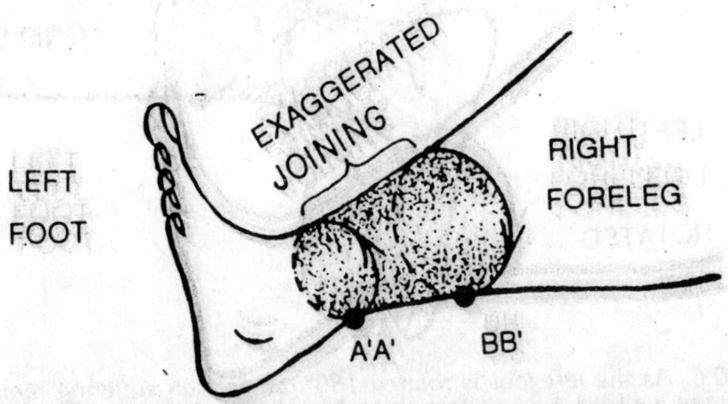

Fig. 10.4. In the stippled cylinder of regenerating and healing tissue, the transition is made from the clockwise orientation of the left-handed graft (foot) to the anticlockwise orientation of the right-handed host (foreleg).

The cylinder of new tissue proliferating in the junction according to Rule 2 is bounded by two complete circles of phase, as indicated. This two – part border has winding number $W = 2$ around the cylinder of skin it encloses. This is shown by slitting

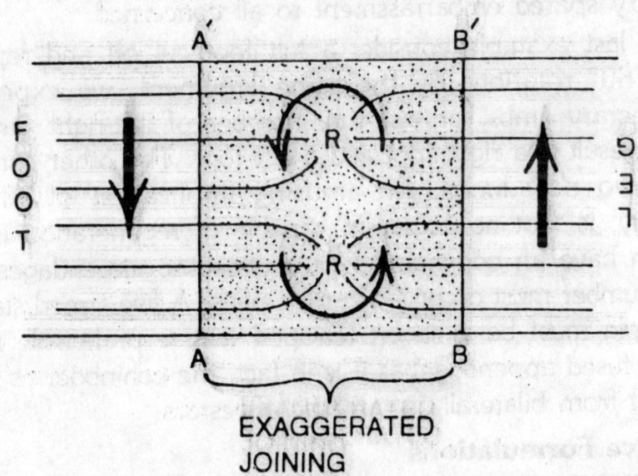

Fig. 10.5. The stippled cylinder is slit open along the line from AA′ to BB′. Lines A′B′ and AB are the same. Contour lines and arrows show the smoothest way to join the oppositely oriented phase circle AA′ and BB′: two phase singularities of right-handed orientation are required.

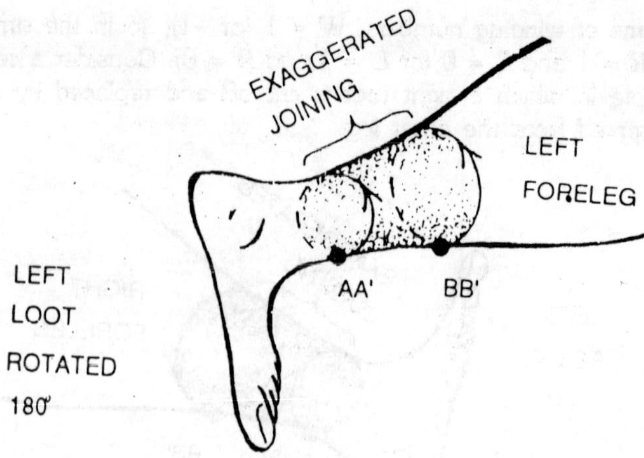

Fig. 10.6. As the left foot is rotated 180° rather than suffering removal to a right stump.

the stippled skin along the line AB and laying it flat: along path ABB 'A'A the phase increases by 0 + 1 + 0 + 1 cycle. Thus we expect two additional new right feet to emerge. In animals capable of regeneration, the actual result is indeed two additional right feet. A Chinese woman suffered this very operation in 1973, following piecemeal destruction of both limbs in a railroad accident (15). The present inability of humans to regenerate whole limbs presumably spared embarrassment to all concerned.

As a last example consider a left foot cut off and replaced with a 180° rotation. By the same argument, we expect no supernumerary limbs, or else any number of left-right pairs. A common result is a right foot and a left foot. The other common result is no new limbs. An amusing implication of all this, incidentally, is that no organism capable of regeneration in this mode can have an odd number of asymmetric appendages, and its even number must occur in left-right pairs. A five-armed starfish, for example must be a covert tetrapod with a prehensile tail (a bilaterally fused appendage)-as it is in fact, the echinoderms being descended from bilaterally symmetric ancestors.

Alternative Formulations

Superficially, the data compiled by French et al., would seem to argue forcibly that phase singularities do play a central role in the morphogenesis of higher animals, much as they do in wave-conducting sheets of oscillators such as the social amoebae, the

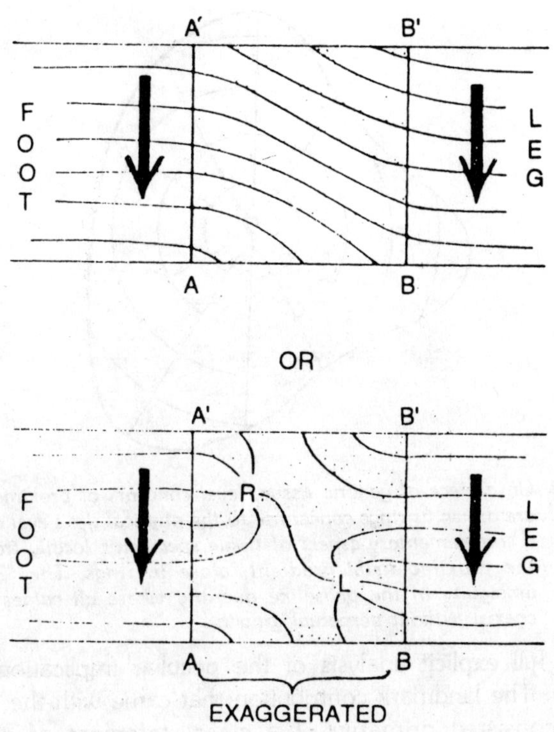

Fig. 10.7. As transition can be made without singularities, by just twisting all contour lines through 180° (top). An alternative construction (bottom) intercalates a superfluous pair of complementary singularities. Both outcomes are common.

ascomycete fungi, and certain chemical reactions. This is a distinct possibility that I personally find very exciting, but it does have difficulties. These arise when we ask, as in the other organisms exhibiting phase singularities, what happens at the singularity? There is no abrupt discontinuity in cell type, no unique structure, and in apparent violation of Rule 2; cells do not keep proliferating indefinitely. All this suggests that the proximo-distal aspect of tissue specificity interacts in an essential way with the azimuthal tissue specificity represented on the phase circle.

Although the original paper does not dwell on this interaction, it is implicit in the polar co-ordinate diagram in which the proximo-distal aspect of tissue specificity is represented radially with the most distal types (making toes, fingers) at the center. Sibatani

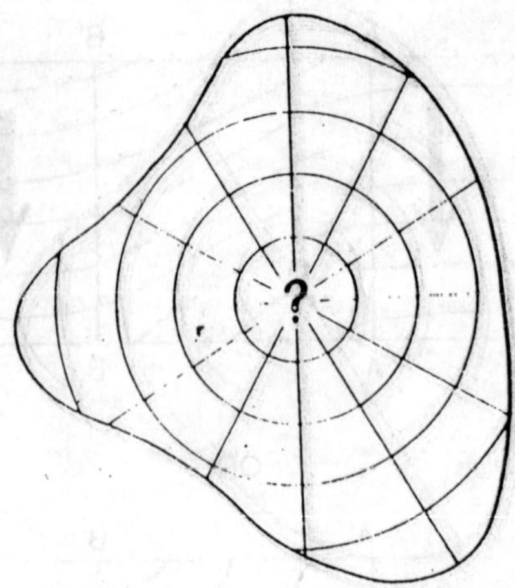

Fig. 10.8. On a piece of generic tissue, level contours of proximo-distalness are drawn as rings concentric to the physical tip. Level contours of a complementary aspect of tissue specificity, locally transverse to the proximo-distal gradient, close in rings. The "?" denotes ambiguity in the phaselike quantity where all values ostensibly coexist without perpetual growth.

gives a full explicit analysis of the peculiar implications of this diagram. The landmark contribution that came with the clockface model consisted primarily of a clear statement of correlated phenomena.

There are some astonishing regularities spanning diverse phyla, never before perceived so lucidly. The model itself had less appeal, its own internal contradictions lying so near the surface. The inevitable flurry of counter-experiments and *ad hoc* amendments was well summarized, but the ostensible elegance of 1976 was irretrievable once the subject had been glimpsed from another perspective. At present it appears that the reported phenomenology is best understood as a consequence of little more than the continuity properties of any topologically suitable representation of "positional information" or (to avoid confusion with physical position) "tissue specificity" underlying visible differentiation. The most tantalizing feature of the clockface model was its use of a *circular* co-ordinate for measuring cell types. It cells have two interdependent aspects of tissue specificity then one must ask exactly what motivates the choice of one co-ordinate system over another on this two-dimensional state space. Of course, one would prefer a co-ordinate system natural to the topology of the state space.

An Inverse Fate Map

A simple co-ordinate-free representation might be constructed as follows. Let us imagine a tissue-specificity space (TSS). We endow it with enough dimensions (two) to distinguish cell types on a two-dimensional surface such as the surface of a leg. We depict as a *place* in TSS the latent tissue specificity of a cell or a patch of neighbouring cells. To each region in TSS there corresponds a type of structure that cells in that region will make when they mature. Map the creature, organ, or tissue into TSS. There is nothing new in this procedure. This is only drawing a fate map, inside out as it were: instead of drawing the organism in the real

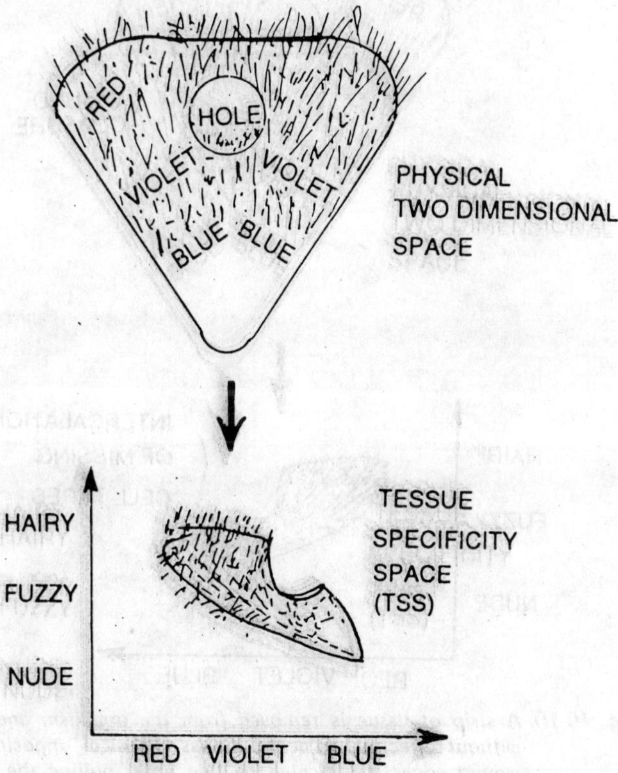

Fig. 10.9. An imaginary two-dimensional organism and its image in tissue specificity space.

world and writing tissue names on an overlay, we write the tissue names at fixed places in tissue-specificity space and then draw the organism as an overlay, distorted as required to put places on their corresponding names. For example, this "inverse fate map" of any bilaterally symmetric organ is folded so that two patches of cells symmetrically disposed to the left and right of a mirror plane both map to the same place in TSS. The mirror axis maps to a fold line. Now consider two pieces of an organism, normally not adjacent in the intact, mature stable organism.

There appear in TSS as two islands of tissue. If they are now physically juxtaposed, tissue specificities are not initially affected,

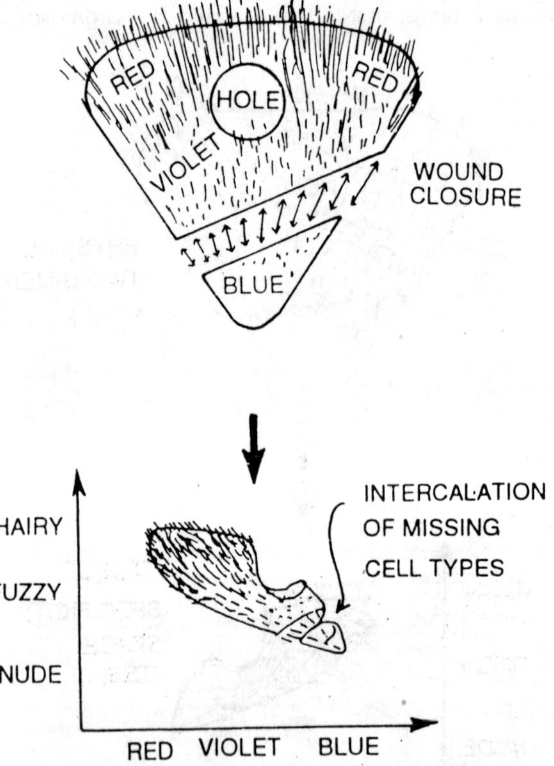

Fig. 10.10. A strip of tissue is removed from the organism and its image without affecting adjacent tissues. Physical apposition of the wound edges is indicated by fine lines, pulling the edges into contact in the upper picture, but not altering tissue specificities in the image.

so fine lines must be used to connect the cells that are quite separate in TSS though physically adjacent along the surface of contact. In experiments on the crayfish, Mittenthal shows that these "in-between-ness" connections *cannot* always be made straight by suitable rubber-sheet rearrangement of TSS, contrary to my conjecture of 1977. This may be a consequence of nothing more exotic than *curvature* of TSS: if TSS were like a steep hill on an otherwise flat plane, then between any two points astride the hill there would be two shortest paths (left and right along the hill's flanks) from one to the other through "in-between" cell states. Any representation that purports to straighten all such curves would thereby inadvertently superimpose these two, though they consist of distinct cell types.

So a two-dimensional TSS cannot always be so defined that the linking curves appear straight. Nonetheless, straight-line connections in two-dimensional-space appear to suffice for the purposes of this Chapter, obviating the need to adopt a special "shortest arc" rule and the implicit metric to decide which path to take in a one-dimensional space. We rewrite the rules now in three parts using this geometric language:

Rule 1: Each little patch of cells is labeled with a state which is part of a smooth gradient of states across the tissue. This state is a point in a two-dimensional space of biochemical specificities topologically equivalent to a plane. Once established, this smooth map does not change. The two components of this "state" specification might be, e.g., the concentrations of two kinds adhesive site on the outer cell membrane.

Rule 2: No exceptional point or state exists.

Rule 3: Cells proliferate at a rate determined by the separation of adjacent cells in TSS (the unlikeness of surface properties, for example). If they are initially quite different in tissue specificity (e.g., cells on the interface, connected by fine lines, then they begin mitosis. The tissue expands in physical space while its image in TSS remains unmoved but becomes denser with cells. Proliferation stops when the density of cells throughout the image has risen to a local norm, i.e., when the local gradient of tissue specificity has diminished to normal for each kind of tissue. There is no problem about endless proliferation at an unremovable phase singularity: no singularities occur. Just as in the original rule, new cells intercalate tissue specificities intermediate between their physical neighbours. (Note that we have yet to specify exactly what is "intermediate" between cell types that are not normally adjacent. This puzzle is pushed much closer to solution in Mittenthal's *Rule of Normal Neighbours*.

Applying the TSS Image Rules

No separate rule is needed for induction of a limb. The image in TSS of any ring of tissue is necessarily a ring, and by Rule 3 the tissue inside the physical ring must acquire (at least) the tissue specificities inside the ring's image. At an appropriate stage in development, cells will develop the structures by which their positions in TSS were named. If those should happen to include the distal parts of a limb (fingers, for example), then such structures (among others) arise when tissue specificities are interpreted biochemically, and we have a "limb". If the most distal structures are normally internal to more proximal structures in TSS, then amputation corresponds to ablation of a disk in TSS, leaving a ring.

Blastema formation followed by wound healing corresponds to stitching across the empty disk fine lines along which new cells take up the missing tissue specificities, restoring the more distal organs. Replacing a left hand (L) by a right hand (R) on a left stump corresponds to cutting out the disk in TSS and replacing it with an identical disk. But remember that because of the inevitable L-R mismatch in physical space, tissue connections from the (R) central disk reach across to the opposite side of the (L) hole. We thus have a three-layered map across the distal core of TSS. The three layers are the existing hand and the cylinder of wound along the wrist that joins hand to forearm. The two new layers (the cylinder of wound proliferation) have the same orientation, so that we get two replicate left hands. Or imagine that the arm is amputated at both ends. Both the proximal and distal blastemas must then span the middle ("hand") region of TSS.

So mirror image hands must emerge (and do) at each end. This is the simple geometric essence of the rule of distal outgrowth. According to this rule, more distal structures regenerate from any stump even at the proximal end where, if regeneration were naively expected, a shoulder should grow. This picture in TSS is the same (not inverted) for right or left limbs. Thus replacing a left hand (L) by a right hand (R) on a left stump corresponds to cutting out the (left-hand, L) disk in TSS and replacing it with

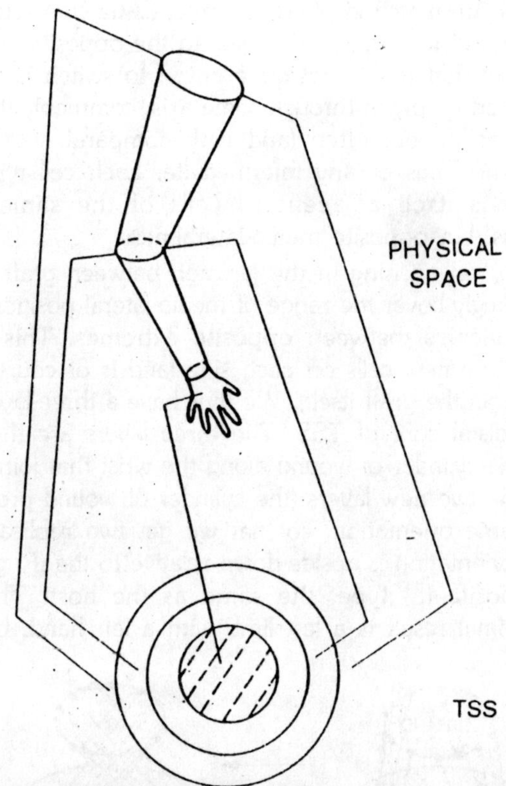

Fig. 10.11. As in a limb maps to its image in tissue specificity space. Most distal structures are believed to map concentrically interior to more proximal structures. Amputation at any level thus deletes a central disk. Contact of "wrist" tissues through the regeneration blastema is indicated by dashed lines: they span the missing (more distal) tissue specificities.

another (right-hand, R) disk. The new disk looks exactly the same as the old in TSS: it covers all the same regions of the "inverse fate map," since it has all the same cell types in the same neighbourly arrangement. But the reversal of handedness corresponds to a reversal in the correspondence between cell types and points in real physical space.

This only becomes important at the boundary where the graft and host are trying to fit together: in real physical space, that ring of cell types in TSS corresponds to a clockwise ring of host tissue, but to an anticlockwise ring of graft tissue. Because of this inevitable L-R mismatch in physical space, tissue connections from the (R) central disk must reach across to the opposite side of the (L) hole. Note that this is not quite crazy: to switch handedness, one must invert an organ through some axis, commonly the medio-lateral axis as in, but often (and with comparable effects) the antero-posterior axis or any intermediate. Each cell type at the junction finds itself adjacent to cells of the same antero-posteriorness, but opposite medio-lateral-ness.

The new cells arising in the junction between graft and host must accordingly cover the range of medio-lateral positional values (tissue specificities) between opposite extremes. This range is spanned by the new cells on each side (and is of course already spanned within the graft itself). We thus have a three-layered map across the distal core of TSS. The three layers are the existing hand and the cylinder of wound along the wrist that joins hand to forearm. The two new layers (the cylinder of wound proliferation) have the same orientation, so that we get two replicate hands. Since that orientation is upside-down relative to the (R) graft, they are of opposite (L) type, the same as the host. The parity-conserving final result is a left limb with a left hand, but with a

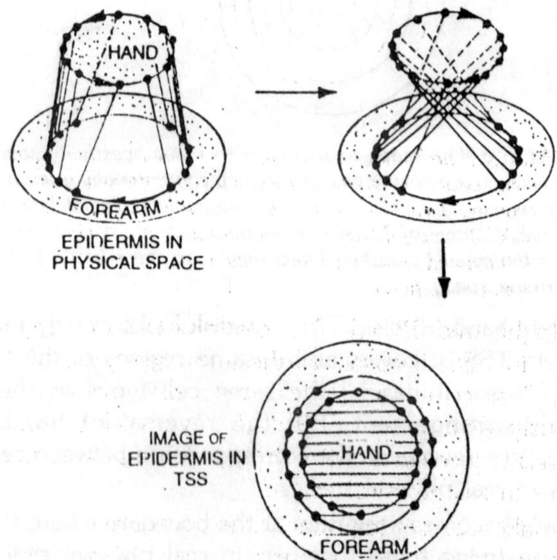

Fig. 10.12. The experiment is represented in the terms, rather than in terms of phase contours.

complementary L-R pair in addition. This strange, and (originally) unexpected experimental result derives from nothing more than an assumption of continuity(!).

Equivalent Metaphors and a Coat of Many Colours

Another way to articulate the TSS rules uses the language of *colour*. The diagrams in every paper about the clockface model would be much simplified, and both design and interpretation of experiments would be made transparent if only we could publish airbrushed colour diagrams as follows. Let the phaselike, circular aspect of tissue specificity be represented not in digits 1- 12, nor in contour lines, but in *hues* taken from a colour wheel. On a colour wheel, red grades into orange grades into yellow grades into green grades into blue grades into indigo grades into violet grades into red grades into orange, and so on: without backtracking, one can progress smoothly and unidirectionally through a cycle of tissue states, as required. Proximo-distal-ness may then be represented by *saturation* of the hue, the most distal tissues being least saturated (gray) so that all hues nearby are compatible without discontinuity. Our labeling of tissue types then consists only of assigning to each tissue type (and so to each place on the organism) a colour: have and its saturation. Every organism, in this view, wears a coat of many colours.

The colour is permanently imbued in each cell at mitosis, in cases of strict epimorphosis. Surgical interventions consist of abutting together colours that were previously separated by gradations of colour. If the colours now "run" and blend (as in morphallaxis), or if new fabric intercalates itself, adopting hybrid colouration compatible with its boundaries (as in epimorphosis), then we end up with arrangements of colour that satisfy all the phenomenology of regenerating and duplicating tissues, and, incidentally, the mathematical phenomenology of phases 1-12 (hues) and winding numbers thereof (complete colour wheels) and singularities (gray spots). The phase circles of Glass may be traced through full cycles of hue, ignoring saturation. They can be constructed by drawing circles on the physical tissue and then following the circles' images (rings) in TSS. The number of singularities corresponds to the winding number of the image about an arbitrarily chosen origin for the angular co-ordinate system.

In the coat-of-many-colours metaphor, the usual choice of origin would be gray, but this is an arbitrarily convenience dependent on ambient lighting. Biologically, the usual choice of origin is the distal-most tissue. From my point of view, the arguments of French *et al.* and of Glass about circles, phase maps, and winding numbers amount to using a circle embossed on the organism as a means of book-keeping the folds and rotations of the regenerating tissue's image in TSS. Glass's calculational procedure is very useful here, because until one has a little practice, it seems awkward to visualize handedness in TSS maps.

The essential feature common to all these metaphors is the mapping from a physically distributed tissue to an abstract state space of two dimensions with continuity properties like ordinary two-dimensional Euclidean space. The most general underpinning for such a mapping is a set of chemical substances arranged in the tissue along transverse concentration gradients. To take but one example, the local concentrations of any three distinct pigments define a three-dimensional state space. Any surface coloured by mixtures of those pigments is thereby mapped into that space. If the total amount of pigmentation (the sum of the three concentrations) is limited to a fixed quantity, then the surface is confined to the two-dimensional triangular plane of standard colorimetric charts.

Colour mixtures happen much as in. If the "pigments" (or cell-surface antigens, etc.) should happen to mutually influence their own rates of synthesis and degradation then, as is well-known from the classical theory of reaction/diffusion structures, the chemical gradients may become self-stabilizing. The only experimentally demonstrable reaction/diffusion structure at present is the "rotor" in Belousov-Zhabotinsky reagent. Its structure, dynamics, and reaction to trauma may be expounded exactly in the manner of this chapter, its development into three dimensions resembles nothing so much as the embryology of simple organisms. From the literature surrounding the clockface model (1), one gains the impression that proper understanding of the phenomena of regeneration somehow repudiates (in favour of the more modern preoccupation with cell surface chemistry) the classical and apparently sterile notion that chemical concentration gradients and reaction/diffusion structures underlie it all. Perhaps it is indeed time for a fresh perspective on these unanswered riddles, as old as the science of biology. But to my mind, nothing more resembles the phenomenology of the (known and nonlinear, as opposed to

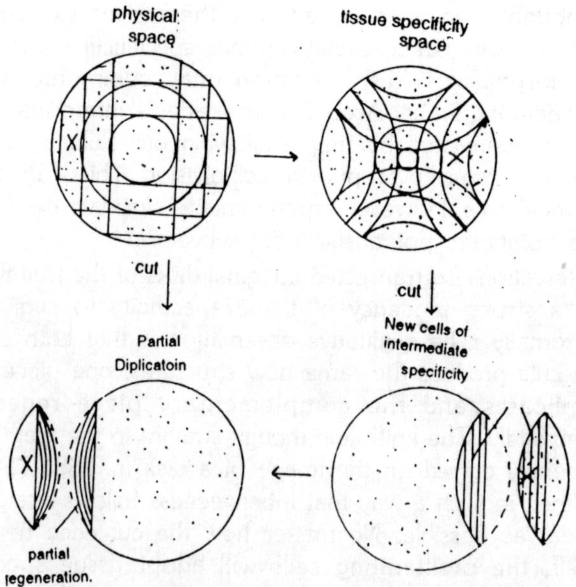

Fig. 10.13. The imaginal disk of a fly appendage.

theorized and linear) reaction/diffusion structure than contemporary reports on regeneration in limb fields of comparable size.

Experiments Needed

Certain implications of the geometric viewpoint adopted above may allow critical testing:

1. This style of mapping physical media or organisms to a state space or a TSS differ its better-established prototypes in that no "dynamic" is postulated in state space. Applied to non-dividing or relatively slowly dividing tissue, this is the assumption of epimorphosis. But dynamics may play an essential role in more rapidly dividing young or regenerating tissue. As Goodwin was first to emphasize, this is a deficiency that needs attention. For example, consider the following:

(a) Every tissue starts as a tiny patch of indistinguishable cells which, as they grow, acquire divergent specificities. Developmental fields originate, like the "big-bang" model of the expanding universe, from a singularity in TSS. At least at that early stage of development a "dynamic" *dominates* TSS. Kauffman *et al*. provide an alluring linear theory of this "dynamic."

(b) Although French *et al.* address themselves exclusively to epimorphosis (old cells retaining their specificity), some amount of morphallaxis (respecification) may commonly occur in regenerating limbs. According to Maden's interpretation, this fact is assimilated by simply allowing our Rule 3 to govern wounded tissues even prior to cell division. This scarcely alters our diagrams. However experiments to elucidate the following two points are potentially more subversive.

(c) Observations on transected imaginal disks of the fruit-fly argue for a strong tendency of tissue specificity to evolve more proximally. The qualitative observation is that both edges of the cuts produce the same new structures: one piece exactly duplicates and the complementary piece regenerates completely. The knife line though straight in the real world, is generally curved on the image of a disk in TSS, TSS being defined in such a way that inbetweeness follows straight lines so far as possible. No matter how the cut edge heals onto itself, the proliferating cells will adopt tissue specificities intermediate (parallel lines) between points on the knife cut's image in TSS. These intermediates cannot include the whole domain occupied by the excised piece if the disk's image is convex in this TSS- and it must be, or else intercalation between protrusive extremes of the image would produce tissue types outside the image, i.e., not normally present. So developmental fields are *convex* and cutting the edge off one leaves each part completely without access to some region of TSS occupied by the other part. Without a *"dynamic,"* one piece must then fail to *fully* regenerate while the other fails to *fully* reduplicate. Transection experiments may thus provide the means to verify the existence of a "dynamic" and to quantitate the rates at which tissue specificity can change during normal growth and during regeneration.

Without a proximalizing "dynamic," there must be parts of any organism that *cannot* be regenerated: the adjacent most proximal parts of neighbouring developmental fields, located on the common rim of their convex images in TSS. By the same token, the kind of regeneration here considered cannot proceed beyond any symmetry plane: a bilaterally symmetric organism, for example, could regenerate at most half of its complete body. How do these limitations contrast with the facts of regeneration in radially symmetric organisms and in morphallactically flexible organisms such as *Planaria*? Recent experiments using artifically constructed bilaterally symmetric limbs are particularly intriguing in this context. An unexpected progressive loss or gain circumferential positional values (hues, clock-phase values, TSS regions occupied). This may betray a slow underlying dynamic such as we inferred previously.

2. According to this coordinate-free representation, ablation of a feature on the symmetry plane of a bilaterally symmetrical organism (a tongue, a nose, a penis, a tail, the genital disk of a fly) corresponds to cutting a hole out of the fold of the two-layered image in TSS. Such a hole can heal in two quite distinct ways.

(a) Closing horizontally (top-to-bottom contacts), the hole in TSS would be filled again. Thus we should anticipate complete regeneration.

(b) In contrast, a vertical closure (side-to-side contacts) would join mirror image tissues. Thus, no proliferation should ensue. Even if it did, the missing tissue specificities would not be recovered.

The results of diagonal closure must vary between the external results a and b according to the angle of the diagonal. This series of experiments may help to distinguish between geometric interpretations of tissues specificity. For discussion of experiments recently undertaken along these lines, consult. One source of ambiguity at present is the lingering uncertainty about the actual (uncontrolled and unobserved) geometry of wound closure.

3. If one had to guess before observing, it might reasonably be supposed that an organism would respond to surgical challenges by intercalating positional values between given boundaries in the smoothest way. This surmise comes from the observation that paired phase singularities (left-right pairs of supernumerary appendages) are uncommon where uniformity of phase might alternatively prevail.

4. It seems not entirely proven by experiment that apposition of dissimilar tissues is required to elicit regeneration. It is possible that the external medium may under some circumstances serve as a "tissue type"? Where is its point in TSS, if so?

5. It would be of interest to treat a strictly non-living chemical "organizing center" as though it were, for example, a limb field, repeating many of the physical manipulations of classical

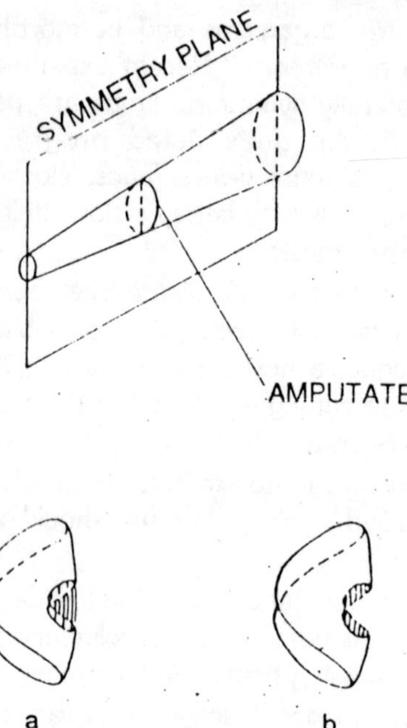

Fig. 10.14. As in a symmetric appendage maps into TSS as a folded disk, its corresponding left and right parts twice covering each occupied specificity. Amputation deletes a central disk straddling the fold line.

embryology. Some of the practical difficulties are similar: Accurate separation and recombination of microscopic chunks of wet gel that tend to adhere through surface tension or to float apart when submerged; accurate observation of outcomes that consist of garbled and sometimes subtle patterns, constantly changing; limited reproducibility due to uncertainty of initial conditions. The outcomes might also bear sufficient formal resemblance to provoke stricter attention to strategic issues: What questions are we asking? How will the answers aid understanding and control? Since the morphogenesis of organizing centers in Belousov-Zhabotinsky reagent can be understood in exact Physical-chemical detail, it might be illuminating to know how nearly it really does resemble cytochemical morphogenesis.

INDEX